Fall Down Nine Times, Get Up Ten

or
Cold Lake Zen Healing

or
Zen And The Art Of Healing
Your Aching Back And Legs
And Numb Or Burning Feet:

Neuropathy Treatment Fact & Fiction
With A Plea For Integrative Health

Martin Avery

This book is dedicated to my cousins, Lyn Philips and Shirley Macdonald, of London and Lac La Biche, and to some great healers in Northern Alberta:
Angela Plaquin at Canadian Health & Sports Rehab Inc., Cold Lake
Suzanne Theckson, acupuncturist, Cold Lake
Dr. Darrel Kopola, The Wellness Centre, Cold Lake
Charmaine and Trudy O'Shaunessy at Cold Lake Community Health Care
The CT Unit at Cold Lake Healthcare Centre
J.F. Bernier at Health First Physiotherapy
Cold Lake Hospital Emergency Unit
Dr. Stander and Dr. Birkill, Cold Lake
J.F. of Health First Physiotherapy, Bonnyville
Dr. Jassar, Edmonton Clinic
Nola, the neurology nurse at the Edmonton Clinic
Jasmine Ouellette, who led me to the gifted reflexologist
and especially to
Mecell Pillon, alternative practitioner, The Wellness Centre, Cold Lake, with a wish for more doctors across Canada, walk-in clinics, too, and even more of an evolution in healing so Eastern healers and Western doctors and alternative healers can work together to help and heal us all.

ISBN 978-1-312-33300-0

Fall Down Nine Times, Get Up Ten

or
Cold Lake Zen Healing

or
Zen And The Art Of Healing
Your Aching Back And Legs
And Numb Or Burning Feet:

Neuropathy Treatment Fact & Fiction
With A Plea For Integrative Health

Martin Avery

"The world breaks everyone, and afterward, some are strong at the broken places." Ernest Hemingway

I guess you never know if the last time you see someone is going to be the last time you ever see someone. -- Eddie Murphy as Jack McCall in A Thousand Words

"Only two things can reveal life's great secrets: suffering and love."
— Paulo Coelho, Aleph

Inside You'll Discover:

The causes of Neuropathy and the little known tricks to reduce the symptoms.The top natural, but highly effective, ways to reduce the pain, numbness and burning.

The RIGHT kind of vitamins to take that can help repair nerve endings.

Small lifestyle changes that can dramatically improve your well-being.

And much more

Activate your short term and long term disability, the neurology specialist said, this could be cancer, AIDS, West Nile, MS, ALS, lyme disease, or Lou Gherig's Disease, or GBS -- Guillain-Barre Syndrome. I'm ordering an EMG and a lumbar puncture, or spinal tap.

I laughed.

Don't laugh, he said. You can't walk. This is no laughing matter.

Good luck dealing with the medical establishment, my ex said. Good luck with the medical/pharmaceutical complex. Good luck dealing with Big Pharm and the hospitals, clinics, and doctors they fund and control.

I'm not as cynical as my friend.

A sudden stab of intense back pain, followed by the loss of the ability to walk, sent me from the emergency department at the closest hospital to the neurological emergency centre in the city, and to a specialist who listed the possible causes as cancer, MS, ALS or Lou Gehrig's disease, HIV/AIDS, Meningitis, diabetes, West Nile, lime tick, but not a chiropractor's adjustment.

He ordered a spinal tap and an EMG and said, Activate your Short Term and Long Term Disability.

Electromyography (EMG) is a technique for evaluating and recording the electrical activity produced by skeletal muscles.

At the end of the summer, I was healthy and happy, getting in shape, again, hiking trails in Northern Alberta, wading in the cold water, and using free weights while I walked. I hiked a 12 km trail in 90 minutes and my speed was getting better every day.

In September, I couldn't walk at all and was reduced to crawling, at home.

In the last week of August, I had intense back pain at work. So, I went to a chiropractor three times.

That took my back pain away, but ....

Now I can't walk.

At work, I used a wheelie chair, as I walked like a spastic.

I went to emerg 4 days ago, got a needle drugs, heat, and a promise I'd be doing the tango in 3 to 5 days.

That was great, I thought, because I never did the tango before.

After 3 days, and no tango, or walking, I went back to emerg.

They gave me new drugs. Stronger. And a prescription of physio plus 7 days off work. Seven work days, that is.

I phoned the physio right away and made an appointment for the earliest date: October 30th.

Although I had a reflexology appointment for today, I couldn't go, because I couldn't walk 100 steps to my car, never mind drive, and get into the building for the appointment. I made it about fifteen steps, felt wobbly, almost fell down, and decided to go back home.

On Tuesday, after Emerg, at the drug store, I couldn't negotiate the curb, so I sat on the ground.

Despite losing the ability to walk, I am oddly cheerful.

Not suicidal.

I feel creative. -- Had a great idea for novel. On Facebook I saw a quote by the American author Anne Lamott about people who get upset when writers describe them in ther books not as warm and fuzzy but as something else. Lamott says, They should have lived better lives.

"You own everything that happened to you. Tell your stories. If people wanted you to write warmly about them, they should have lived better lives." -- Anne Lamotto.

That made me laugh out loud.

I have wanted to write about this person I met a decade ago and got to know over a five year period.

X-rays revealed that one disc has thinned, a little, over my lower back area, the doc says. Not enough to cause pain and trouble walking.

Dr. Decker said, It looks like a chiropractor hurt your sciatic nerve and that's why you can't walk.

Oh, I said. I was thinking, How would I kill myself?

I made a short list.

1. A long walk in Northern Saskatchewan.

-- But now I can't walk.

2. Do I have enough drugs for an overdose?

-- All I've got is ibuprofen. So, no .... I don't think so!

I am laughing as I write about suicide.

Okay, it entered my mind.

But, really, I'm not the type.

At the end of the school year, I was exhausted. And then I caught a cold.

Summer colds, or the flu, ain't what they used to be. I rarely get sick, seldom catch cold, but the last couple were real killers. They turn into bronchitis, fast, and last for weeks. You feel like death would be better. But eventually you recover.

I went for two or three months without a day off, due to weekend conferences, the drama festival, and other school-related events. And I had no prep time, at school, plus drama and basketball after school. And one of the drama groups was, shall we say, difficult. So, by the time exams were over and report cards were in, at the end of my first year as a teacher in Alberta, I was so exhausted I was facing burnout. And then I got sick with a cold or flu.

For the first couple of weeks of summer, I was flat on my back, only got out of bed for the basics, like little trips to the bathroom or kitchen, and thought death would be better than the well-known symptoms of pneumonia.

Before I got sick, I was working out at the gym three times a week and hiking 12 k in the park on the weekends. When I finally crawled out of my sick bed, I tried to get in shape, again, by hiking a bit more every day. By the end of the summer I was hiking 12 k again, faster and faster, with a backpack on my back and eight pound weights in both hands.

When I went back to work, feeling fit and strong, looking a little leaner, I suddenly got hit with intense back pain.

The lower back area hurt like never before.

It was so intense, I had to do something.

It's impossible to get a family doctor, and I didn't want to go to the hospital, so I went to a chiropractor.

I have a brother who believes in going to the chiropractor once a month for an adjustment and he has done that for decades, with no ill effects. Even so, I didn't like the idea of having somebody adjust my bones, violently, and painfully.

In the summer, I went for acupuncture a few times, and it helped me feel better, generally, but it wasn't so dynamic that I thought of going there for back pain relief.

I was afraid the chiropractor would hit a nerve and hurt me seriously.

And my fears turned into reality.

At first, I was happy, because my back pain went away.

But then I lost the use of my legs.

First, I walked funny. And then I avoided walking. And then I couldn't walk at all.

I went to emerg at the hospital twice and got two different assessments.

Dr. Dekker, a guy, prescribed a needle, for pain, and pills for inflamation and pain, and told me to take a hot bath with a shot of whiskey.

As a non-drinker, I didn't know how to respond to the doctor's advice. -- I thought it was very odd to have a doctor recommend mixing drugs and alcohol.

But I tried it anyway.

A little Jack Daniels with a hot bath had no more effect than the drugs or heating pad. Ice packs on the back helped a bit.

At work, I staggered from my car to my classroom, then sit in a wheelie chair all day.

At first I could stagger from one classroom to another but by the end of the week I was wheeling from class to class.

The second time I went to emerg, the doctor was a woman and she was more severe. She prescribed stronger pain drugs, anti-inflammatories, physiotherapy, and almost two weeks off school.

I thought that was a bit much but after one day in bed I realized she was right.

My symptoms got worse and worse.

When I went to the drug store, to get her prescriptions filled, I found I could not negotiate the curb. I sat on it and two people came to my rescue. A rough looking guy in his twenties and a grey haired woman in her eighties helped me up and into the store. -- How embarrassing is that?

Funny .... When I recall that moment, think back, and see it in my mind's eye, they guy looked like Jesus, as he had a beard, was thin, calm, peaceful, and kind. And the woman looked like Mother Theresa, out of uniform.

At home, I had to crawl around my apartment, to get from the bedroom to the kitchen and bathroom and back again.

After a few days of bed rest and a lot of sleep, I phoned telehealth, the telephone helpline, to talk to a nurse, and she said, Go to emergency!

My cousin and my girlfriend said the same thing. Call an ambulance or a taxi and demand stronger drugs or the next step, cortisone shots or surgery, to fix your back.

I got up on Sunday morning with the intention of calling an ambulance but, after taking a shower and getting dressed, realized I could walk a bit better than the day before. I decided to walk to my car and call a taxi, to go to the hospital, if I felt like it when I got there.

Surprise! I walked the short distance to my car without any problem. The last time I tried that, my legs gave way and I wound up on the ground, again.

So, I drove to the drug store, bought two canes, two hundred dollars worth of supplies, then drove to the gas station, where I got some chicken and used the cash machine, then drove around town looking for drive-thrus, and finally went to the drug store. I didn't fall down this time, since I was using two canes.

A week ago, I was completely disenchanted with the idea of using a cane. Today I was thrilled to be able to walk with canes.

After I did all those errands, I drove home. It was a challenge getting my groceries from my car into my apartment, but I did it.

And then I wrote for about three hours, all about the experience of losing the use of my legs.

I am feeling quite emotional right now. Maybe that's what I was supposed to learn from all of this.

I feel like proposing to my girlfriend.

We broke up, rather dramatically, just before the end of the school year.

She came up for a visit, for a week, just before exams, and I was a mess. She was pretty stressed, too, trying to meet deadlines for her job as a newspaper reporter. She is a multi-media reporter for a big daily newspaper in Edmonton.

We went out for dinner every night. On her final night, I took her out to dinner, again, and we talked about the future. She wanted to take a job as the editor of the local newspaper and move up north so we could be together. I told her that would be a terrible career move for her and I only had a one year contract so the timing did not appear to be right.

So she left.

We split up.

After a couple of weeks, we became Facebook friends, again, and she started phoning me, just about every day.

She wanted to get married. I thought it was too soon. -- How well did we know each other?

For nine months, we spent just about every weekend together.

Those weekends were great, but how well do you know somebody after nine months of weekends?

I thought we should do the same thing for another year.

She said, No, if you don't know after that much time, then it's a no-go.

Today, I feel like phoning her and proposing.

She has been my most useful friend during this episode that started with back pain and has led to the loss of the use of my legs. She has been very bossy, as she says, telling me to take an ambulance to emergency at the hospital and demanding stronger drugs or the next step, which is cortisone shots or surgery.

I haven't taken her advice, and I told her she was scaring me, instead of inspiring hope, so she apologized, but I know it comes from her heart, which I think is big and beautiful, so .... There you go.

Maybe I was thinking that my contract here would not be renewed and I would go to China.

It was time to lie down, now, and follow doctor's orders.

The first doctor said the chiropractor hurt my sciatic nerve.

The second doctor ordered x-rays and said I had one disc that has thinned a little bit.

Louise L. Hayes has other advice.

That's it for now! Twent-three pages, today, in about four hours. That's pretty good for one day, one run, one writing session for a guy with a bad back!

Emerg

Alberta's health care system, like Ontario's, and the health care systems of other places, is in a bit of a crisis. There aren't enough doctors, apparently, so it's almost impossible to get a family doctor, you have to wait half a year in Ontario, and forever in Alberta.

Where I was living, in Northern Alberta, there were no doctors taking new patients and there was no walk-in clinic, so people used the Emergency wing of the hospital as though it was a walk-in clinic or made use of nurse practitioners, working in the hospital, or went to various wellness centres, where you could find chiropractors, acupuncture, reflexology, and so on.

At the end of my first year as a teacher in Alberta, as I was telling you, I was exhausted, facing burn-out, and then I caught a cold, so I spent the first couple of weeks of summer in bed, flat on my back, coughing and sneezing, and then recovering. I spent the rest of the summer getting in shape, again, by hiking on a trail in the nearby provincial park and wading in the cold water of Cold Lake. It was great.

I started slowly but built up so that by the end of the summer I was hiking a 12 k trail with a knapsack on my back and holding free weights in my hands, and my time was improving each time out.

I was losing weight and getting in shape, feeling strong and happy, and looking forward to going back to work.

I also wrote a couple of books. Two and a half books, to be more precise. One for kids, one for teenagers, and one for adults. And I found that work to be quite healing, too.

When I went back to work, the week before classes started, I suddenly felt a sharp and severe pain in my lower back. The pain made me double over and go to a chiropractor.

I would have gone to a doctor, if I had one, and you needed an appointment for to see a nurse practitioner in the so-called walk-in clinic, and I thought Emergency at the hospital was for people who were close to death. So I went to a wellness centre.

The first one I went to was booked up for months so I went to one down the road from them, that took me right away, and offered acupuncture as well as adjustments by a chiropractor.

With a couple of adjustments, my back pain went away.

First it hurt more, but then it went away.

I felt ecstatic.

But a couple of days later, I couldn't walk.

My lower back no longer hurt, and my legs and feet did not hurt, but they felt funny. It's hard to describe. I walked like a drunk, or a spastic, and I couldn't walk very far before my knees just gave way and I found myself on the floor, or the ground, laughing at my predicament.

I went back to the chiropractor and after a third adjustment, he suggested I go to the hospital for pain relief.

The first doctor I saw prescribed a pain killer, an anti-inflammatory, and ice packs for my back. He also suggested taking a hot bath and drinking whiskey.

He said I would be doing the tango in three to five days.

He said the chiropractor probably hit my sciatic nerve and after some bed rest I'd be back in the game.

He said my leg muscles were alternately contracting and spazzing, and that's what made me walk like I was swimming.

The second doctor prescribed a strong pain killer, anti-inflammatories, a hot pack for my back, plus anti-depressants, an wrote me a note so I could take seven days off work, and recommended physiotherapy at a place downtown, not at the hospital.

The place downtown booked me in for a month and a half later. The hospital was booking two months down the road and taking only post-operative cases.

She said I should stay flat on my back, except for going to physiotherapy, and I would be able to go back to work after some serious bed rest.

She sent me down the hall for x-rays and reported that I had one disc that was thinning slightly, but it was above the small of my back, and was not likely the cause of my pain or inability to walk.

The third doctor said, There's something going on down there, but I don't know what it is. He got a lab technician to take some blood, sent it to a neurologist in Edmonton, and said the neurologist would contact me sometime in the future, and then I would have to go to Edmonton for an appointment.

The prospect of driving to Edmonton, and through Edmonton, appeared to be impossible to me, at that moment.

I booked an appointment for acupuncture and one for reflexology.

I had seen the acupuncture person a few times in the summer, as my legs were a little sore, and my feet felt funny. The balls of my feet did not feel numb or like pins and needles, as they say, but somewhere or something in the middle of those two feelings.

Three acupuncture treatments made my legs feel better.

When I went to the acupuncturist in mid-September, I could not make it up the stairs to the room where she operated. So, she brought her needles and things down and looked after me on the couch in her livingroom.

She stuck needles in my thumbs, my forearms, under my nose, and doubled up on my thumbs. She got me to stand up for this, and then sit down.

We talked about Chinese medicine and the concept of being hot and moist. She said that was my issue.

She was disappointed that her treatment did not have a dramatic effect.

In fact, I felt a little bit worse.

My legs hurt, a bit, and walking was more difficult.

I was able to drive home, though, and lurch down the hall from the elevator to my apartment.

It was a challenge and I felt as though I was going to fall, or collapse, and wind up crawling on the floor, again, but I made it to the couch and my computer, where I typed these last four pages.

It's eight thirty, now, and I feel tired. I think I'll check my e-mail and then watch or movie, but I might just go to sleep.

The acupuncturist said she wanted me to call her in the morning, to tell her how I was feeling.

I had an appointment with a reflexologist people said had a real gift for healing that morning, too.

Qigong

I know what you're thinking: If this guy has done all this spiritual healing, qigong, Zen meditation, new age energy healing, and on and on .... How come he's sick?!

I ask myself that question sometimes, of course.

My environmentalist friends blame the environment. They point out that the planet is polluted, Mother Earth is under attack, and a lot of people are ill with immunity problems and other diseases.

Somebody suggested that if you do qigong in the tar sands, send your roots down into trouble ground, where sandy oil is extracted by heating up the earth and pumping in toxic chemicals, while war planes fly by overhead, it's going to burn your feet and hurt your whole etheric system.

My new age guru says there might be some truth in that, but the answer is not to cut off your roots, it's to get more sensitive. Keep sending your roots deep down underground to the magma core and embrace it, to send down stale energy and pump up fresh healing energy.

In qigong, that's what you do. You turn your body into a pump by doing various exercises while sending your roots down through the floor and the ground and what's under the ground all the way to the middle of the planet. You imagine it, visualize it, believe it, and you might feel it happening.

In Cold Lake, we create and celebrate global warming, with a highly productive oil sands industry, while jet fighter planes fly by in the sky, going up north to test weapons in a wilderness area that looks like Russia. We are surrounded by Reserves and there is a Metis community nearby, too. It's a little like living on the set of Avatar, the movie about a military airforce backing up open pit miners looking for unobtanium and fighting with the aboriginal people who get in the way.

The difference is that the aboriginal people here do not get in the way. They work in the oil fields.

Some of them.

They run a casino, too, at the edge of town.

Some of the big reserves south of here have oil wells on their land. In situ oil extraction operations are surrounded by fields of wheat or grazing areas for cattle.

Maybe I should teach Zen meditation with massage and qigong here. -- Call it Cold Lake Zen. -- Open the Cold Lake Zen Centre.

-- I'll get right on it, as soon as I can walk again.

Reflexology

After a chiropractor, a few trips to emerg, a visit to an acupuncturist, attempts to get into physiotherapy, and xrays, blood work, drugs, a week off work, lots of rest, a little exercise, needles, and canes, I went to see a reflexologist who came highly recommended. "Gifted," people said. "Really gifted."

One of my high school students, who is regarded as a "sensitive", or someone with a gift for healing, and who senses auras, energy, and so on, recommended her. She worked at the Wellness Centre, which was headed up by a chiropractor. The acupuncture person I went to used to work in the same Wellness Centre, but had moved out so she could work out of her house, after she had a few children.

When I phoned the Wellness Centre to book an appointment, the receptionist, who happened to be a double amputee, said, You are making an appointment with a goddess of healing.

Oh really, I said. That sounds great.

She really is very gifted, she said.

I had to cancel my first appointment, as I couldn't walk or drive, but we talked on the phone and she encouraged me to take it easy and come in when I could.

So I did.

She arrived a little late. The receptionist, the double amputee, who had artificial feet, ankles, and shins, said, She's running a bit late because, well, that's the way she goes, sometimes.

She stuck her head through the door and smiled at me.

Martin! she said.

Mecelle! I said.

Sorry I'm late, she said. I have to do a couple of things and then I'll be right out to get you.

Take your time, I said. No hurries, no worries.

She laughed and hurried up the hallway to talk to the receptionist, and then came back.

You're the guy from Saskatchewan, she said.

Nope, I said.

Avery, from Saskatchewan, she said. No?

I shook my head, no. I live in Cold Lake, I told her. I come from Ontario.

Have you been here before? she said. You have. Right?

I made and appointment but couldn't make it and we talked on the phone for a bit, I said.

Oh, yes, she said. That's it.

Well, she added. Come on in.

I followed her down the hall.

The receptionist had given me a form to fill out with the usual information plus a couple of unusual questions.

It asked what I had already done to deal with the problem I was having. And it asked what I expected from this session and how long I thought it would take.

I said I expected a miracle to happen in about one hour.

It asked what my major source of stress was.

I filled in the blank with, The loss of the ability to walk.

There was a place to indicate your goal for healing.

So I wrote that I wanted to walk well, again, as I did a few weeks earlier.

She watched me walk as she showed me to her room.

Nice room, I said, as we walked in. Smells great.

It was a twelve by twelve room with a healing table, a big comfy chair, two other chairs, several certificates on the wall, new agey music playing, and an aromatherapy candle.

Thanks, she said.

Which chair? I asked.

The big comfy one, she said.

After I sat down, she pushed it back and a leg rest popped up.

She lifted my feet, put them on a pillow, and took off my socks.

She talked a lot, happily, said she loved her job, noticed I had written that I wanted a miracle, to be able to walk again, and promised to deliver.

We'll get you relaxed, sedate your nerves, find out what's going on, and improve your range of motion.

She held my bare feet lightly, looked at them, and told me my life story, in terms of injuries, checking in with me, for confirmation.

She described my dental work, told me my hips were about three inches out of alignment, which would pinch my sciatic nerve, asked me if I got headaches, if I had trouble sleeping, and said, You had an accident and broke your leg, you got hit from behind, and you were in a car accident.

I told her I rarely get headaches and sleep well but she was right about everything else.

I told her I lost a few teeth playing soccer, shattered my leg skiing, got hit from behind in hockey, and got hit from behind by a truck, too.

You had a concussion, she said.

Yes.

Memory loss?

Temporary, I said. It all came back except for the details about the crash. And that came back on a reiki table when I was in a course for spiritual healing.

She wanted to hear about that.

She said, You were wearing your seatbelt when the truck hit you, and it put your hips out of alignment.

Oh! I said. So that's what did it.

I told her about wearing orthotics, how they didn't work for me, and about having plantar faciitis for years, and finally healing it with medical acupuncture.

She nodded her head knowingly.

I told her I had acupuncture the night before but it didn't do much.

Don't worry, she said. I'll get you fixed up.

She coached me to give her lots of feedback as she did her work.

She lightly massaged my toes, touched other parts of my foot, and then asked if I had had bronchitis.

Yes, I said. And the last time I caught a cold, at the start of the summer ....

It was like getting hit by a truck, she said, finishing the sentence for me.

Next time you get a cold, she said, tell the doctor to listen to your lungs through your back. Once you've had bronchitis, it changes the bronchial tubes, and if the doctors catch on to that, they will prescribe different medicine, so you'll be back on your feet in a couple of days or a couple of weeks.

You're amazing, I said.

What about the pins and needles on the bottom of my feet? I asked her.

That's a circulation problem, she said. A lot of healing is happening and that often leads to circulation problems. We'll clear that up, too.

She said, as well as the hip problem, the pinched nerve which is affecting the way you walk, or don't walk, your old injuries are acting up again, and getting healed. You have a high tolerance for pain, or else you would really be feeling these things and complaining about them. The hip injury, from the car crash, the hit from behind in hockey, the broken or shattered leg, from skiing, all healed, at the time, but now they need to heal again, or some more.

She blamed it on the weather.

It has been humid, here, she said. And warm. A lot of people are going through the same thing. The weather has triggered this kind of healing.

She told me some funny stories about her life in Cold Lake, how people thought she was an aboriginal person, because of her dark skin, and then assumed she was from Haiti, when they heard her name, since Haiti is French and so is her last name.

I'm from Trinidad, she said. Trinidad and Tobago.

Why would they think I'm from Haiti? she said.

I told her that people from Haiti had a reputation for being spiritual and for alternative healing, especially voodoo, so maybe people were putting two and two together and getting the wrong answer.

Ah, she said.

I told her a bit about my background working with a Zenmaster monk at a Buddhist retreat centre as well as getting a diploma in spiritual healing, and doing some work teaching Zen meditation with massage like Reiki self-massage plus qigong, as I was a Reikimaster and a certified qigong instructor and loved doing that work but was happy teaching high school and some elementary classes, too, these days.

Math or Science, she said.

English, I said.

Oh, she said. I'd like to have your male energy in one of my reflexology workshops.

I nodded my head.

Her next client was late, so she kept working on my feet.

At the end of the session, she asked me to take a few deep breaths and hold one. She counted the seconds.

That's good, she said. People in pain generally cannot hold their breath for more than five seconds. So, you're doing well.

I held it for 27 seconds, she said.

She asked me to stand up, which I did rather smoothly, for a change, and then she asked me to put my hands up in the air, which I hadn't done for a few weeks, and then to bend backwards, which would have hurt so much I would have screamed the day before.

I told her it had hurt to stand up and take a drink from a glass or a can, for the past few weeks.

She got me to walk a bit, back and forth, in her little healing room.

I told her my balance was better, I felt loss wobbly, and I could walk without my canes for the first time in weeks.

It's a miracle, I said. You rock!

She said, You're wise to get two canes, instead of one, as it gives you better balance, and won't put you out of alignment.

She gave me a bottle of water and encouraged me to take a drink and recommended going home to have a sleep.

I told her I had a date to meet colleagues for dinner and I had little appetite recently but now I was feeling hungry and thirsty.

Good signs, she said. Sleep a bit, this afternoon, and then go out, and then get back into bed, and come back here tomorrow and the next day.

I want to go back to work on Monday, I told her.

Okay, she said. We'll fix you up over the next two days, so you can have a nice weekend, and then you'll be ready to go back to work on Monday.

She said, I programmed your feet so you could drive home, so that should go okay.

Okay, I said.

She walked me to the door and noticed I was walking much more smoothly. She walked me outside and watched me get into my car.

Look at you go, she said, jumping into that car!

I said, Thanks a million. I can't thank you enough. That was reflexology plus, and it was great. I'll look forward to seeing you tomorrow.

As I drove home, effortlessly, I reflected on the fact that we had a real good conversation. At the hospital, the doctors were distracted by other patients, who were throwing up, or yelling in pain, so you didn't feel listened to, or heard, and they didn't have much to tell you, and said it in a way you could barely understand. We had talked non-stop for over an hour and laughed a lot.

I told her my eyes were watering, on the outside, and that meant happiness. She liked learning that, she said. Tears from the inside corners of the eyes mean pain or sadness, sorrow, or grief. Tears from the outside corners mean pleasure, joy, happiness, or bliss, even ecstasy.

The last time I felt ecstatic, I was in the big comfy chair getting reflexology. The last time I felt ecstatic before that .... I had to think about for a few seconds.

When I discovered I could walk and drive well enough to get myself to the drug store and buy a couple of canes, and walk with the canes, I felt ecstatic.

A week before that, the idea of walking with a cane horrified me. But then it thrilled me.

My back pain was gone, for a while, I could walk a bit better, and I felt very hopeful that the gifted healer doing reflexology was going to get me going again and soon I would be walking the way I did before.

CATscan

Just before I left for the appointment, I got a call from the hospital, saying I had an appointment for a CAT scan on Friday. And when I got home, I got a call from the neurologist's office in Edmonton, saying I could have an appointment on Thursday.

I told them there was no way I could drive to Edmonton, or walk when I got there, so I postponed it for a week.

Okay, they said.

Why would a neurologist want you to go to Edmonton, a three hour drive away, when you couldn't drive or walk?

Did Alberta have some sort of medical transport?

I decided I would research that possibility in the morning.

I went to bed at ten but couldn't sleep, so I got up and wrote this.

It's midnight now.

Goodnight! Drive-Thru

The second session of reflexology was also very good, but quite different.

I had a good afternoon, evening, and night, after the first session, went out to dinner with a few colleagues, had an appetite, walked across a busy street twice, got home, did some work on the computer, slept in the next morning, which is unusual, had a bath, did some work for school on the computer and e-mailed it in, had lunch, and then ....

Downhill.

It was difficult walking to my car, but driving was okay, and walking down the hall at the Wellness Centre was a challenge, but the session went well.

Mecell pointed out that I came in saying my pain level was 9 out of 10 and on day 2 reported it was down to zero, and that's quite a change.

She said my hips were three inches out of alignment the first day but only half an inch the second day.

I asked her about exercises, neurologists, physiotherapy, sleep positions, and she worked away on my feet.

At one point, I noticed a tender spot in the middle of my foot, at the start of my arch, and she said, That's digestion, so she worked on it for quite some time, until my stomach rumbled, and then she suggested I have a drink of water.

I asked her if I would need a wheelchair when I went back to work on Monday, and she said, No.

But she suggested I shop for one, anyway, as you have to be fitted, she said, and at the same time she said the goal was to keep me out of a wheelchair.

At the end of the session, I held my breath for 35 seconds, instead of 25, the day before, she said, but felt wobbly when I stood up and tried to walk.

It was a challenge walking out and getting in my car.

I told her my feet had that pins and needles feeling in a smaller area, at the start, and it was gone by the end of the session.

That was a relief.

She said she did a lot of work and my body was doing a lot of healing while dealing with pain and it was tiring, so I should go home and get into bed.

You might want to watch a movie on the couch, she said, but you need to eat something first.

Promise me you will eat something, she said.

Go to a drive-thru, on your way home.

I got a grilled turkey sandwhich and some fries at Dairy Queen and carried it home.

Getting from the elevator to my door, about ten feet, was a chore, and walking across the room, once I got inside, was painful.

But I did it.

As soon as I sat down, ate, drank some water, and got on Facebook, I felt better.

I feel worried about the future, though, and wonder what it will be like when I go back to work.

Will I get fired?

Will I get re-hired?

I don't want to say anything negative, but I had a few dark thoughts, I have to admit.

On Facebook, I found an old interview with my mentor, W.O. Mitchell, from the Banff Centre archives, and I reposted it.

That made me happy.

I did the math. If he was born in 1914, as it said, he was a little older than I am now, when I studied with him at Banff. And he put in several more years there, while working at the University of Windsor, so .... I should be good to go for a few more years!

My brother is sixteen years older than me and still rockin' on.

I want to recover from this and get back in the game.

The reflexologist said I have had a rather severe injury with a great deal of dammage and I should have the CAT scan and go to the neurologist and the physiotherapist, as well as continue working with her.

Happy, I'm not.

On the other hand, I felt close to death, ready to die, years ago, and could go at any moment without feeling I hadn't written 100 books, taught for two and a half decades with tons of success, experienced a lot of love, had enough adventures, loved my life, and would not likely be coming back.

It's 2012 and I thought that a percentage of the population would survive the apocalyptic weather that has been hitting us, but now I'm not so sure.

December 21, the Mayans say. That's the day after my mother's birthday. -- What would have been her birthday.

I'll be in heaven with her, by then, I guess.

Note: As Apple says, this statement is subject to radical change.

Wheels

I got wheels, man!
The Evolution Series is the latest and the greatest, they say.

The Evolution Series, available in four sizes, are the strongest and one of the most stable walkers on the market today.

Every Evolution model uses 1-1/8" cold rolled steel tubing, the same material used in mountain bike fabrication. Strong 1" tubing is used on the handles, every fastener is stainless steel or plated against corrosion and the frame is powder coated for UV protection against fading. All plastic components are molded utilizing the toughest plastics, capable of withstanding temperatures down to –40°C.

The Evolution model offers unmatched stability for users suffering from neurological conditions such as Multiple Sclerosis and Parkinson's, in addition to arthritic and stroke conditions and users with wide gates. The stable steel frame and 400 lbs seating capacity make this walker ideal for bariatric users.

The Evolution series also include the largest baskets available anywhere.
The rigid steel frame gives this walker unmatched stability for users suffering from neurological conditions such as Multiple Sclerosis, Parkinson's and stroke conditions.
Available with standard rigid back support or flexible back straps
Wide models have a larger seating area with a bigger soft seat cushion for added comfort.
Soft seat cushion in every model.
Plastic tray or basket installed tote bag included.
Now with stronger and wider wheels for ease of use on gravel roads, wet grass, deep carpet, etc.

Three weeks ago, I hiked the 12 k trail in Cold Lake Provincial Park in 90 minutes. Three days ago, I could barely take twelve steps, I was in so much trouble. First I had intense back pain and then my legs shut down. After three consecutive one hour appointments with a reflexologist, the pain was down to zero, or one, or zero, or one, I had a lot of trouble walking, and the reflexologist phoned a new health centre, where two young nurses hooked me up with a walker.

The first time somebody said, you might need a cane, I shook my head, no, as I felt horrified. A week later, I was thrilled to be able to get around with the help of two canes.

The walker was way better than the canes. It had a seat, so you could sit down, when your legs got tired. It had a basket, so you could carry things. It was well-engineered, low-tech, and paid for by Alberta Health plus my insurance plan at work.

"We'll give it to you for six months," the home care nurse said. "And then we'll renew it over the phone."

They were incredibly helpful. One nurse even walked me out to my car, to make sure the walker fit into the back seat, and gave me the paperwork needed to get a HANDICAPPED permit to stick in my window, so I got all the best parking spots.

I made a vow to myself that I would go back there, when I got better, and do volunteer work with them, as they had been so helpful, useful, fast, even fun.

"You can get attachments for it," Charmaine said charmingly. "A cup holder, a flag holder .... All kinds of things."

I have to take the paperwork over to the registry office in the morning and I have to go to the hospital in the afternoon for a CAT scan.

Session three with the gifted reflexologist was as great as the first two. I told her I had had a rough night, couldn't sleep, worried about a lot of things, then slept during the day. She said, I'm not surprised, after what you went through yesterday.

What were you worried about.

Work, I said. Walking, getting to Edmonton for the neurology appointment, the CAT scan, losing the ability to walk, getting fired, not being able to walk well enough to work .... Stuff like that.

Wheelchairs, I added.

I told her I had been looking for a wheelchair and Walmart could order one or a drug store could sell me one. They rented chairs at $25 per day, so it was cheaper to buy one. But the guy who sold them had to fit you for one, first, and he wasn't there, the three times I phoned. And they said he would be there, or call me, and he would be there later, and it was a painful process.

She pulled out her cell phone and talked to a couple of people at the pharmacy. They told her I would have to go to the new healthcare facility at the south end of town.

They never told me that.

Mecelle, the reflexologist, called the healthcare place, but the person she was told to ask for wasn't in at the moment.

I'll drive over anyway, I said.

The home care nurse and the physiotherapy nurse were both there and came out quickly when I walked in, using two canes, and asked for one of them by name.

It was the opposite of the hospital experience. They competed with each other to see who could do more and do it faster.

Here, let me get that.

Have a seat over here.

Let me get that for you.

They asked me a few questions that were appropriate and listened to my answers, nodding their heads and smiling. They were compassionate and friendly, empathic and kind, supportive and encouraging.

I told them I was a teacher and you could tell by the look in their eyes they thought that was great.

I want to get back to work, I said. Basketball season will start soon. I've got two drama groups rehearsing plays. And I have great classes.

So, you want to get back to work, one of them said.

Monday, I said. I've got a doctor's note to be off seven work days, so I go back on Monday.

One of them shook her head.

When you're at the hospital for a CATscan, she said, talk to a doctor about getting a note so you can be off at least one more week, so you can see a physiotherapist in the next town and arrange to see the neurologist in Edmonton.

We'll give you the walker for six months, the other one said. And a sign for your car, so you can park in all the best places.

I left the new healthcare building feeling as though I had just met two angels.

The Evolution Walker folded up, the basket popped up, and it was light enough so I could lift it with one hand and, somewhat awkwardly, put it in the back seat of my car.

I drove home feeling happy.

I couldn't walk, I had to go from canes to a walker, my lower back hurt, my legs weren't working well at all, I had upcoming appointments for a CATscan and to see a neurologist, I found out there was a physiotherapist in the next town, I had been given this idea that I might not be at work on Monday, or the next week, and I might be using a walker for six months or a year.

Despite all that, I felt happy.

-- Go figure.

I phoned my brother. He said, Get down to Edmonton somehow! Good luck with the CATscan. This is incredible. You were hiking a few weeks ago and wading in the cold water of Cold Lake.

My cousin sent me a message on Facebook saying, I'll drive you to Edmonton.

My cousin.

She had needed a lot of help over the years and could no longer work. Her therapist had recently told her she should do some volunteer work, to get her out of the house and out of herself, and into the community, and so on, it would be good for her.

She was on a waiting list for moving into an assisted living facility. And she was going to help me.

Strange things happen, as I have always said. And people change.

When I couldn't sleep, the night before, I read half of the book called Aleph, by Paulo Coehlo, about a guy who stops time and travels in the past, briefly, and takes a train ride across Siberia in order to get in touch with himself, again, reconnect with his spiritual side.

I enjoyed reading Paulo Coelho's novel. It was entertaining and a little inspirational, in the first half. It felt like food for the soul. Aleph it is quite a page turner, which is not always the case of Paulo Coelho's books. He is a philosophical writer.

The plot and theme of "Aleph" can easily be related to by many. Routines wear us down. Great losses in life may dip us into an emotional abyss. We may find faith and then lose it along the way. Or regrets in our past and worries in our future can weigh us down. How can we get out of this? A pilgrimage, as suggested by the author, may have the answer we need.

In "Aleph", the 59 years old Paulo Coelho has taken a trip on the Trans-Siberian Railway to meet his readers across Russia. During his journey, a 21 years old Turkish girl named Hilal has insisted on traveling with Paulo and his team of editor, distributor, and translator. Coelho (who is married) and Hilal became physically, emotionally, and spiritually close.

Two people performed rituals together in an intimate setting. And if you buy into the concept of reincarnation, that two lifetimes ago, Coelho and Hilal were lovers, that past and present are one, I suppose it is OK to be that close. A bit confusing if you are an outsider. But say for a moment that reincarnation does exist and we do happen to meet with the same people through time and space, perhaps what Coelho and Hilal have done is beautiful. It is certainly romantic to read. I love you like a river, she said him to her. That is probably the most artistic thing a married man can say to his admirer (and lover from another lifetime).

There are quite a few memorable moments and lines in the novel. -- Here's a few:

1. It's what you do in the present that will redeem the past and thereby change the future.

2. When faced by any loss, there's no point in trying to recover what has been, it's best to take advantage of the large space that opens up before us and fill it with something new.

3. People never leave. We are always here in our past and future lives.

He says, We are travellers on a cosmic journey. Stardust, swirling and dancing in the eddies and whirlpools of infinity. Life is eternal. We have stopped for a moment to encounter each other. To meet, to love, to share. This is a precious momoent. A little parenthesis in eternity.

Reading Coehlo makes me think of lost loves, and one woman, in particular, who wants to travel across Russia by train.

Handicapped

The term handicapped was considered politically incorrect where I used to live, but it's used out west, in Alberta.

I don't like it.

Disabled, or person/people with disability/ies.

Handicapped is offensive- it's a limiting term. Challenged is just sugar coating, as is impaired or any other word that attempts to 'dance around' the subject matter.
A blind person is either blind or they have low vision. They are not visually challenged or visually impaired.

A deaf person is either deaf or hard of hearing. Hearing impaired, you might get away with calling an 80 year old.. but not a younger deaf person. The word 'Deafie' is not offensive to most culturally Deaf people.

A physically disabled person is physically disabled. In this context, it is appropriate to use mobility impaired to signify the person's limitations.
A wheelchair user is not wheelchair bound or confined to a wheelchair. You can either say that the person is a wheelchair user or you can identify them by their specific disability- x has paraplegia.

If a person is disabled due to chronic illness, chronically ill is fine, or just disabled.

A person with a cognitive disability is just that: cognitively disabled.

A person with Autism or Asperger's syndrome doesn't want to be called retarded. Calling an autistic a savant is almost never acceptable, even if they have a particular talent. A person with autism is either neurodiverse, autistic, or an 'autie' within the autism community. A person with asperger's is either neurodiverse, autistic, or an 'aspie' within the autism/asperger's community. A person who is not on the pervasive developmental disorder spectrum is usually referred to as someone who is 'neuro-typical'

Between the disabled community, there are many more 'appropriate' terms- we throw around cripple and gimp without offence, and so do many of our able-bodied friends.

This morning, I got up at 7, called the hospital at 8, confirmed my CATscan appointment for 2, drove over to the registry office for 9, got a handicapped sign for my car and a handicapped sticker to go with my licence, and got on with my handicapped day.

I drove over to my school, parked in the handicapped spot, and our head secretary, Daniele Desjardins, came out to meet me and greet me and hold the heavy door at the front of the school. She was working with Carol, her teacher for computer applications, and they took a break to talk with me about backs, legs, injuries, chiropractors, doctors, healers, reflexologists, Reiki, acupuncture, the good, the bad, the who's who, locally, personal experiences, and I told them I'd be back on Monday, but they said, I don't think so!

After I went to my classroom, to get assignments to mark, my book bag, and tidy up a bit, I ran into Fern, our custodian, also moose hunter, a former biker, and a great guy, from Quebec, and he said, Looks like you could have this for a long time.

Don't even say that, I said.

On my way home, I stopped at Shopper's Drug Mart and put a poster in my trunk, for safekeeping, but decided not to go in the store as I felt wobbly. I told Daniele the feeling in my legs was like the way you feel when you're in love. I think she liked that. It feels the way your legs do when they fall asleep, as they do, once in a while, right, but they've always woken up when you put your feet on the ground, or stomp on them, or wiggle them a little. Well, my legs haven't woken up for about three weeks.

I wanted to stop and shop for something to give Trudy and Charmaine, the angels, I mean nurses, at the healthcare centre, but my legs wouldn't cooperate. -- I would have to do that another time.

Thought I should head home and rest a bit before going to the hospital for the CATscan in the afternoon.

In two hours, I got a lot done.

Just before heading home, I went to the drive-thru and got the turkey combo again.

Each stop means getting the walker out of the car and putting it back in the car. The walk from my car to the side door is 100 steps. Another 25 takes me to the elevator. Ten more to my door. -- Home.

I feel fine, in a funny way, despite the fact nobody who sees me thinks I'll be back at work for a long time. While I was at school, I sat in my wheelie chair, to see how well I could get around.

-- No problem.

-- Unless there was a problem.

Thank god I'm at a good, small, school, where the students love me, with one or two exceptions, so there aren't stairs or long halls or teenagers who would push you over or little kids who would do something to your chair so you fell on the floor so the bad kids could kick you in the back.

Thank God for that.

-- What a thought!

Heres's what the Service Alberta website has to say about handicapped parking:

Parking Placards for Persons With Disabilities

IMPORTANT INFORMATION: If parking privileges are abused, your parking placard may be cancelled and your parking privileges revoked. Do not allow anyone else to use your parking placard and do not park in marked stalls unless you are exiting the vehicle.

There is a ZERO tolerance on parking placard abuse.

A parking placard and/or disabled licence plate enables those with the greatest needs to use specially designated parking facilities. The placards are issued to individuals who provide proof of eligibility under this program. An applicant requesting plates must either have the vehicle(s) registered in the name of the person with the disability or be a joint owner on the vehicle registration.

To apply for a placard, an application form (pdf): must be completed by your physician, physiotherapist or occupational therapist and submitted to a registry agent office.

To qualify, an individual must be unable to walk more than 50 metres. There are three types of placards:

Blue placards, which can be issued to those with long-term disabilities and are valid for five years

Blue placards, which can be issued to those with Permanent disabilities that can be self-declared every five years without the medical personnel signature

Red placards, which can be issued to those with disabilities that are temporary in nature, but will affect them for a period of three to twelve months (Note: A visitor in need may use their valid (non-expired) parking placard from their home jurisdiction during their vacation or visit to Alberta.)

A visitor in need, without an existing placard, may apply for a temporary red placard for the duration of their visit. Visitors will be considered for a parking placard and should contact a registry agent for more information.

-- My gimp sticker isn't red, it's blue.

"Live the present moment wisely and earnestly." ~Buddha

Of all the funny things flying around Facebook these days, the one that caught my eye today is by Mr. Positive, Jerry Criss, that says, This is a hold-up. Give me all your fear, your doubt, your pain, and your shame, and nobody will get hurt.

I also noticed something cool about global warming, some hot news from the Arctic, some good news from a group called 350.org, that said 350.org
Good (?) news: it appears that the Arctic has stopped melting, at least for 2012.

The new record for lowest sea ice cover in human history is 3.41 million square KM - 760,000 square KM below the last record.

This is some of the scariest news of 2012. But click LIKE if you're ready to organize to stop the melt for good.
a global movement that's inspiring the world to rise to the challenge of the climate crisis.

350.org is a grassroots global movement working to unite the world around solutions to the climate crisis—the solutions that science & justice demand. 350 is you, and the thousands of people like you who take it upon themselves to inspire your communities to action.

It was one of the groups I watched in the summer, when I was writing a thriller about global warming. The novel is actually about a guy writing a novel about global warming. This guy goes to a thriller writing workshop and comes to the conclusion, again and again, Why are we writing thrillers when we are living in one.

After CTscan, went to emerg, to get a note for another week off work, and had a better communicator, who said my x-rays showed a little deterioration in 2 discs, not 1, plus some sign of arthritis, and he concludes my spine is putting pressure on my sciatic nerve, but they will send the x-ray and ct scan to the neurologist to interpret and he'd discuss it with me on Friday. and I should go to physiotherapy AFTER that. AND he said there's a new doctor coming to town in November so call the clinic and get on his list.

The Morning After

There's got to be a morning after, they say.

There's got to be a morning after
If we can hold on through the night
We have a chance to find the sunshine
Let's keep on looking for the light

"The Morning After" , also known as "The Song from 'The Poseidon Adventure'" is a song first released in May 1973. It was the first success for singer Maureen McGovern and used as the love theme for the film The Poseidon Adventure. Many singers have covered it, including Joan Baez, my favourite.

Okay, I just Googled Joan Baez and The Morning After and looked for it on Google and couldn't find it, so I must have made it up.

If she hasn't already covered it, Joan Baez should sing "There's got to be a morning after ...."

Bob Dylan has written some songs about coal, but not about oil, I noticed while I was looking for that information. There are a lot of songs about coal and coal mining, but not many songs about oil.

But this chapter is not about Bob Dylan and Joan Baez or songs and music or oil and coal, it's about the morning after my first day with a walker.

At the end of my first day with a walker, I was exhausted but happy. I got a lot done but by the end of the day I was very tired. I was out in the morning and in the afternoon, had a CATscan, went to Emerg, found out I had deteriorating disc disease with a touch of arthritis, and would be off work for another week, probably longer, talked to people at work, talked my supply teacher into coming back by promising to plan lessons and do the marking, and by the time I got home and checked e-mail, it was too late to sit in the sun, as a few people had suggested.

It was a beautiful early autumn day in Northern Alberta, warm and sunny, with clear blue skies, not a cloud to be seen, and that made my first day with a walker a lot better. -- If I had to pull that walker out of my car and walk slowly into the hospital and the school and then load it in again on a dark, cold, rainy day, it might have been a different story.

How about the winter?

Instead of reading some more of the novel I started, Aleph, by Coehla, I did a little bit of writing and then I watched a movie.

On my computer, I watched The Other Guys, on Netflix, a cop movie starring Will Ferrell, which I have now seen half a dozen times. What can I say? It's funny!

I woke up at four in the morning, as usual. That has been my habit for decades. I check e-mail, write, wander, eat, and go back to bed. Usually.

In my e-mail was an announcement from Ezine @rticles, awarding me a Certificate of Achievement for Expert Authors. There was another award, from Bleacher Report, saying they were giving me a gold medal for an article I wrote about the future of the NHL.

The first thing I saw on Facebook was a bright little poster from a group called Attitude Of Gratitude that said May You Wake With Gratitude.

And I did wake with gratitude, I realized. Even though I was stiff and sore, and had to walk very slowly with the walker, it was gratitude that was on my mind. I was grateful for the walker, which meant I could get around a bit, since the canes weren't working for me anymore. I felt appreciative of the fact the little city where I live has a good hospital, a CATscan, and I met a good doctor who communicated clearly, even if I did not like what he had to say.

I was impressed with the CT unit, the people working there, the technology. I got there early, since I was moving slowly, and they took me in as soon as I got there.

I said a little prayer expressing gratitude for my little school and the people who work there, and the benefits program of our school board, and the health care program of Alberta.

To be sure, I felt a little cranky because I was stiff and sore, but I was glad it was because I now had a walker and that meant I could still get around and do things.

It reminded me of an old story about a couple with a crowded little house. The complained about how crowded their house was after they had a baby. And the advice they got was unusual but they followed it. First they were advised to get a chicken and keep it in the house. And then a cat and a dog. And a cow. And so on. And one day they were advised to get rid of all the animals, so it was just the couple with their baby and all of a sudden the little house seemed to be enormous, to them.

How happy was I when I could walk, before I had back pain.

I had other worries, other complaints.

How happy am I now that my back pain has gone away, except for the odd jab or stab, now and then. How happy am I that this walker will let me look after myself and get things done, like going to the grocery story and the neurologist in Edmonton.

I woke up thinking that maybe I should take a little trip sometime this week before going to Edmonton, to find that old Chinese guy who does acupuncture in Lloydminister.

He is famous, around here; a legend. I keep hearing stories about him.

At first, the stories sounded sexist, racist, and ageist. My acupuncture person was a young, white, woman from Vancouver, who had studied in China. Why did so many people appear to be prejudiced and believe acupuncture had to be done by an old Chinese man?

But now I'm thinking .... If he is a legend and I am in pain with severely limited mobility, why not go see the living legend?

Why not go to Martyr's Shrine? Why not go to Lourdes?

Night after night, I pray to the Lord.

And you know what they say: The Lord works in mysterious ways.

Sometimes you wake up at four in the morning with an odd idea in your head, like .... Maybe I should go to Lloydminister to get acupuncture from the old Chinese guy.

I Googled him. I found him on the website of the College of Physicians and Surgeons of Alberta. It said that said Dr. Shiu Kee Yen had a general practice in Lloyd, spoke English, Chinese, and Japanese. And it said this doctor is only accepting patients that meet the following criteria: ACUPUNCTURE, BACK PAIN, ORTHOPEDIC MEDICINE, Family members of existing patients, Friends of existing patients.

His phone number was (780) 875-2221.

Was this the legend?

How many Chinese men in Lloydminister could I find who did acupuncture? -- One.

I decided to call him in the morning.

Second Breakfast

There is a voice that doesn't use words - Listen!
~ Rumi

Up at ten, again, for a second breakfast, like a hobbit.

A little back pain, a little stiffness, my feet feel funny. Aside from that, I'd say a good night's sleep had a healing effect.

Even so, my plan for the day is to follow doctor's orders and "take it easy".

What's new on Facebook?

Only 60 days until The Rise Of The Guardians.

I should create something for my Grade 3s about that.

A friend of a Facebook friend posted an odd quote about death:

Death is awkward.

Despite all the planning, all the talking,

no-one really knows what to do.

The hush is stunning.

I have not made any plans to kill myself or to die. I've planned to teach until the retirement age, 65 or 67, another decade or so, and then go back to Ontario, build a log cabin in the Zen Forest, live there for another couple of decades, and when it is time to go, I'd like to be cremated and have my ashes poured into the wild lake at the north end, where visitors can see them, as a reminder that life is temporary, it's dust to dust.

The monk and Zen master at the Zen Forest likes to say, "This is the truth: We are born, we grow older, health fades, and we die. So .... wake up!"

It's a simple philosophy, not very comforting or encouraging, but inspiring, nonetheless. I think it's Forest Zen, the simplest form of Zen, that you get at the Zen Forest.

Facebook, this morning, also says, Happiness is the highest level of success.

That was posted by a place called Love Tree, which is a store in Huntsville, the town in Muskoka where, today, they are holding the awards ceremony for the Muskoka Novel Marathon. -- My pet project for a decade or so.

Get a bunch of writers together for a long week, write as much as you can in three days, and do it as a fundraiser for local literacy. Writers have said, This is the best writing event of my life. And, This is the best life event for this writer.

Get them together again in the fall when the autumn leaves are at their peak and have an awards event.

You can check out Ontario Tourism's fall colour report to find out where to go to see the amazing display of changing leaves. Right now, the MNMers are lucky, because they are close to Algonquin Park and the report says not many leaves have fallen, orange and red are the dominant colours, and about 70 per cent of the leaves have gone from green to one glorious colour or another.

At Arrowhead, even closer to town, the leaves are yellow and orange, but not quite as many have changed colours.

You can follow the fall colour report as the change makes its way south and see some of it or the best of it or a lot of it. -- I did that every year I lived in Ontario.

I've exchanged messages with some of the writers who have traveled to Huntsville today and are getting together for lunch, before the big event, and complaining because it is raining. "Just a drizzle," says Dawn Huddlestone, a freelance writer who went to the University of Calgary. "Don't believe the weather reports."

Overcast with rain. High of 15C. Breezy. Winds from the WSW at 20 to 25 km/h. Chance of rain 90% with rainfall amounts near 5.8 mm possible.

That's their forecast for today.

Partly cloudy. High of 23C. Winds from the SSE at 10 to 15 km/h. Clear this evening.

That's our forecast for today, in Cold Lake, Alberta.

Before I left Toronto for Alberta, I thought the weather in Northern Alberta would be a lot worse than the weather in central Ontario, where I spent half my life. But it's not.

Muskoka gets a lot of snow. It gets more and more freezing rain. Summers are hotter, over 30, so the water in the lakes is like soup.

Cold Lake had a few cold days last winter when the schools were closed but everybody thought they should have just waited an hour or so. And there was just enough snow for skiing.

I'm happy the novel marathoners are getting together and having a good time at the little event I invented over a dozen years ago. I'm happy I live in a place that has better weather.

If happiness is the highest level of success, as they say, then I should feel successful.

The Love Tree posts something like that every day and I usually "like" it. The store on the main street of that hilly town is cool. They sell books and things for people who want to develop their spiritual lives. There's a new yoga studio downstairs. They sell my books and I've done a few workshops there on Zen meditation, massage, and qigong.

Today, for a change, they have posted a few things. One of them is a colourful mini poster that says, Always believe that something wonderful is about to happen.

I "liked" that, too.

Yesterday, when I found out I had to be off work for another week, I was devastated, for a minute. I was looking forward to going back to work. I miss my students. I love teaching. However, I cycled through the stages of grief quickly and came up with a great idea. -- I could use the time to make something wonderful happen.

While the novel marathoners were getting together in Muskoka, I could write. And while my students had a supply teacher, I could write. And in the autumn or at the start of winter, before Christmas, during book season, I could launch a few books at the bookstore in town.

A colleague and friend of mine said she would set up a book signing at Lotsabooks and discussed it with them, without mentioning any of it to me, and they said, Good idea.

Over the summer, I wrote a kids book, a teen book, and started an adult book.

The YA is about Santa Claus. -- Last year, some of my younger students said I looked like Santa Claus, so I wrote a book about my connections with the big elf.

The teen book was about a summer at an international arts camp near Niagara Falls and a kid who gets incredibly inspired to turn his life around.

The adult book is a thriller set in the oil patch.

I also have another book ready to go. I put together the short stories from my first three books, published by Oberon, with the rave reviews those books got across the country, so I could have a book of my collected stories.

So, I could have a book launch for five books at once, including this one. -- That might get a little bit of attention, maybe an article in the local paper, The Cold Lake Sun, and sell a few books. It could be fun. It could count as 'community involvement', at work. And it would be a good time to announce that I'm running a writing workshop in town for Portage College.

If I am.

If I can walk.

If I am back in the game.

I believe that something wonderful is going to happen!

The Wellness Centre

Last summer, I went to the Wellness Centre, since they had a big sign outside, by the highway, saying, New Patients Welcome, but I was told they were not taking new patients right away. You had to rent a video, or leave a twenty dollar deposit for a dvd, and then go to a group meeting, and after all that you could book an appointment.

Although I took the dvd home with me, I did not feel as though I could wait around for an appointment, so I went elsewhere.

My back was sore. It was summer, and I had a little time to look after things like that, so I wanted to get moving on it.

A month and a half later, I watched the video.

Since that time, I've had four adjustments by a different chiropractor, four visits to emerg at the hospital, four acupuncture treatments, and four kinds of drugs, plus a needle, x-rays, and a CT scan, or CATscan, and now I walk only with the help of a walker.

The big back pain is gone, but I still get a few little stabs when I walk. The problem is no longer my back. It's my legs. They don't work.

They aren't numb. But they don't cooperate. It is as though I have almost no control over them.

Muscles contract and spaz. Sometimes I go weak in the knees. It feels as though the ligaments in my ankles are stretched and useless.

Starting tomorrow, I'm off work for the third week.

Last night, after a day of bed rest, following the doctors' orders, I felt like watching a movie, so I loaded the dvd into my computer.

Dr. Kopala's Health Report #3, made in 2009, explains chiropracty a bit and his approach to it. He tells some stories, asks some questions, shows some slides, and leaves you with the message that he doesn't do miracle cures, but he sees miracles every day, and he doesn't want clients to get one adjustment, he wants them to commit to regular treatments combined with diet and exercise, so they can return to the highest level of health. And he doesn't want to put down the mainstream medical establishment, which he says is great at dealing with emergencies, but he does think hospitals should be called sickness centres instead of healthcare facilities, because of their approach to sickness and wellness.

There is an information session on Tuesday, apparently, so I guess I will go, since I don't have to be at work. I'll be the big guy with the big walker at the back of the room.

It's all about nerves, Doctor Darren says. -- That's what people call him.

If there is something pressing on a nerve, blocking it, health problems can result.

By adjusting bones, pressure can be taken off nerves, so the nervous system can function smoothly, and then the body heals itself.

That's the theory he subscribes to.

And there is a promise of gentleness. Some chiropractors are rough, and they used to be rougher, he says, but now there are gentle techniques and instruments suitable for babies and old ladies, I mean elderly women.

So, sign me up, I say. Adjust my bones, starting with my hip bones. I'll commit to regular sessions as well as a nutritional diet and exercise.

I've always been athletic and active and I've tried to eat right.

I played hockey, skied, hiked, coached soccer and basketball, walked, jogged, coached swimming, and love to swim. I've joined gyms and lifted weights and tried all the machines. I enjoy all of that stuff.

I've been a vegetarian for decades, or a chickenetarian, until last year, when I moved to Alberta, and decided to eat beef, as well. Steak and eggs was my favourite meal, for a while. After eating roast beef and big beef ribs and steak, I felt stronger than I had in years.

Three years ago, suffering from plantar fasciitis, I could not ski, any more, or hike, so I did yoga for a year, and I loved that, too.

Working as a teacher means you are on your feet on cement floors a lot, but when classes were over I walked in the woods or went swimming, and I spent weekends at the Zen Forest, with the monk and Zen master, building roads and cabins, doing landscaping work and forest management, as well as leading tours through the property and teaching Zen meditation, massage, and qigong.

An old friend from my undergrad days suggests another theory: We're getting older, she says.

The truth is, I've been in denial about getting old. I like the sayings we've all heard the past few years about 50 being the new 40 and 60 being the new 50, and on and on.

I don't feel 57. I feel 17.

When I was 17, I was the fullback on my high school soccer team, and we won our league, district, region, and went to the all-Ontarios. I was the captain of my hockey team. The next year, I would play for the Bracebridge Bears and win the provincial championship, and I would be named Athlete Of The Year, at school.

Right now, I'm a guy who can't walk across a small room, without a walker.

But I'm planning to make a comeback.

How long will it take?

Three to five days, they said, at first. Then seven more days. Then another week and possibly more. That's what doctors said.

When I met the reflexologist they call the goddess, she asked me what I wanted, and I said, I want a miracle. In one hour, I want to walk well again.

She said she'd have me walking back to work in a week.

Later, she said she'd make a couple of phone calls to help me get a walker.

So it goes, as Kurt Vonnegut Jr. used to say.

Informally, people at work take a quick look at me and say, You're going to be off for a long time, buddy.

Dr. Internet, or all the info I can find with Google, says six weeks for most cases like mine, or three months, and then it's time for surgery.

My brother had surgery, twice, and it didn't work. First they removed a disc. Later they fused his spine. Both times, the pain remained the same.

He can walk, but he has to take morphine for the pain.

My other brother has gone to a chiropractor regularly for decades and spent this summer swimming in the warm waters of Georgian Bay just about every day.

My sister moved to Australia, where she walks a lot and swims sometimes, and has other aches and pains, but never complains about her back.

Our parents, long gone, complained about each other, but never their backs.

A lot of North American men get back pains. Millions of them.

What are we going to do?

A nutritional diet, good exercise, and regular chiropractic treatments, the doc says.

Meditation, qigong, energy from nature, the Zenmaster monk says.

Qigong and etheric healing, my new age guru says.

And what do I say?

All of Thee above.

The Love Tree says:
Nothing is inevitable.
But all things are possible.

A Sunny Sunday

Cold Lake is quiet on Sunday mornings, so I waited until then to make my second outing with my walker. On Saturday, I slept and wrote and slept and wrote, and that was about it. I slept a lot and wrote a little. That's what I did on Sunday, too.

In the morning, I made it to Shoppers Drug Mart, just up the highway, a few minutes, parked in the gimp spot, unloaded my walker, and dropped a hundred bucks on supplies so I won't starve to death any time soon. I got some cleaning supplies, too, and some B vitamins. B complex and B12. -- Good for rebuilding the nerves, I read somewhere, online, and I always take them anyway, as they re-energize me.

By the time I got home, I was a tired, sweaty, guy and I didn't feel like doing much more than having a shower and lying in the sun like a big cat.

I watched a comedy, Dinner With Schmucks, on Netflix, and then phoned my brother.

He's obsessing over prostrate cancer, so found some stats on the Canadian Cancer Association website. Ten in one hundred thousand men die from prostrate cancer, it says.

I had that checked out just before I moved out west. The doc said, a lot of men die with prostate cancer, but not many men die of prostate cancer.

Then I phoned my friend and booking agent, Jackie, in Parry Sound, who has gone through everything I'm going through. She has had chronic arthritis for a long time, dealt with a lot of pain, has had hip replacements and knee replacements, learned a lot about healthcare, drugs, mainstream and alternative therapies, and lived to tell the story. And she recently became a grandmother, a little sooner than she planned, but it appears to be working out well.

On Saturday, she drove over to Huntsville for the Wrap-Up party and awards presentations for the Muskoka Novel Marathon. They raised fifteen thousand dollars for literacy.

I started this fundraiser over a decade ago and it just grows and grows.

Many people have told me that the MNM would make a lot more money if it was for cancer than for literacy.

Maybe we should have one for prostate cancer.

How about spinal injuries, nerve damage, or lumbar pain.

Have a bunch of writers sitting around hurting their backs while raising money for lower back pain.

-- Maybe not.

In the evening, I finished reading Aleph, by Paolo Coehlo, and found out Aleph means qui, or chi, and also a moment when everything becomes one thing. There is a lot of description of time travel, in the second half of the book, as the author and narrator go back to the Inquisition and a few other times and places.

The novel takes readers on an adventurous journey that spans all 9,288 kilometers of the Trans-Siberian railroad from Moscow to Vladivostok and a parallel mystical journey that transports its narrator through space and time. Coelho presents himself as a pilgrim seeking to regain his spiritual fire, much like Santiago, the main character of his runaway bestseller The Alchemist.

Coelho's books have sold more than 130 million copies and have been translated into 72 languages. Besides The Alchemist, his international bestsellers include Eleven Minutes, The Pilgrimage, and many other books whose characters grapple with seemingly simple spiritual themes: light and darkness, good and evil, temptation and redemption.

In Aleph, Coelho writes in the first person, as a character and a man wrestling with his own spiritual stagnation. He's 59 years old, a successful but discontented writer, a man who has traveled all over the world and become widely acclaimed for his work. However, he can't shake the sense that he's lost-and deeply dissatisfied. Through the leadership of his mentor "J.," Coelho comes to the conclusion that he must "change everything and move forward," but he doesn't quite know what that means until he reads an article about Chinese bamboo.

He commits to a journey through Russia to meet with his readers and to realize his lifelong dream of traveling the entire length of the Trans-Siberian railroad. He arrives in Moscow to begin the journey and meets more than what he's expecting in a young woman and violin virtuoso named Hilal, who shows up at his hotel and announces that she's there to accompany him for the duration of the trip.

Hilal won't take no for an answer, Coelho lets her tag along, and together the two embark on a journey of much greater significance. By sharing deeply profound moments lost in "the Aleph," Coelho begins to realize that Hilal can unlock the secrets of a parallel spiritual universe in which he had betrayed her five hundred years earlier.

In the language of technical mathematics, Aleph means "the number that contains all numbers," but in this story it represents a mystical voyage wherein two people experience a spiritual unleashing that has a profound impact on their present lives.

Coelho's tendency to describe spiritual concepts in simple terms sometimes borders on cliché. "A life without cause is a life without effect," he repeats, along with other pithy sayings such as "Life is the train, not the station." These sayings take on greater depth, however, as this story's narrator travels back in time and returns to the present with experiences that give them new meaning. `

The tension in Aleph builds as the train nears its destination at Vladivostok, the final stop on the Trans-Siberian railroad. The narrator Coelho and Hilal have become entangled in a spiritual web that must be broken if they are to continue on in their separate lives.

Through their delicate negotiations, readers will come to understand the interconnectedness of people throughout time and find inspiration in this story of love and forgiveness.

It was the best book I've read in a long time.

Because of Aleph, and my back pain, the loss of the use of my legs, being off work, and so on, I've been thinking of my brothers a lot. They have both had big issues with sore backs.

My oldest brother hurt his back when he was 25. He was a university student driving a beer truck for the summer and he had to stop for a tree that had fallen on the road. He was a strong guy, so he picked the tree up and pushed it off the road. However, he heard some odd sounds in his back and discovered he could not straighten up.

So, he went to a chiropractor. He told me they were pretty rough in those days. It was like getting tackled in a football game, he told me, or getting hit from behind in hockey.

Our other brother had a lot of back pain and had a disc removed. He was still in pain, so he had the spinal fusion operation.

He was told, beforehand, that the operations could take away his pain or do nothing at all or leave him without the use of his legs.

What do you do?

He had the surgery and came out of it the same way he went in: with a lot of pain.

My legs are feeling a little different after two full days of rest, and two weeks with a lot of rest. I still stagger like Frankenstein, but it is not as spastic as it used to be. I'm getting a stab in the lumbar region and occasionally in the butt, when I stand up, but it isn't much.

I have two plans for the coming week. One is to follow doctors' orders and rest flat on my back. The other is to track down a legendary local acupuncturist in Lloydminister, sign up with a different chiropractor, and go back to the reflexologist known as the goddess.

I'll sleep on it.

The Alchemist

Well, after sleeping away another day, but getting lesson plans in for my new supply teacher, and marking assignments online for my students at school, I had a little vision. Okay, maybe it was just a dream. This Christmas, I want to be healthy and ambulatory and take a trip to Ontario.

If I don't move, and just sleep in the sun, I feel great.

My feet feel funny, like they are full of tar sand, or something. And they are a little sore.

If I could walk, I would be at work in the morning!

Oh, Brother

I am a man of constant sorrow
I've seen trouble all my day.
I bid farewell to old Muskoka
The place where I was born and raised.
(The place where he was born and raised)
-- traditional folksong, made famous in the movie called O Brother, Where Art Thou? by a band called The Soggy Bottom Boys, featuring George Clooney.

After three weeks of back pain and the loss of the ability to walk, plus three days flat on my back in bed, following doctors' orders, things are shifting.

A bit.

My feet feel like they are full of beans, like bean bags.

My legs feel alright and I can even take a few steps. If I use my walker, I can take hundreds of steps. I get stabs of pain in the small of my back and in my butt. That's it.

I'm not dying. -- I'm going to beat this thing.

The reflexologist says, You have a high threshold of pain, but that's not always a good thing.

No? I say.

It means you can walk around for years with the kind of pain that would send someone else to the hospital. And that's not a compliment, because sometimes it makes sense to pay attention to your aches and pains and get them looked after.

Did this really hit you out of the blue at the end of August? she said. You've never had a pain in the back before, or trouble walking?

I didn't answer right away. I thought about it.

I've never had trouble walking or a pain like that, I said. But I have had little back pains, especially this year, and sometimes I walked stiffly, now that you mention it.

For several years, I had to deal with the pain of plantar fasciitis. Before that, I had a little back pain, but I got orthotics, and the pain went away. After plantar fasciitis, I went to a very good orthotics person, but the shoe inserts didn't do it for me anymore.

No, the reflexologist said. It wouldn't. -- Your back was too far out of whack. It had to be brought back into alignment.

When were you in that car crash, hit from behind by a big truck.

That was several years ago, I said. My car was totaled, but I walked away without a scratch.

No scratches, she said, but in a crash like that, if you were wearing your seatbelt, it can twist you hips out of place. -- Were you wearing a seatbelt.

Of course, I said. It's the law. But I didn't feel any back pain back then.

Your body goes into shock, she reminded me, whenever there is an injury that causes pain you can't handle.

Did you ever get hit again, after that?

Now that you mention it, I said, there were a few times when somebody jumped on me, playfully, but caught me by surprise, and I thought of going to the hospital, but the pain went away after a night flat on my back.

She shook her head.

You have a high threshold of pain, she said. But it hasn't always worked for you.

My brothers are polar opposites in this regard. One pays attention to every ache and pain and likes to tell you about it in great detail. The other is the strong, silent type. Or used to be.

I've been thinking about my other brother a lot, lately. Not that he is ever too far from my mind.

He is six years older than me and I looked up to him as though he was a god when I was a little kid. But that phase did not last too long.

After our father left, or was chased out, however that happened, when I was eight years old, my brother tried to take his place as the centre of attention, the wild card, we had to tiptoe around. The raging lunatic, we called him. He got into drugs in high school, but never got caught, and started drinking after he dropped out of university.

My brother became an alcoholic, my father was an alcoholic, after World War Two, and my grandfather was an alcoholic, after World War One.

My father's father fought in the trenches and came back home shell-shocked as well as cross-addicted to alcohol and nicotene, booze and fags, he said, and with TB, as well.

War is hell, he said.

My father fought in England, part of the top secret radar project, and saw some action at the end of the war, joining the Canadian Forces that liberated The Netherlands, cleared out concentration camps, and chased Jerry back to Germany. And he came back with undiagnosed PTSD as well as cross-addicted to alcohol and nicotene, or booze and fags, and with TB as well.

War sure as hell is hell, he said.

My oldest brother was in army cadets, my other brother was in air cadets, and they said I was in space cadets. When I was in Wolf Cubs, I also had a huge interest in rockets, NASA, the space race, from Sputnik to the first man on the moon.

It wasn't until I was in my early thirties that I looked at my life and realized that I came from a long line of alcoholics. I knew that alcohol affected me differently than my friends, and I didn't like it, so I stayed away from it. Oh, I tried different things in high school, but it was easy to see it was not for me.

My brother, on the other hand, got addicted.

When I learned what it meant to be an enabler, to enable someone else to drink, to allow and even encourage someone with a drinking problem to keep drinking, well, I did what they recommend in ACA. I cut him off.

You say, If you don't stop drinking, we are cutting you off. We won't help you, we won't even talk to you.

It is hard to say something like that to your own brother.

On the other hand, if he was a bully from the time you were a little kid, and it got worse the more he drank, it was not so hard to put an end to all of that.

Decades of abuse of all kinds were cut out of my life.

That's the ACA way.

Adult Children of Alcoholics.

My sister and brother stayed in touch with him, but grew more distant as he drank more and more. He was a violent drunk, and nobody wanted to be around him when he went to that dark place.

He was a bully at the best of times.

That's not true.

I've been told that he can be very entertaining, when drinking, tell jokes, make puns, be funny, be the life of the party. I've never seen that side of him, but I've been told it exists.

He was the athlete of the year and best all 'round boy in Grade Thirteen of high school in Ontario and he became an elementary teacher, specializing in phys. ed.

He had a back pain that became so bad he had to set up chairs in the corners of his classroom so he could lurch from one to the other as he made his way around the room to work with different students.

He did not notice the pain for a while because he drank a lot. And when the pain got bad, he drank some more.

Eventually, he went to a doctor, who sent him to a specialist, who recommended surgery. First he had a disc removed. It was damaged and causing pain, so they took it out. Later, he had the spinal fusion operation. -- Neither operation ended the pain.

He was given morphine. And he drank a lot.

I have missed my brother for about twenty five years.

In that time, my sister has cut him off and so have all our other relatives, except our oldest brother. He visits, sometimes, and then complains about it to my sister and I. He says, Why do I go there, to get abused, over and over again?

Well, lately I have been feeling his pain more directly, as my back pain is something like his.

He never lost the ability to walk.

But he has had chronic severe back pain for decades.

Imagine that.

That twinge a lot of people feel, from time to time, when they lift something the wrong way, or twist into a certain position, taken to the maximum, and never going away.

That's a lot of pain.

True, my brother caused a lot of pain. He was a pain in the neck at home and then he started playing rugger, or rugby, and turned out to be quite good at it. He was named the captain of a central Ontario team that toured England and Wales and never lost a game.

Their secret? They got my brother angry at the start of the game, pushing him around and calling him names, and then pointed him at the biggest guy on the other team. In the first play of the game, my brother would break the big guy's ribs.

Not only did it take him out of the game, it also made the rest of the team tremble in their boots, because he had been taken out and would no longer be there to protect them.

In game after game, the ambulance was called, and came, and carried the biggest rugger player on the other team to hospital, with broken ribs.

Some call it karma, some say payback, and others say, God will get you for that!

I took a different path, exploring one spiritual path after another, but always returning to my roots.

And now I wish I could call my brother, commiserate, get the benefit of his experience, teach him what I have learned about Zen meditation, massage, and qigong, to help him deal with his pain.

I've been planning to go to the store and get him a card, one of those greeting cards for brothers, and add a little note, then send it to him. But it is such a chore to load and unload the walker and make my way through a store.

I will find a way to do it.

But what will I write in the card?

Dear Rob,

Oh, brother! I've missed you. More importantly, I am writing to say I feel your pain, bro. Literally, as well as metaphorically. I'm sorry you've had to deal with decades of back pain, severe and chronic, without any relief from surgery, and the mixed blessing of prescription drugs like morphine, which come with problems of their own. Or so I've heard.

Anyway, this is to say .... Hang in there, man. I'm thinking about you. Everybody loves you and hopes for the best for you. In case you aren't getting that message, I thought I'd send you this card as a token of the way I feel.
Love,
Your brother.

Love?

Let's not go crazy!

I do love my brother. I love him like a brother.

Forgiveness

"In your light I learn how to love. In your beauty, how to make poems. You dance inside my chest where no-one sees you, but sometimes I do, and that sight becomes this art."
— Rumi

While we are in the forgiving mood, there is somebody else I should send a little letter to.
Maybe a card would do.
Oddly enough, it's another redhead with freckles.
Aside from that, she has nothing in common with my brother.
Believe me.
And she looks nothing like my sister, either, despite the red hair and freckles.
She is a woman my age who I met when we were in our late forties and we had a great relationship as we turned fifty and entered our early fifties.
It was one of the greatest times of my life.
It ended the way a lot of relationships do, and I don't want to go into that here, I just want to express forgiveness.
In the movie called Eat, Pray, Love, based on the novel of the same name, by Elizabeth Gilchrist, the actress Julia Roberts, playing the main role, tells her ex, "Whenever you think of me, just send me light and love, and drop it."
My ex looked a lot like Julia Roberts.
But that's beside the point.
The point is, I want her to know that I have thought of her often, and I have always sent her light and love, and tried to drop it.
I find her unforgettable.
I wish her well in her life without me.
Last night I saw a funny movie called Dinner With Schmucks, and Steve Carrol, in the role of a gifted goofball, says, You CAN die of a broken heart. I know. It almost happened to me.
His buddy in the movie has a bad back.
I'm getting off track.
Anyway, I think I'll get another one of those corny greeting cards, and send it to this woman, just to say, in the words of Sarah McLaughlin, I will remember you, and in the words of Julia Roberts as Elizabeth Gilchrist, whenever I think of you, I send you light and love, so you must be bathing in it by now.

I found her on Facebook and sent her this little message:

In your light I learned how to love.
In your beauty, how to make poems.
You dance inside my chest,

where no one sees you.
- Rumi

I re-wrote Rumi, a little bit, to make it fit.
      We both liked Rumi a lot and, when we were together, felt connected to the famous old Sufi poet.

Hmmmm .... Anybody else out there who I have cut off and feel a need to express some empathy or send some love?
      -- Nope!

Well, actually, I could send that letter to a few heart-breakers!
      This goes out to all the heart-breakers: I forgive you!
      I'm sure you don't care! But I forgive you anyway.

One Month

It is about one month since the day I got zapped with that big pain in the back, and I think I may have turned a corner, or at least entered a curve.

When I follow doctor's orders and rest flat on my back, I feel fine.

After a lot of rest, I can walk a few steps. After walking or just standing for five minutes, I feel wobbly, tired, and feel as though I am going to fall over.

My feet have that needles and pins feeling.

Yesterday, I couldn't move my big toe, on either foot, by itself, but today I can.

My legs feel fine. -- If only they could hold me up!

Sometimes I get a little shot of back pain.

If I sit in an odd position, my legs fall asleep even more.

That doesn't sound so bad. Does it.

Could I go to work?

Could I drive myself to Edmonton?

Could I do that in a few days?

This is Tuesday, September 4th.

I have an appointment with Mecell the magic reflexologist tonight. And an appointment with her chiropractor tomorrow.

Should I postpone the appointment with the chiropractor?

What if I get hurt again?

What if he can put my bones back in place and help my nervous system?

The most positive thing I can say is that after getting worse for a month, I may be slightly better.

30 Quotes on Healing

My ex sent me a funny quote about healing that she found on a wrapper for cough drops. It said, "You've survived tougher. You can do it and you know it. March forward." - Wrapper for Halls No Sugar Added Cough Drops.

It inspired me to look for better quotes, some quotable quotes, and I found a good list online on the Psychology Today website. They suggest putting on some healing music while reading and re-reading these famous and not-so-famous quotes.

Published on February 17, 2011 by Stephanie Sarkis, Ph.D. in Here, There, and Everywhere

Healing, like forgiveness, is a gradual process. Give yourself all the time you need... and play some good music while you're at it. Music is a big part of the healing process, as you will see from some of these famous quotes on healing. - Stephanie Sarkis

"All healing is first a healing of the heart." - Carl Townsend

"Although the world is full of suffering, it is also full of the overcoming of it." - Helen Keller

"Healing takes courage, and we all have courage, even if we have to dig a little to find it." - Tori Amos

"Eventually you will come to understand that love heals everything, and love is all there is." - Gary Zukav

"I'm touched by the idea that when we do things that are useful and helpful - collecting these shards of spirituality - that we may be helping to bring about a healing." - Leonard Nimoy

"Healing is a matter of time, but it is sometimes also a matter of opportunity." - Hippocrates

"I think music in itself is healing. It's an explosive expression of humanity." - Billy Joel

"There is something beautiful about all scars of whatever nature. A scar means the hurt is over, the wound is closed and healed, done with." - Harry Crews

"Healing may not be so much about getting better, as about letting go of everything that isn't you - all of the expectations, all of the beliefs - and becoming who you are." - Rachel Naomi Remen

"Humor is healing." - Brad Garrett

"The practice of forgiveness is our most important contribution to the healing of the world." - Marianne Williamson

"Live your life from your heart. Share from your heart. And your story will touch and heal people's souls." - Melody Beattie

"Our sorrows and wounds are healed only when we touch them with compassion." - Buddha

"When you hear the music ringin' in your soul
And you feel it in your heart and it grows and grows
And it comes from the backstreet rock & roll and the healing has begun" - Van Morrison, "And the Healing Has Begun"

"If there's no breaking then there's no healing, and if there's no healing then there's no learning." - One Tree Hill

"Healing does not mean going back to the way things were before, but rather allowing what is now to move us closer to God." - Ram Dass

"The wish for healing has always been half of health." - Lucius Annaeus Seneca

"I've experienced several different healing methodologies over the years - counseling, self-help seminars, and I've read a lot - but none of them will work unless you really want to heal." - Lindsay Wagner

"It's when we start working together that the real healing takes place... it's when we start spilling our sweat, and not our blood." - David Hume

"Of one thing I am certain, the body is not the measure of healing, peace is the measure." - Phyllis McGinley

"Healing yourself is connected with healing others." - Yoko Ono

"There are so many ways to heal. Arrogance may have a place in technology, but not in healing. I need to get out of my own way if I am to heal." - Anne Wilson Schaef

"For me, singing sad songs often has a way of healing a situation. It gets the hurt out in the open into the light, out of the darkness." - Reba McEntire

"Gracious words are a honeycomb, sweet to the soul and healing to the bones." - Proverbs 16:23-25

"Love one another and help others to rise to the higher levels, simply by pouring out love. Love is infectious and the greatest healing energy." - Sai Baba

"Music is such a great healing balm and a great way to forget your troubles." - Ricky Skaggs

"The soul is healed by being with children." - Fyodor Dostoyevsky

"It is reasonable to expect the doctor to recognize that science may not have all the answers to problems of health and healing." - Norman Cousins

"Healing requires from us to stop struggling, but to enjoy life more and endure it less." - Darina Stoyanova

"When I stand before thee at the day's end, thou shalt see my scars and know that I had my wounds and also my healing." - Rabindranath Tagore

A "Rocky" Moment

Try bearing this in mind today: All beings want to avoid suffering, but keep stumbling into suffering.

After a reflexology treatment, I felt so much better, I took a letter to the post office. When you can't walk, little jobs like that become big obstacles. The thought of parking, unpacking the walker, staggering to the post office, and back, knowing how much energy that takes, feels intimidating. But not after reflexology treatment #3.

I parked in front of the post office, left my walker in the back seat, grabbed my canes, and climbed up the three stairs leading to the post office. The curb, the sidewalk, slightly bigger than average stairs .... No problem.

I popped my letter into the mail slot.

Income tax.

And then I turned around for the little return trip. But first, I stopped, looked around, and there it was: I heard the theme music from the movie Rocky and thought of that scene where the boxer in training runs up the long flight of stairs in Philly and does his little dance, arms up, fists clenched, in celebration and anticipation.

The next day, I had a meeting with Dr. Darren, who runs The Wellness Centre. I tried to get in to see him in the spring, for acupuncture, but he was taking patients only at certain hours during the day, and I was working, so it never happened. But now I had made it through his video and intake interview and he gave me a very thorough exam.

He was a pharmacist who became a chiropractor and is now studying Chinese medicine, so his exam was unusually thorough. We talked, he heard my story, he told me what he had to offer, then he got me to stand and stretch, checked my reflexes, and then he looked at my tongue and at my ears. -- Not in my ears. AT my ears.

It appeared as though he got the most information just be looking at my ears.

He said, Your balance is just about zero, you have a lot of internal heat and a busy mind.

In the end, he said, As a chiropractor, I wouldn't touch your back with a ten foot pole.

However, your case intrigues me, as it is unusual to lose the ability to walk without getting a lot of pain at the same time.

He said he wanted to get a copy of the neurologist's report and then, perhaps, try fire cupping.

Not acupuncture, this time, he said. Fire cupping.

He also said he wanted to rule out MS and ALS.

-- That scared me quite a bit.

He said the neurologist would likely tell me to take another couple of weeks off work.

After the examination, I went for reflexology, one more time, and we talked about past lives in Egypt. I told my story about my Rocky moment. She thanked me for the book I gave her, as a present, the last trip, and about a friend of her who had recently mellowed after getting involved in Buddhism, even though she seemed to be the most unlikely Buddhist. I told her my brother had recently run into my childhood best friend, and he had mellowed, too, apparently. He coached a juvenile hockey team and set a new record for penalties in minutes, which is really saying something, because juvenile hockey in Ontario is pretty tough. But now, my brother said, the guy was calm, polite, even self-effacing.

Things happen, I said. People change. -- That's what I've been writing about since the beginning.

She said she wanted to take my writing workshop. She had a book in mind for a long time.

After that treatment, I felt good, again, despite the scare of MS and ALS, so I drove over to the drug store and found a card for my brother, and a few other things.

I went home and wrote in the card, as I had planned, put it in an envelope, and then looked for his address. Google and the internet could not find him.

I sent my sibs an e, to see if they could provide a mailing address.

That was a long day with no sleeping in the sun like a big cat. The phone rang at 8:30, with a question from my supply teacher, at work, and there was one thing after another, including the two appointments and the little trip to the drug store. The day went by quickly until 8:30 at night.

I played a little Scrabble, online, won a few games, and decided to go to bed.

But I couldn't sleep.

My back hurt a bit.

I took some ibuprofen and wrote up lesson plans for my five classes, marked everything my students had handed in, returned it, online, and then went to bed.

First, I Googled MS and ALS, and decided they did not describe me.

According to Wikipedia, Multiple sclerosis (MS), also known as "disseminated sclerosis" or "encephalomyelitis disseminata", is an inflammatory disease in which the fatty myelin sheaths around the axons of the brain and spinal cord are damaged, leading to demyelination and scarring as well as a broad spectrum of signs and symptoms. Disease onset usually occurs in young adults, and it is more common in women. It has a prevalence that ranges between 2 and 150 per 100,000 .

Amyotrophic lateral sclerosis (ALS) – also referred to as motor neurone disease in some British Commonwealth countries and as Lou Gehrig's disease in North America – is a debilitating disease with varied etiology characterized by rapidly progressive weakness, muscle atrophy and fasciculations, muscle spasticity, difficulty speaking (dysarthria), difficulty swallowing (dysphagia), and decline in breathing ability. ALS is the most common of the five motor neuron diseases.

Where no family history of the disease is present – i.e., in around 95% of cases – there is no known cause for ALS. Potential causes for which there is inconclusive evidence includes head trauma, military service, and participation in contact sports. Many other potential causes, including chemical exposure, electromagnetic field exposure, occupation, physical trauma, and electric shock, have been investigated but without consistent findings.

There is a known hereditary factor in familial ALS (FALS), where the condition is known to run in families.

There was no ALS in my family. I have no trouble speaking, swallowing, or breathing.

In the morning, my girlfriend phoned to say, What about sacroiliitis?

Oh my aching sacroiliac, I said. Isn't that a line from a tv show from the Sixties?

I don't know, she said.

She's several years younger than me.

According to the website of the Mayo clinic, Sacroiliitis (sa-kro-il-ee-EYE-tis) is an inflammation of one or both of your sacroiliac joints, which connect your lower spine and pelvis. Sacroiliitis can cause pain in your buttocks or lower back, and may even extend down one or both legs. The pain associated with sacroiliitis is often aggravated by prolonged standing or by stair climbing.

Sacroiliitis has been linked to a group of diseases called spondyloarthropathies, which cause inflammatory arthritis of the spine. Sacroiliitis can be difficult to diagnose, because it may be mistaken for other causes of low back pain. Treatment of sacroiliitis may involve a combination of rest, physical therapy and medications.

The night before, we chatted on Facebook, and she told me she was praying for me.

She said, Give yourself lots of time with travel and remember chargers etc., and your computer. Sleep well Moto. I will be praying for you.

And then she added, Not like I have any special pull with the Almighty, but it's what I do.

Frankly, I think she does have special pull with the Almighty, as she says.

She has had a full life, as they say, with a lot of experience praying for family and many others.

But that's her story, and I'll let her tell it.

The Edmonton Clinic

Opening in late 2012, the Edmonton Clinic will support the emphasis on putting patients at the centre of care at the University Hospital site by consolidating specialized, outpatient clinical care and streamlining access to numerous services and specialists.
Unique to Alberta, this nine-storey, 670,000 square foot facility will bring multi-disciplinary teams of specialists together under one roof including surgical, medical, family and seniors' clinics, as well as orthopedic and neurosciences clinics. This integration will provide seamless ambulatory care for up to one million patient visits per year.

The Edmonton Clinic will allow rural Albertans to stay in their communities while keeping them connected, via new technologies, to the latest advances in diagnosis and treatment. In addition, this tremendous new out-patient facility will enhance the partnership between Alberta Health Services and the University of Alberta by strategically aligning state-of-the-art technology, education and research.

The Edmonton Clinic, formerly known as the Health Sciences Ambulatory Learning Centre (HSALC), will be an interdisciplinary health science facility located on the University of Alberta main campus, in Edmonton, Alberta, Canada.

The estimated cost was just under one billion. The Edmonton Clinic Health Academy opened its doors for classes in fall 2011, and the Edmonton Clinic South is slated for completion Fall 2012. The joint venture between the University of Alberta and Alberta Health Services is funded by the Province of Alberta.

The pain ended a month ago. The issue is walking
Lyn and Shirley invited me for a month or whatever it takes
Why did you drive all that way instead of staying home and resting? Were you tired of your own company or what?
I had to go to Edmonton to a neurology specialist in metro emerg
I couldn't do it solo do l and s offered.
They have been GREAT.

Went to emerg, told them the specialist didn't have xrays and catscan, yet, and the nurse SWORE.
Bullshit! she said. He should learn how to use the internet and AlbertaHealthCa.!!! she said." "WTF! How could he dismiss anything without these? I don't get it."
He's a cowboy! she said. "I make my own decisions!" he said. "I do this all day, every day, and I've seen thousands."

First "his" nurse got family history and the basics. his questions were more specific about me, pain, what happened, and walking. And then he did a very thorough and unusual physical exam. not unusual in a bad way. Just things nobody else has done when examining me. head to toe. he asked me to stick my tongue into my cheek and push as he pushed his fingers against it. he got me to push my feet in every direction and test my strength. he said, you have a lot of upper body strength, it looks like your knees are back, and your foot muscles are not working well, in some areas."

Sounds like he was checking checking body strength and mobility. Normal neuro stuff. Haven't heard the tongue test before. Muscular and neural test I guess.

How are you feeling now Marty? My sister asks, in an e-mail from Australia. Any better? I hope so!

No. I can't walk. drove to lac la biche, stayed with 2 cousins, they drove me to edmonton in the a.m., pushed my wheelchair through the fancy new Edmonton Clinic, Lyn stayed with me during meeting with God, i mean the neurological emergency doctor who thinks he's God, and it was a good thing because after he said cancer, ALS, MS, AIDS, syphillis, my brain froze for a second, but then he said You DONT have cancer, ALS, MS, we'll do blood work for other stuff.

His suggestion was more tests, EMG and NCS in Edmonton, physiotherapy to work on balance, std then ltd for disability, and bloodwork.

So, I'm off indefinitely.

Well, they ruled out cancer, ALS, MS, and anything spinal -- said it has nothing to do with chiropractic adjustment

After Edmonton

New beginnings are
often disguised as
painful endings
-- Lao TZu

When I got back home, after going to the Edmonton Clinic, I went to emerg, as directed,
but they said, We don't do lab work on weekends. So I went on Monday.

Emerg said they would sign a form for school, if I took it in.

I told them the doc in Edmonton clinic didn't have x-rays or CATscan and the doc
here got mad and said, BULLSHIT! She said, Tell him he has to learn how to use the
internet and the Alberta Health website.

Uncle John

My cousins still talk about their father a good deal. He was my uncle. They call him John,
rather than dad or father or daddy or anything like that. I call him Uncle John.

He was blind and very strong-minded.

He wasn't born blind. He went blind slowly over many years.

First he list his peripheral vision, then he had tunnel vision, and eventually he was
stone blind.

As Shakespeare said, he was sand blind, then gravel blind, and then stone blind.

There is a genetic disease in my extended family, and I missed it "by that much".

Corrhoideremia is carried by women and shows up in men.

There's a lot of folklore attached to this genetic disease in my family, and I have
written about it elsewhere, but this is not the place to get into all that Irish mythology.

John Macdonald and his brother Will both inherited the disease from their mother.
And my mother and her sisters did not pass it on.

The boys in my family grew up worrying about it, but worried about it a little less
each year as the tell-tale signs did not show up. -- We kept checking our peripheral vision.
John was a cook, a chef, a baker, a woodworker, a father, a husband, an uncle, and a
character. He had a big personality, a strong presence, an indominatable will.

My cousin just told me that when he went for his annual medical check-up, when he was
old, the doctor said, Why are you walking?

And he said, What?

The doc said, Well, all the muscles in your legs have atrophied. -- You should not be
walking.

And he said, "Oh, I'm walking, alright."

He had no intention of sitting in a wheelchair.

If he could have stopped the blindness by willpower, he would have had 20/20
vision.

My brothers and I saw him as a father figure. Our father left our family when I was eight years old and my brothers were fourteen and twenty-two. Before that, he was a wild card, to say the least. He came back from World War Two with TB and was cross-addicted to alcohol and nicotene. Our uncle John was stationed in Iceland with Canadian troops for most of the war and that was where he learned how to cook meals for hundreds of people at a time.

My oldest brother lived with him for four or five years while he went to university and figured out what he wanted to do with his life. Our other brother spent two full summers with him, cooking at a summer camp. I spent one summer with him at that camp, and that was enough for me.

My brothers loved him. So did I. But in a different way.

My father left when I was eight and I worked with my Uncle John when I was sixteen, so I had half a lifetime without a father, by then. I had some good, strong, role models outside of our family. My hockey coaches and the male teachers at our elementary school were great with me, I thought. They had the right blend of interest and disinterest, closeness and distance, fatherly concern and cheerleader.

Uncle John was tough and straight but he had a huge sense of humour.

His humour was sometimes used as a weapon, but it was never aimed at me.

He often included me in his funny plots of revenge and vindication.

We had a blast, working together, that summer.

I loved every minute of it.

My cousin was at the camp that summer and every summer of her childhood and adolescence. What a way to spend your summers. It cost thousands of dollars to send your kid to that camp. Uncle John's two youngest daughters went to camp for free every year their father was the head cook.

I wasn't just the second cook, Uncle John said. I was his eyes.
I enjoyed that job but I have to admit there were a few times when I failed and his eyes worked. He ran his kitchen and most of his life by memory: he memorized where everything in his world was located and insisted that the people in his life always put things where they belonged. Precisely. Accurately. Unfailingly.

So I did. And when I couldn't find something, looking at boxes and bottles, reading labels, there were times when he said, Well then let me have a look.

He would go to the place where he knew the item should be, pretend to read the label, and say, Here it is, Mart; right where it's supposed to be.

Was he pretending to read, or could he see a tiny area right in front of him? It was hard to tell.

He loved puns, which Freud said was a sign of hostility, but he used them good humouredly. On a trip to town, I found a goofy placque that said, "I'm not a slow cook, I'm not a fast cook, I'm a half-fast cook," and I bought it for him. -- He laughed for days. Uncle John had five girls, three daughters close to the same age, my cousin who was my age, and one more, to look after him in old age.

He made his youngest daughter this strange deal: Stick around to look after me and your mom in old age, and I'll give you the house.

And she did it.

My cousin my age had a different experience with John than her sisters or my brothers.

My experience was something like hers, but I only spent one summer with him. When I was sixteen, I was old for my age, everybody said, and I was longing to be free, itching to be independent, aching to be out of school and on my own, with nobody to tell me what to do. It was the end of the Sixties, the start of the Seventies, and I liked the hippies, so my uncle and I were not the greatest fit in the world. But, even so, we got along well, laughed a lot, and had a good summer.

He hurt his back one year, working on his roof, and it affected his legs.

What was a blind man doing on his roof? you might ask.

The answer is: He was fixing it. Doing some roofing work.

So, the summer I worked with him, he was blind and walked the walk of a man with a sore back and not much sensation in his toes. -- Maybe it was his feet, or his legs. He would never tell you.

Never complain, never explain: that was his motto.

And now two of his daughters are helping me get through a period in which I have back pain and walking problems. They pushed me in a wheelchair when we went to a fancy clinic in the city.

I look something like my grandfather, now, the way he looked when I was a kid, and his son, my uncle, looked a lot like him, so it must be a strange experience for my cousins, like getting your father back, for a while, decades younger, but now they are decades older, so we are the same age he was when we were teenagers.

My cousin told me the story about his legs, how his leg muscles atrophied, in old age, and how the doctor told him he shouldn't be walking, but he refused to get in a wheelchair. He had a strong mind, she said, and it was mind over matter. He walked with willpower when he had no leg muscles left.

I tried that approach.

And it didn't get me very far.

Not the first time.

But I started walking. -- Just ten steps, the first time, without canes or a walker. Then twenty steps. Then one hundred, two hundred, five hundred, and more every day.

I've seen several doctors and a few alternative healers, but I think it's the spirit of Uncle John that is getting me going, getting me walking, once again.

The Comeback

This is Day Six of The Comeback. I just walked 100 steps without the walker or crutches or holding onto walls or counters or anything. First I visualized it, then I just did it.

One hundred steps. It's a short walk for a man but I giant leap ahead for they guy who was getting attached to his walker.

On Day Five of The Comeback, I got up, got out of bed, dragged a comb across my head, as The Beatles said, got dressed, packed, picked up my suitcase with one hand and leaned on my crutches, clutched in my other hand, and walked from my cousin's bedroom to the front door. One hundred steps.

I drove one hundred kilometres, used my walker from the parking lot at the hospital into Emerg and back, went home, did laundry, went to the Base for more quarters, stopped at the drug store to get 'thank you' cards, came home again, watched The Secret Twice, and walked from the couch to bed, without crutches or anything.

In total, I walked around 1000 steps, almost all with canes.

On Day Four, I went to the Edmonton Clinic, was driven by my cousins, wheeled around the new hospital facility by my cousin, sat in a car for four hours, stayed up late to yak but didn't walk much.

On Day Three, I drove 100 km, used canes to walk into a restaurant and back, and later went for a walk, outside, around a circular driveway, with canes, around 1000 steps.

On Day Two of The Comeback, I went to The Wellness Centre, met the guy who runs the place, had my fifth session with the gifted reflexologist, and then ....

Let's look at this again.

Day One was really last Sunday. On Saturday and Sunday, I rested a lot. On Saturday, I did not feel so hot but on Sunday my spirits lifted, my back felt find, my hips weren't wobbly, my knees were more like knees, but I was still walking with my walker.

On Monday, I showed off a little bit for the reflexologist, hurrying up the hallway.

Let's strive for accuracy.

This turn-around started with the first session with the reflexologist, which was last Tuesday. I went three days in a row. On Thursday, I had that "Rocky" moment, leaving the walker in the car, using canes to climb three steps to the post office to mail a letter.

Let's call that Day One.

Friday, Saturday, Sunday, I rested a lot.

Monday, I showed off a bit for the reflexologist, hurrying down the hallway.

Tuesday I met Dr. Darren at the Wellness Centre and had my fifth session with the reflexologist. After that, I felt like going to the drug store and getting an empathy card for my brother.

Thursday I drove to my cousins' place.

Friday was the trip to the Edmonton Clinic.

Yesterday I walked 1,000 steps with canes, and did laundry with the walker.

This morning, I walked 100 steps in the morning, getting breakfast.

So, let's call this Day 9 of The Comeback.

After three weeks of going downhill, first staggering and holding on to things, and then a week using two canes, and a week with the walker, about 21 days, I've had 9 days in a row, feeling better and better.

Call me The Comeback Kid.

Okay, call me The Comeback King.

The Rocky of Teachers.

I feel like watching Rocky.

I feel like taking five or so trip to the garbage disposal, outside, to get rid of the big green bags of garbage that have accumulated in my place!

Also, this morning, I had an idea that might work, a plot that might serve many purposes.

I want to thank my cousins. I want to see my brother get some work. I'd like my brother to bring out some of my stuff.

My cousins want my brother to do some painting for them. He would like to do it, but said he needed some help with his trip out here.

He thought he would fly.

I wonder if he would drive.

A plane ticket costs about six hundred bucks.

What if I gave him six hundred to drive out here, bring me my Reiki table, which is worth six hundred bucks, and then he did the painting for my cousins.

They would feed him and look after him for a few days or a week or however long it took to do the painting, and pay him something, which would cover his trip back.

-- We aren't millionaires, so we have to think of things like this!

-- Maybe it would make sense to just let it be, as The Beatles said.

"Let it be, let it be, let it be, let it be. Speaking words of wisdom: Let it be, let it be."

In the movie called Dinner For Schmucks, they misquote John Lennon, on purpose, saying, "They say that I'm a dreamer, but I'm not."

There are several more funnies like that in the movie. They say Jesus wrote the Bible. They claim the Wright brothers said plywood is lighter than air. And Sir Francis Bacon helped the Earl of Sandwhich create the BLT. The movie ends with with another funny misquote: The mind is a terrible thing."

That movie made me laugh when I was feeling down.

Maybe I'll watch some comeback movies to help inspire my comeback.

The Secret was inspirational, last night, and I might watch it again.

What else?

How about From Here To Eternity, with Frank Sinatra.

Anastasia, with Ingrid Bergman.

Ulee's Gold, with Peter Fonda.

Iron Man.
The Wrestler, with Mickey Rourke.
How about Rocky!
Hoosiers, Major League, the other Rocky movies.
I've never seen Caddyshack ....

Positive Affirmations

Love Tree posted this: How empowering to become a witness/observer of this third dimensional reality. Recognizing that all conflict and beauty in the outside world is a direct reflection of myself... an opportunity to learn, love and undo. I release myself from being a victim. ~Mantra...

'As anger arises and the unexpected occurs, I remain a calm witness knowing this too shall pass.' ~Joshua David Stone

Zenda, a friend of mine in Vancouver, who I met in the Zen Forest, had this to say, chatting online, via Facebook:

OK - I can totally relate because I've had almost the same problem. Louise Hay says the foot problem is about fear of the future and moving forward, the back problem has to do with lacking support and the nerves are related to communication. I believe the Holistic approach to healing works well, i.e. mental, physical, emotional, spiritual - they're all connected and it doesn't matter which level you address, all levels will be affected. You are taking care of yourself right now and that's great. Keep up the good work!!

While I was chatting online with Zenda, I noticed a posting in my newsfeed on Facebook that was all about positive affirmations. It took me to a website, www.vitalaffirmations.com, and an article explaining how affirmations work, with several free samples.

It said Positive Affirmations work. This page explains how and why positive affirmations can be used to manifest your needs and bring positive and permanent change to your life. You will also learn how to Create and supercharge your own affirmations.

Dr. Ho's Decompression Belt

At four in the morning, on Monday morning, I woke up, feeling hungry, but remembered I was fasting for a blood test at the hospital, so I walked over to my computer to check e-mail and Facebook.

A friend in Toronto had posted pictures from Nuit Blanche, the all-night arts event with installations all over the downtown area. It looked very cool -- but I thought you would have to do a lot of walking!

Somehow that made me think of sending an e to my ex about the Dr. Ho Decompression Belt. She had sent me an e saying she lost a couple of months to back pain, recently and it had changed her life, or the way she lives it, to avoid that kind of pain.

The Attitude of Gratitude

Our lives are a direct relation to how we view it. Everything we think becomes our reality. As Brian Tracy likes to say, "We are what we think about most of the time." There are many people surrounded by negativity that start to believe that their life has limited value and potential, because someone told them so. The only person who truly creates value and purpose in your life is you. It doesn't matter what anyone says or does to you, because you determine your reality. You have the choice to listen to allow their insidious opinions to form your opinions if you choose to. Having faith and believing in yourself has more power and strength than any negativity thrown your way. Remember what we learned in math class. A positive plus a positive equals a positive. A positive plus a negative equals a positive, a negative plus a negative equals a negative. If you choose to see the positives in your life, there isn't anything than can bring you down.

The Attitude of Gratitude is a Facebook page that goes with a website of the same name and both are filled with positive pictures and sayings. There's an update from them every day in my Facebook newsfeed. And yesterday it was good to get one because I lost the attitude of gratitude in the afternoon.

The morning was spent at the hospital, again, from nine thirty in the morning until one o'clock or so. First, I had to fast for ten hours, then they took some blood, at the lab, got me to drink a chubby little bottle of some gunk, wait two hours, and then they took more blood. And after that, I was feeling hungry, so I checked out the hospital cafeteria, had an egg sandwich on brown bread, some pumpkin bread, and a V8. I sat by myself and looked out the window into a courtyard with trees and picnic tables but no people as it was cold and spitting rain. The worst part was the two hour wait in the lab waiting room as sad looking people came and went, getting blood work done, and a TV was on, and hard to ignore.

I stopped watching TV over a dozen years ago, so when I see one, these days, I find it mesmerizing.

CTV's all-news channel was on. And you know what that means. Nothing but bad news 24 hours a day.

Barbara Ann Scott, Canada's Olympic, gold-winning, skater died at age 84.

Omar Khadr was flown to Canada by an American military plane and sent to Millhaven prison.

Justin Bieber threw up on stage but did not stop his show.

Allegations about kickbacks from construction projects hit the mayor's office in Montreal.

Jack Layton's widow Olivia Chow says she accepts the apology of a Conservative MP who said Thomas Mulcair hastened her husband's death. Rob Anders, a backbench MP representing Calgary West, suggested in an interview that Mulcair pushed Layton to work too hard during the last federal election.

Investigators in Longlac, Ont., a community near Thunder Bay, say they received a report on Sunday of a suspicious male in distress, and identified the suspect as Leblanc.

I lost my attitude of gratitude.

When I got home, I got some great news: My cousin Lyn in Lac La Biche sent me an e saying that her sister, Shirley, found me a doctor taking new patients, and I had an appointment for October 22nd.

At the hospital, the receptionist asked me if I had a family doctor, as usual, and I said, "Trying." And she said, "Me, too." She added that she was new to the area and had been told that getting a family doctor around here was like finding gold.

Later in the afternoon, my cousin sent me another "e" saying her sister found me a physiotherapist, too, and I had an appointment in Lac La Biche on October 19.

My cousins told me that the situation is even worse in Fort MacMurray, two hours north of them. Waiting times in Emerg at the hospital can be four to seven hours long, so some people drive down to Lac La Biche, about two hours, and go to the Emerg at the hospital there, because it's faster, despite the long trip.

I thanked my cousins but, really, it felt kind of flat. I phoned the specialist's nurse at the Edmonton Clinic to give them the message that my x-rays and CTscans were online, and that my employer said they needed a letter from the specialist in order for me to get Short Term Disability. But all I got was an answering machine that said they took calls between 9 and 5 from Monday to Friday. I called on Monday at 2:00 in the afternoon. But all I got was their answering machine. -- I left a message.

My phone rang a few times and I talked to the head secretary at work and to our Science teacher, who told me she felt as though she was coming down with something and was getting work ready for the next day, just in case.

I got a nice note from my new supply teacher, saying that my students all missed me a lot.

That cheered me up!

The other good news of the day was that I walked without canes or the walker twice as far as the day before. And walking with the walker from the parking lot to the hospital and through the hospital was no problem.

However, by the end of the afternoon, around 3:30, I was "done". My back was sore, I felt tired, but not exhausted, and I felt very frustrated with the medical system of Alberta.

My friend Jackie in Ontario chatted online with me and I told her about my frustrations.

She said, In Ontario, you would be lucky to get an appointment to see a doc or a physio in six months.

She lives in Parry Sound.

What happened to the medical system in Alberta and Ontario. Canada is famous for its medical system.

To cheer myself up, I put on the Dr. Ho decompression belt and I watched The Secret, again, but stopped it after ten or fifteen minutes when somebody said that listening to music and singing along is a great way to turn your feelings around. Instead of watching The Secret, I watched some videos on Youtube. Walk Off The Earth, my new favourite, were in the news for a cover of a Bob Dylan song. I watched that one and a dozen others by the same band. I sang along with their megahit, Somebody That I Used To Know, and noticed it was up to well over 136,400,000 views. -- Incredible.

I found a few new videos by the band, including one I really liked called PayPhone. I cut and pasted and posted PayPhone on my Facebook page.
That made me feel happy again.

Tired but happy, I went to bed early, around nine o'clock, and fell asleep fast.

I was wondering why I hadn't heard from my friend Jackie in Edmonton and then I got a chat message on Facebook from her saying that the daughter of a friend of hers had died.

I sent her a note saying, "Sorry" and pasted a link to the Bob Dylan song called Death Is Not The End.

When you're sad and when you're lonely
And you haven't got a friend
Just remember that death is not the end
And all that you held sacred
Falls down and does not mend
Just remember that death is not the end
Not the end, not the end
Just remember that death is not the end

I fell asleep thinking that Day 10 of The Comeback felt like a step back but I turned that around with the help of music and singing and I realized that walking more, getting a doctor and a physiotherapist, even if they were two hours away, by car, was like finding gold.

Day 11 Of The Comeback

It has been over a month since I got that big back pain but this is day 11 of my comeback. The bad news is that my back pain has come back. The good news is that I read the info on the Dr. Ho belt and it says you can wear it all the time. And right now I am trying that. It's seven thirty in the morning and I'm still tired.

Today I am supposed to see the reflexologist again.

My ears are bugging me.

I have this thought in my head: What if something else happens .... Say I catch cold or something.

Getting around with the walker and dealing with doctors around that feels like just about all I can do.

Well, I've been doing lesson plans and marking for work, too.

I just put the belt on, tightened it up, and .... We'll see how it goes!

Meanwhile, on Facebook, I got an update from Sue Kenney, from South Muskoka, who I met at a few of my novel marathons. She had walked the Camino -- barefoot -- and was now walking The Bruce Trail the same way.

She said The Bruce Trail was a lot more challenging.

Sue Kenney posted to Explore The Bruce: I just finished walking over 400 kms on the Bruce Trail, mostly barefoot. Incredible journey. Most gorgeous trails I've ever walked in the world. Even heard/saw a rattlesnake (about 20 inches from my foot) and then saw another one! I walked for a cause, a foundation to support artists. If anyone has questions about walking BAREFOOT feel free to ask. You can find pics, videos, and stories about my experiences on www.indiegogo.com/visiononewalk.

She walked 400 km of The Bruce Trail, from the Bruce Peninsula to Collingwood, barefoot.

I've walked that section of the Bruce Trail, but I was wearing hiking boots.

I'd love to walk it again!

I'd love to walk The Camino, the famous spiritual hiking trail across Spain, someday. -- Next summer!

This morning I walked from my bed to my couch, but it hurt my back so much I had to sit down right away.

It's not the Bruce Trail.

It's the Marty Trail!

On the website called Explore The Bruce, I read that the Grey/Bruce area, where I used to leave, had a hard year, starting with an extreme weather event in the springtime. There was an odd combination of high and low temperatures in March and April, in the area, and that did a lot of damage to the early budding apple trees.

Generally, apple trees don't grow that far north; however, the Beaver Valley has a microclimate of its own, so apple trees can be grown. But this year, for example, an apple farm that usually harvests 12,000 bushels of apples this year harvested just 12 bushels.

Every fall the artists of the Autumn Leaves Studio Tour invite the public into their homes and studios. This year on September 28th - 30th explore the diversity of our seasoned artists and the artworks of a number of first time studios. The range of mediums include; glass, watercolours, pottery and ceramics, acrylics and oils, fiber art, wood, preserves, print and metal. Discover the talents of 51 artists exhibiting at 23 studio locations. Attend the 2012 tour and make it a weekend get-away by using our Advertiser links to plan your stay. You'll be warmed by the experience of country roads, inspiring studios and this amazing network of Canadian Art.

Always held the weekend before Thanksgiving and always FREE.
http://www.autumnleavesstudiotour.ca/

Other exciting things on my list for today: Go to Staples, get my health cards laminated, and get photocopies of the x-ray, CT scan, and specialist reports.

At eight thirty, wearing the Dr. Ho belt, I walked to the kitchen and back, got myself a bowl of cereal, and now I am feeling hopeful again because it didn't hurt my back!

An hour earlier, walking that far hurt quite a lot.

-- I'm thinking about going back to work soon!

-- I DO get carried away!

-- My walk is still zombie-like, but not as spastic as a month ago.

Well, at the very least, maybe I will leave my walker in the car, when I go to see the reflexologist, and walk into her office. -- That will make her squeal with delight!

My little plan worked! Mecelle walked by me and said "Hello" as she arrived and got ready for work. But then she came back with a funny look on her face. She said, "Where's your ... walker ... or your ... canes?"

I had a big grin on my face and she smiled, too.

"In the car," I said calmly.

Just a minute, she said. I'll be right with you.

She got her room ready for me and came back to get me. I was standing up, waiting for her. She broke into a great big grin and said, "Okay! Well, you made my day! I just got here but I'm ready to go home now! That's enough for one day! You can walk again!"

I asked her if she'd like to dance.

We waltzed into her room in the clinic.

As she worked on my feet, doing reflexology, I told her the story of my trip to Edmonton, how great my cousins were, what the specialist was like, going to Emerg when I got back, taking the lab test, the frustration of finding a doctor and getting into physiotherapy, and how my cousin came through for me -- bigtime -- with a doctor's appointment and a physio appointment on the 19th and 22nd.

Alright! she said. The doctor will be able to get your Short Term Disability activated and I don't think you'll need the Long Term because you will be back at work.

You might want to take your walker with you, she said, in case you get tired, but I think you will be back there before too long.

I'm glad you aren't going to physio right away, she said. It can be exhausting. -- I'll help get you ready for it.

While we were talking about healing, the arts, books, and so on, Mecell asked me if I would be willing to teach Zen meditation at the Wellness Centre.

I gave her one of my books and she was showing it to her friends and they all said they'd like to try Zen meditation with me. She suggested a Saturday at the Wellness Centre.

Sure, I said. I'd love to.

And, I added, tell your friends you don't have to sit in some weird position to meditate. -- That scares some people off. Let them know you can sit on a chair to meditate.

Oh, she said. Good. -- We have lots of chairs here!

She walked me out, when the hour and a half session was over, and watched me walk to my car and climb in.

She smiled broadly and gave me the "thumbs up".

I drove home but decided I was up for a bit more so I drove over to Staples to get copies of my medical documents, and to get my health cards laminated, and I took one cane but I was able to walk in and around the store and back to the car without using it.

I'm still wearing the Dr. Ho belt.

My walk, by the time I got home, and left my walker by the door, so I could stroll over to the couch, was a little wobbly, as though my hips were not connected quite right, or something. But I was happy I had just walked about 1,000 steps!

Day 12 Of The Comeback

After reflexology, yesterday, and Staples, I drove home and went to bed. At first, I thought I would sleep for an hour and then get up to go out and meet my colleagues for wing night, as we do every Tuesday night. The high school teachers at my school get together to talk about school and related issues for an hour or two at whatever restaurant has the best deal on wings. Sometimes we go to OJs, sometimes it's the place on the base, and once in a while we go to BP. On Monday, Julie phoned me to see how I was doing and to tell me wing night was on.

After sleeping for a couple of hours, I looked at the clock and decided rest would be better than going out.

In the evening, I moved from the bedroom to the livingroom, but was flat on my back, again. I watched a great movie called A Thousand Words, with Eddie Murphy as a high rolling book agent who rips off a big guru and has to pay for it, seriously, when a bodhi tree appears in his fancy back yard and he realizes that every time he talks the tree loses leaves. The tree drops one leaf for every word he says and the guru tells him that when all the leaves are gone, so will his life. He will be dead.

I watched it twice, with The Secret in the middle, plus some Scrabble time with an old friend down south, chatting online with Jackie, back east, and a phone call from my former girlfriend in Edmonton, which was interrupted by a call from one of the teachers at my school, phoning to discuss work and offer any kind of help I needed. She was a former nurse, then a librarian, now working as an EA. We talked for quite a while and I thought that was great.

I couldn't sleep, so I watched A Thousand Words, or dozed while it was playing, and finally fell asleep, after exchanging a few e-mails, at three in the morning.

-- Very unusual. -- I've been keeping fairly regular hours.

Everybody said the same thing: Don't push it, don't try to come back too fast, keep following doctors' orders, rest a lot ....

A Thousand Words was a very good movie, I thought, like Click, with Adam Sandler.

Spoiler alert!

With his life falling apart and the tree running out of leaves, Jack goes to Dr. Sinja -- the fictional guru -- and asks how to end the curse.

The guru tells him he has talked to a lot of other gurus and nobody knew what to do. The guru thought it was just a story, like a parable, and not something that really happened. He tells Eddie, I mean Jack, to make peace in all of his relationships.

With just one branch of leaves left, Jack tries to work things out with his wife, he visits his mother in an assisted-living center, and he visits his father's grave.

His wife throws him out. His mother, who suffers from dementia, tells Jack, who she thinks is Jack's late father Raymond, that she wishes Jack would stop being angry at his father for walking out on them when he was a kid.

Jack realizes that this is the relationship that needs the most mending, so he goes to visit his father's grave. Jack expends the last three leaves of the tree with the words, "I forgive you".

With no leaves remaining on the bodhi tree in his back yard, Jack suffers a heart attack and appears to have died.

The bodhi tree, by the way, just in case you never heard of it, is the tree The Buddha sat under, famously, when he reached enlightenment. It is also called the Sacred Fig tree.

The Bodhi tree is recognizable by its heart-shaped leaves.

There is a Bodhi Tree Spa in Canmore, Alberta; a Bodhi Tree Eco Boutique, up in Peace River; and a Bodhi Tree Yogo studio in Regina, Saskatchewan.

The tree is from India, apparently, but has been planted all over south-east Asia and also in the U.S.A. and some places in Canada.

The movie made me think about making peace in all my relationships.

But I believe I've already done that!

My former girlfriend in Edmonton came to mind. She phoned while I was in reflexology and again at night but that call was interrupted. She told me she had a bad night and cried a lot after I told her the specialist said HIV/AIDS could be a cause of what was bothering me.

Why? I said. I had an AIDS test before I left Ontario and got the all-clear.

What about that guy you lived with down in Texas? I said. The guy who worked on oil rigs around the world?

She had gone on a lot of dates after she got divorced, trying to find somebody to marry, for the second half of her life. She went online and set up a series of dates. She called it 40 Dates In 40 Nights.

But they weren't 40 one-night stands.

She said she cried because she would feel horrible if she had caused my problem.

I told her I doubted it and she shouldn't worry about it, but if she was worried about it, it's easy to get an AIDS test. Just go to the doc and tell him you think you should have one, they take some blood, and you get the results fast.

Ya, she said.

In the movie, the guru says, Pain is the touchstone of growth.

The book agent says, That would make a badass t-shirt!

Earlier, the guru was described as the leader of a spiritual movement.

The book agent says, I have a spiritual movement every time I eat a bran muffin.

This morning, I slept in until the phone rang a couple of times. A nurse asked me if I had a family doctor in town, yet, and I said, No, but I'm working on it, so she said, Oh, and that Dr. Sander thought I should have a lumbar puncture.

Oh? I said.

She called back and told me Dr. Stander talked to the neurologist at the t and they decided I should have a lumbar puncture in Emerg on Saturday, in town, at nine in the morning.

Okay, I said.

The nurse suggested I get to the hospital a bit before that, to register, so I can be in Emerg for nine.

"I'll be there!" I said.

The call was completely unexpected. The two calls. At nine and ten. And what was a lumbar puncture?

Google that!

Wikipedia said it's another term for a spinal tap.

A lumbar puncture or spinal tap is performed in order to collect a sample of cerebrospinal fluid (CSF) for biochemical, microbiological, and cytological analysis.

Sometimes a spinal tap is done for therapeutic purposes, to relieve intracranial pressure.

Usually the spinal tap is done to get spinal fluid to test for meningitis, apparently.

And what is meningitis?

Google says meningitis is inflammation of the protective membranes covering the brain and spinal cord, known collectively as the meninges.

The inflammation may be caused by infection with viruses, bacteria, or other microorganisms, and less commonly by certain drugs. Meningitis can be life-threatening because of the inflammation's proximity to the brain and spinal cord; therefore the condition is classified as a medical emergency.

That freaked me out for a nano-second!

The most common symptoms of meningitis are headache and neck stiffness associated with fever, confusion or altered consciousness, vomiting, and an inability to tolerate light (photophobia) or loud noises (phonophobia).

-- I didn't have any of those symptoms.

Sometimes, especially in small children, only nonspecific symptoms may be present, such as irritability and drowsiness. If a rash is present, it may indicate a particular cause of meningitis.

A lumbar puncture may be used to diagnose or exclude meningitis.

But that's not all.

Here's how they do it: The patient is usually placed in a fetal position, or sits on a stool and bend his or her head and shoulders forward. The area around the lower back is prepared using aseptic technique. Once the appropriate location is palpated, local anaesthetic is infiltrated under the skin and then injected along the intended path of the spinal needle. A spinal needle is inserted between the lumbar vertebrae L3/L4 or L4/L5 and pushed in until there is a "give" that indicates the needle is past the ligamentum flavum. The needle is again pushed until there is a second 'give' that indicates the needle is now past the dura mater.

Doesn't that sound like a whole lot of fun?

The procedure is ended by withdrawing the needle while placing pressure on the puncture site. In the past, the patient would often be asked to lie on his back for at least six hours and be monitored for signs of neurological problems, though there is no scientific evidence that this provides any benefit.

Patient anxiety during the procedure can lead to increased CSF pressure, especially if the person holds their breath, tenses their muscles or flexes their knees too tightly against their chest.

Reinsertion of the stylet may decrease the rate of post lumbar puncture headaches.

I don't think I want this procedure!

A headache is a common side effect of a lumbar puncture, with a third of people who have the procedure developing a headache within 24 to 48 hours.

I found some more info on a website called buzzle.

It is by and large, a safe procedure; serious side effects being very rare. The most commonly encountered side effect is headache, that usually lasts for two to three days. This can be relieved by regular analgesics or pain killers. But, if the headache persists for several days, and occurs when sitting up, then it can be an indicator of cerebrospinal fluid leak. If the condition does go away with enough bed rest, then physicians can treat it with epidural blood patch. In this method, patient's own blood is injected into the site of the leakage, so that the blood clot formed could seal off the leakage.

Headaches caused by lumbar puncture are generally felt either at the front of the head or near the base of the skull and is experienced by about 40% of people who have gone through the procedure. Apart from headaches, a few patients (approximately 1 in a population of 1000) can get minor nerve injury. As has been mentioned above, serious side effects are very rare.

Read more at Buzzle: http://www.buzzle.com/articles/lumbar-puncture-side-effects.html

Hmmm .... Just when I was thinking I was out of the woods, with a dozen days of improvement, the neurologist called for a lumbar puncture.

This morning, I have no back pain, or anything else, and I walked from bed to kitchen to livingroom with little more than the odd twinge of back pain.

Do I really need this stupid spinal tap?

Wasn't there a movie or a band called Spinal Tap.

I Googled that.

Rotten Tomatoes said the movie was largely improvised by director Rob Reiner and his cast and looks like a "real" documentary about a going-nowhere British heavy metal band called Spinal Tap.

One reviewer said it is really about spiritual exhaustion.

Hail Mary

That reminds me .... If had the same dream a few times reently. It is like that vision I had of being close to the light and seeing some of my mentors and some of the greatest avatars looking at me. I see it a little differently now. It looks like a huddle, as in a football huddle, and at first I thought I was the ball, but later I realized I was the quarterback.

I've never played football, except touch football, informally, and in phys ed class at high school, but I've seen CFL and NFL games on TV, of course, like any red-blooded North American male.

Well, I don't know if I've ever watched a full game.

I've seen a few American movies featuring football, too, of course. Waterboy comes to mind.

I like the Saskatchewan Roughriders and the Edmonton Eskimos.

So, picture this: You are the quarterback, down on one knee, looking up at your team, in silhouette with bright stadium lights behind them, but you can see who's who, mostly, and there's Jesus with his arms around Buddha and Mohammed, with Dr. Usui and the Dalai Lama, and Bodhidharma is there as linebacker or tackle.

The crowd is screaming, Go Team!

Okay, the screaming crowd is not part of the dream. -- I was just getting into the fantasy.

When I have this dream, I feel good. It makes me happy. I get the feeling I have assembled this fantastic team and I get to call the plays but how can anything go wrong with a team like this?!

On Facebook, Love Tree shared Angels Dancing the Cosmic Rainbow's photo, with these words:

Everything really does happen for a reason, and even if you can't understand it all, everything that happens to you is for the progression of your soul.

A Facebook friend, a woman I met at Centauri, the international arts camp, near Niagara Falls, first as a camper and years later as a program director, posted this update:

To all my friends who are going through some things right now--Let's start an intention avalanche. We all need positive intentions right now. If I don't see your name, I'll understand. May I ask my friends wherever you might be, to kindly copy, paste, and share this status for one hour to give a moment of support to all those who have family problems; health struggles, job issues, worries of any kind and just need to know that someone cares. Do it for all of us, for nobody is immune. I hope to see this on the walls of all my friends just for moral support. I know some will!! I did it for a friend and you can too. You have to copy & paste this one, no share button!

So, I posted it for an hour or so.

An old friend now living in Nashville has been a faithful correspondent via e-mail over the years and especially during the past month and a half. She has had numerous health issues over the years and wrote to say, again, how shocked she is to realize we are 57, as she usually thinks of herself as 25, until a sore back slows her down.

I told her about my dream.

PS New recurring dream, related to a little vision I had a few years ago .... I'm in a football huddle, the QB, calling the play, looking up at a team
in silhouette with bright stadium lights (or something!) behind them. And the team includes Jesus, Buddha, Bodhidharma, the Dalai Lama, and on and on. The Dalai Lama's just a blocker. He wants us all to meditate, doesn't "get" football. Bodhidharma could take on the other time single-handed.
What do you think?
Just had an idea: Call for the Hail Mary!
lol

BTW: A Hail Mary pass or Hail Mary route in American football refers to any very long forward pass made in desperation with only a small chance of success, especially at or near the end of a half.

The expression goes back at least to the 1930s, being used publicly in that decade by two former members of Notre Dame's Four Horsemen, Elmer Layden and Jim Crowley. Originally meaning any sort of desperation play, a "Hail Mary" gradually came to denote a long, low-percentage pass. For more than forty years its use was largely confined to Notre Dame and other Catholic universities.

The term became widespread after Dallas Cowboys quarterback Roger Staubach (a Roman Catholic) said about his game-winning touchdown pass to wide receiver Drew Pearson in a 1975 playoff game against the Minnesota Vikings, "I closed my eyes and said a Hail Mary."

The Angelic Salutation, Hail Mary, or Ave Maria (Latin) is a traditional Catholic prayer asking for the intercession of the Virgin Mary, the mother of Jesus. The Hail Mary is used within the Catholic Church, and it forms the basis of the Rosary. The prayer is also used by some Anglicans as well as by many other groups within the Western Catholic tradition of Christianity. A somewhat different form of the prayer is used in the Eastern Orthodox and Oriental Orthodox churches and other groups of Eastern Christianity. Some Protestant denominations, such as Lutherans, also make use of some form of the prayer. Most of the text of the Hail Mary can be found within the Gospel of Luke.

It goes like this: Hail Mary, full of grace, the Lord is with thee; blessed art thou amongst women, and blessed is the fruit of thy womb, Jesus. Holy Mary, Mother of God, pray for us sinners, now and at the hour of our death. Amen.

Update: It's almost noon and I just did a set of qigong exercises. It felt good.

At the Zen Forest, where the Zenmaster follows Forest Zen, he recommends a daily morning routine of meditation, twenty minutes, followed by massage, like Reiki self-healing solo massage, and then qigong. -- It's great.

I've written books about it and criss-crossed Ontario, teaching it, and everybody loved it. It's energizing and makes you feel good, happy, ready for anything.

I started with the basic qigong stance, one foot ahead of the other, at ninety degrees. All you do is shift your weight from one foot to the other. You can move your upper body, too, so your shoulders point in the same direction of each foot, in turn.

That exercise is turned into one called The Prayer Wheel by adding coordinated arm movements. Pretend you are turning a big wheel. Imagine and envision that you are drawing in energy from nature and placing it in your heart.

After that I changed feet positions to stand in the cross-country skiing stance and swung my hands back. When your hands are at the front, form fists, and as you push them back, like doing the double pole thrust in cross-country skiing, you open your hands.

They say this gets rid of stale energy.

You can rock your legs while you do this.

It's good to follow that one with its opposite. You throw your hands forward and up, again and again. This warms you up and pumps up energy.

I did sixty of both.

I did a warm-up and cool-down exercise that is a lot easier to do than to describe. You let your arms hang loose and flop around as you turn your shoulders so your hands hit your back and chest. It requires very little movement of the shoulders and you feel as though the exercise takes over so you are doing nothing but your arms keep moving and your hands keep hitting your back and front for a light, thumping, massage.

And I followed all that with walking meditation, but I walked faster and faster, not that I was moving very fast at all, around and around my place. You are supposed to focus on your breathing and count breaths but I counted steps. I did 500, which is five times more than I've done before, since I lost the use of my lefts.

At the end of this little exercise routine, I felt good. -- A little warm, a little wobbly, but energized and happy.

As I sat down, I got a little shot of back pain.

Last night, I thought about doing yoga, but I got up and did qigong.

-- Maybe I'll do some yoga later on.

The specialist said to work on balance, and yoga is all about balance.

Qigong is all about pumping up energy.

I used to do tai chi, which has a lot of movements you have to do just so an in the right order, like a martial arts ballet. Qigong is a little bit like that but without the martial arts or the memorization. You are supposed to do the various movements in random order and that is supposed to help your brain deal with chaos.

That's what I teach in a workshop I call Zen Power Hour.

B6 Causes Peripheral Neuropathy

After talking to the nurse at the neurology clinic, I had an idea, and did some online research.

The doc said, No B6. He said the other B vitamins were good but I should check my B complex and if it had B6 in it, to get rid of it. Don't take it.

I looked at my B complex bottle and it said there was B6 in it, so I stopped taking it.

Just now, I looked up B6 in energy drinks and discovered it's there.

There's 40 mg of B6 in 5 Hour Energy. B6 was an ingredient in Rockstar, too.

I drank a lot of Rockstar last year, or from January to June, and then I switched to 5 Hour Energy.

There's 2 mg of B6 in a can of Monster

And I found out it causes problems.

So ....

I hate to admit it, but I was drinking a couple of cans of Rockstar or Monster energy drink, coffee flavoured, for almost six months. On the odd day, I had three.

Why?

I felt I needed the energy.

I had no prep period at work, no lunch, because of basketball and drama, and didn't have a day off for a few months, because of festivals, conferences, and tournaments.

I don't thank that those drinks, alone, could cause my symptoms. I've just read that too much B6 can cause peripheral neuropathy, which means tingling in the hands and feet. I had a funny feeling in the balls of my feet, then a great big back pain, taking away by a chiropractor, followed by the loss of my legs. My knees were week and feet felt unattached.

Maybe it was a combination of B6 and bone alignment and chiropractic adjustment.

-- Just checked the nutribar package. I had a diet bar for lunch just about every day, at work, last year, as I was trying to lose weight and didn't have much time for anything else. And nutribar has B6 added, too.

Day 13 Of The Comeback

Up early, chatted online with my sister in Australia, who has some serious health concerns of her own, these days, and to call the Cold Lake Primary Care Network. The CLPCN phone line had a recording saying they were closed today because they are booked up for weeks but they're still taking appointments for a few things, such as ear irrigation, so I left my name and number.

My Facebook newsfeed a link to an interesting article about self-compassion.
Lots of people talk about self-esteem, but who talks about self-compassion.
There's a new book out with that title.
The article was called "How 'self-compassion' trumps 'self-esteem'".
It describes the 1970s as a time when adults were looking for a way to raise confident, go-getter children, who would celebrate the person they were to become.

So parents and teachers started showering them with praise, creating a pop movement of self-esteem that played up their worth. Those youngsters grew up with grand aspirations of becoming celebrities, astronauts — anything they wanted to be.

Children of the self-esteem movement had their identities shaped by I Am Special songs and "Princess" t-shirts. They became entitled, confused, and self-critical youth and adults. They were raised to believe they can do anything and got frustrated, sometimes devastated, when they found out they can't.

Now, decades since the praise began, psychologists and researchers say they've found a way to ease the mental self-battery that has become prominent in North American culture.

A new wave of research on self-compassion — the ability to treat yourself the way you'd treat a friend or a loved one — has been creeping into the mainstream, aiming to rescue people from the depths of narcissism and unreasonable standards they will never meet.

Borrowing principles from Buddhism and mindfulness, the practice demands people be kinder to themselves instead of sizing themselves up against others and beating themselves down.

Fascinating, I thought, as that describes a lot of high school students I taught.

I was in university in the Seventies and when I started teaching high school, after that, it was obvious to me that something big had happened because kids had changed.

The sense of entitlement was huge, in that generation, and teachers could see where it was leading.

It's all explained in Meet The Parents and Meet The Fokkers -- the movie with Dustin Hoffman, Barbra Streisand, Robert De Niro, Blythe Dannar, and Ben Stiller.

I haven't seen the sequel, yet, called Little Fokkers.

-- Maybe I'll watch it today!

Prof. Neff published her first book on the topic this month, entitled Self-Compassion: Stop Beating Yourself Up and Leave Insecurity Behind. In December, American psychotherapist Jean Fain released The Self-Compassion Diet, a book that applies the practice to weight loss.

After reading that, I decided to activate self-compassion and take the day off. After breakfast, I played the movie A Thousand Words, again, and went back to bed. And after sleeping, on and off, until almost three in the afternoon, the phone rang just as I was getting up. It was Nola, the nurse who works with the specialist.

She told him that I was walking and he left her a note saying that maybe I didn't need disability.

She said she didn't agree with that and she would be talking to him about that and get back to me.

She also said to go on a website for physicians in Alberta to find a doctor because there were a lot of forms to fill out for short and long term disability and the specialist didn't do that because he sees a patient once and that's it. A family doctor can fill out all those forms.

She said she would call me again early next week.

Here's what I heard: It's up to me.

I heard fireworks exploding. I can go back to work?!

Woo-hoo!

But then I asked myself: Am I ready?

Should I go through with the spinal tap on Saturday, which was scaring the heck out of me?

Was I really walking?

So far, I had walked from the bed to the couch and that was it. I crawled into the nest I made on the floor between the glass door to the balcony and the coffee table that holds my computer. I had crawled under the covers and slept the sleep of the exhausted.

Suddenly, I had a surge of energy, like Jack after he falls to the ground, in the rain, in front of his father's grave, but his cell phone rings, he answers it, and he finds out the bodhi tree has come to life, so he can talk again and his life is not over.

A little while later, he says that a part of him died, the fake part, but the real "him" was still alive and very happy about it.

He writes a book called A Thousand Words, his former assistant sells it for him, the great guru guy writes the foreward, he buys the house his wife wanted, the tree reappears in his yard, and it's a Hollywood ending.

This is not a Hollywood ending. I did not have an epiphany, a profound moment of revelation, get a sign to change my life or mend my relationships or anything like that.

After getting that good but confusing news, I got up and walked around. I took 500 steps and felt I could walk more, and better, than yesterday.

My back hurts a bit, my feet feel very funny, and I'm not exactly rock solid on my feet. I have been bags in the balls of my feet and my balance is not the best.

Could I go to work like this? Or next week, after the long weekend?

I want to say, "Hell YES!"

But let me go for another walk, first ....

But first I checked Facebook and I got this reminder from a meditation group: Remember to pause and give gratitude for the ten thousand things that are going right in your life right now.

-- That sounded like a good idea.

As a result of this episode in my life, I have the attitude of gratitude for several new things: I felt my brother's pain and wrote him a note so we have a connection again. I got closer to my cousins. I talked to my other brother more in the past three weeks than the past year. I met the magic reflexologist. I heard from a few colleagues -- one in particular -- and feel closer to them. I heard from some students and feel loved. My former girlfriend phoned a lot and last night she said, Maybe you're right and it does take more than a year to get to know someone.

I got invited to teach Zen meditation at the Wellness Centre in town.

And then there were the dreams relating to that vision I had a year and a half ago. First, I saw a group of great healers gathered around me, with a light behind them, so some of them were impossible to identify, but I could see Jesus and Buddha and Bodhidharma and Dr. Usui and the Dalai Lama and others. But now I saw it as a huddle, the kind they have in football games, and at first I thought I was the ball, but then it got bigger and it hit me that I was not the ball, I was the quarterback, and I had the greatest team around me. All I had to do was call the plays, throw the ball, and try not to get hit. Or I could run with the ball. Whatever. And the rest of my team would make things happen.

My journey had led me to the books of Paolo Coehlo, Aleph and The Alchemist, and the movie called A Thousand Words.

I learned the term "self-compassion".

Math is not my subject, but I would say there are more than a dozen huge things on that list.

What else?

I've learned that B6 causes neuropathy and to look for it in vitamin supplements, energy drinks, diet bars, and everywhere.

I've learned that the health care system is great, but not perfect, as it's stretched to the max and under a lot of pressure, and it's evolving to include some alternative practitioners to the doctors, nurses, and hospitals designed to handle emergencies and disease.

While I was writing that list, the sun came up.

Also, I felt inspired to try a piece of Nola's advice. She said to go on the website of the College of Physicians & Surgeons for Alberta and search beyond Cold Lake for a doctor. I had an appointment with a doc in Lac La Biche on the 22nd, but I didn't want to wait that long.

I found a doctor in Bonnyville, the closest town, and the website said he was taking new patients, so I phoned, but I found out that listing was out of date and they had no doctors taking new patients.

I tried the website for Lac La Biche and found the names of two doctors who were listed as taking new patients. I phoned the first one and the secretary said I could get in before the end of the month. We talked a bit and she said there was another doctor I could see sooner. -- On Tuesday.

She made me an appointment with Dr. Birkill on Tuesday at 2:00.

Woo-hoo!

Now what?

Objectively speaking, at a moment when it is difficult to remain objective, here's what I'm thinking. Seriously. I'm almost ready to go back to work.

Today is Thursday. Forget about tomorrow, that's too soon. And Monday is a holiday, for Thanksgiving. And Tuesday, I have an appointment with a doctor.

How about after that?

That gives me five days to get into better shape for a return to work.

What about physiotherapy?

I have an appointment in town for the end of the month.

What will a physiotherapist tell me?

What about Mecelle? I have an appointment with her on Tuesday. Maybe I can move it to Friday!

Called. Left a message.

It's 3:30, the sun is shining, I want to go outside and celebrate -- and see how far I can walk. Really.

After 200 steps on the balcony, I came inside, sat down, and the phone rang.

I love it when that happens.

Mecelle said she was phoning to see if she could move my appointment.

Synchronicity.

So, we talked.

She wanted to go ahead with the Zen meditation event at the Wellness Centre, so we talked about details, including money. I said, Do you want to do it for charity, for free, or to make some money?

She said, You should make some money.

I said, You should make some money.

She said, How about doing it as a fundraiser for kids at your school, so a poor kid can go on a trip?

Good idea, I said. How about doing it as a fundraiser so somebody who can't afford it can go and see the gifted reflexologist?

She said, I could add something to that.

Do you guys at the Wellness Centre have a favourite charity? I asked.

Yes, she said. We've been talking about that, thinking about Christmas coming, since it snowed today, and we usually make a short list of charities to support at this time of year.

Let's talk about it when I come to see you for my next appointment, I suggested.

She booked me in for Wednesday at noon.

How about doing it as a fundraiser so my cousin can get in there to see the reflexologist?

When I told her about it, she gave me a blank stare, which seemed to say, I'm sick of going to doctors of all kinds and the way you described reflexology kind of freaks me out.

So. We shall see.

But it feels great to get things happening.

She said to get a second opinion about the spinal tap as it can be painful and lead to complications.

Now I really don't want to go through with it.

-- Maybe I'll walk some more and think about it.

She said we had some snow so I should take my walker if I'm going out. And I said, My walker is locked in a closet and I never want to use it again!

She laughed a lot but said, Use the walker or I will use my special radar and hunt you down!

And that made me laugh a lot.

Amazing how things can turn around in a short space of time.

After taking another 200 steps, for a total of 700, I felt like doing something different. -- Not sure what.

Day 14 Of The Comeback

This morning I started the day with a TED video featuring "the happiest man in the world". He was a French biochemist who moved to the Himilayas to become a Buddhist monk, in search of happiness.

The "happiest man in the world" is Matthieu Ricard. He is now a Buddhist monk, author, and photographer.

After training in biochemistry at the Institute Pasteur, Ricard left science to pursue happiness, both at a basic human level and as a subject of inquiry.

Achieving happiness, he has come to believe, requires the same kind of effort and mind training that any other serious pursuit involves.

He was in North America to speak and he had a slide show to go with it as he was also a photographer. He had beautiful pictures of snow-capped mountains, turquise lakes, and people meditating in the mountains. The pictures were taken all around Tibet.

He started by saying, So, I guess it is a result of globalization that you can find Coca-Cola tins on top of Everest and a Buddhist monk in Monterey.

He talked about meditation, developing the ability to control your thoughts and your mind, how happiness is related to compassion and well-being, and human nature, and he ended with pictures of leaping monks and flying monks, or happy monks expressing their feelings.

As for me .... Right now, happiness would be getting out of that spinal tap scheduled for tomorrow morning.

In the morning e-mail, a note from a friend in Port Perry, Ontario, who I told about B6. She says: If spinal tap is not a vibrational match for you then perhaps continue your search. I am firm believer there is no dis-ease just ask function of cells.
Look up Raymond Francis work on the malfunction of cells and there is no disease.
Also with eating beef you would have had lots of b6 already in your body.

That was from Heidi, an alternative healer, in southern Ontario. She lives in the country outside of Port Perry, between Lake Scugog and Peterborough.

I felt like going out and walking so I headed out for the drug store and I left my walker behind.

That felt very liberating!

I walked to the elevator, out to my car, drove to Shoppers, parked in the blue spot, walked in and around the store, and ordered the right size of the Dr. Ho belt. The pharmacist said it would be in by Tuesday, as Monday was the Thanksgiving holiday.

I picked up a few other things and when the cashier said would you like to make a donation to our Tree Of Life, I said, "Sure!"

How many steps was that? I counted 500. But I didn't feel as though I was finished.

Next stop: the post office. I returned to the place where I felt my comeback started, after reflexology, when I left my walker in the car, grabbed my canes, again, and struggled up the curb and steps to mail a letter. This time, I left the canes in the car and just walked up.

It made me laugh to see there were only two stairs, not three, as I described earlier, and they were small steps, not the big ones I believed I had climbed. The curb and a couple of stairs were no problem for me now.

Did I hear the Rocky theme song this time?

No. -- That seemed silly, all of a sudden. After all, it was just a couple of stairs. The letter I mailed was the card with the note I wrote to my brother, who has had decades of back pain to deal with. Chronic, severe, pain in the back.

After mailing that letter, I drove over to Walmart to walk around. It's a big store with a lot of flat ground to cover. And I walked another 500 steps, so I was up to 1000.

While I was wandering around the book section, my cell phone rang and I was surprised it was Julie, our Science teacher, and her class. She said they wanted to say hello.

That made me smile, too.

Mostly it was just one of the Grade 10 students who did the talking. He asked me if it was a pinched nerve, or what, and if I would be back by the start of next semester, when he would be in my English class. I told him the doctor said no but I said, "Yes!"

He told me he was playing football and I said I would love to see a game.

He asked me what I'd been up to, so I told him I'd seen a lot of doctors, including a specialist at the new Edmonton Clinic, on the grounds of the University of Alberta, and they all said the same thing: There's something going on down there, with your feet and legs, but I don't know what it is, and they ordered more tests.

I didn't mention the spinal tap scheduled for the morning.

After Shoppers and Walmart, plus the post office, I felt hungry, as I skipped breakfast, so I went to Humpty's for lunch and had their pan scrambler, which is scrambled eggs with hash browns, cheese, green pepper, and mushrooms, with a side order of rye toast, and a tea.

I never order tea in a restaurant, but I felt like something hot with a little caffeine.

And then I drove home, walked from my car to the elevator, and back to my computer, for a total of, perhaps, 1500 steps.

Next? The bookstore had called the school to say some books I had ordered were in, a while ago, and I hadn't been able to pick them up, so I thought I would go over there.

I read a few health testimonies about having a spinal tap.

One said, When the doctor says spinal tap I think fear automatically goes through a person's mind. Fear of the unknown, fear of needles, fear of a big gauged needle going in your back, needles breaking, paralysis. One person may say it was the most painful procedure they have ever had to go through in their entire life. While another person will say, it was not that bad. I can personally say I have had to undergo three spinal taps in my lifetime and I had an epidural too when I was in labor with my daughter which is very similar except they are giving you medication and not taking fluid out and it really isn't as bad as some people say.

I found an explanation and description of the spinal tap online. A spinal tap (also known as a lumbar puncture) is an important tool for doctors to have in diagnosing certain medical disorders such as multiple sclerosis, and Gillian Bar Syndrome. During a spinal tap, the doctor removes a small amount of spinal fluid, which is then taken to the lab for examination.

A little humour from Jeff Foxworthy helped, too. He said, If the Mayans are right, now would be a great time to buy all the stuff with "No payments until 2013..." If they are wrong, the repo man sure is going to be busy!

While I was contemplating a trip to the bookstore, to pick up another book by Paolo Coehlo, as well as books I'd ordered, I looked online for a suggestion about which of his books to read next, and that led me to Pirate Bay, the big Swedish website that is anti-copyrite, and I found Coehlo's book, Aleph, with this description: Transform your life. Rewrite your destiny, it said. In his most personal novel to date, internationally best-selling author Paulo Coelho returns with a remarkable journey of self-discovery. Like the main character in his much-beloved The Alchemist, Paulo is facing a grave crisis of faith. As he seeks a path of spiritual renewal and growth, he decides to begin again: to travel, to experiment, to reconnect with people and the landscapes around him.

And I thought, well, isn't that my story, exactly.

The phone woke me up at noon. The Glen Rose Hospital in Edmonton had an appointment for me on Wednesday at 2:00. 780-735-8210.
The Glenrose Rehabilitation Hospital is a hospital located in downtown Edmonton, Alberta, Canada. The Glenrose Rehabilitation Hospital is a health care facility unique to Alberta, devoted primarily to high-level rehabilitation care of both adults (including the elderly) and children. The Glenrose is also the academic centre for several University of Alberta medical education programs and plays a strong role in the development and advancement of leading edge research and training opportunities in rehabilitation fields.
The Glenrose is the largest freestanding comprehensive tertiary rehabilitation centre in North America and offers services to children and adults on an inpatient, outpatient and

outreach basis. Opened in 1964, the 244-bed facility has developed an international reputation for excellence in key areas of complex rehabilitation and specialized geriatrics.[citation needed] The Glenrose offers highly specialized assessment, treatment, consultation and technology services, as well as education for patients and families through more than 120 clinics and services.

In addition to rehabilitation services for all age groups, areas of focus also include mental health and psychiatric services for children and seniors, as well as cardiac rehabilitation for adults. Specialized technology enhances patient care in programs such as the Syncrude Centre for Motion and Balance, I CAN Centre for Augmentative and Assistive Communication, Alberta Caregiver College, Cochlear Implant Service, Telehealth Seating Service, Prosthetics and Orthotics, and Scoliosis Clinic.

Glenrose rehabilitation research aims to improve the quality of life for people of all ages. Themes of research include:

Assistive technology to assess and treat people with disabilities

Helping children with developmental disorders through better understanding, assessment and care

Improving function in persons with chronic conditions

Anticipating the needs of an aging population

Did I want to go there?

No!

Do I want to drive to Edmonton and back?

No!

Oh well.

Day 15 Of The Big Comeback

Got up early and found this note on Facebook from The Attitude of Gratitude
If you're feeling frightened about what comes next, embrace the uncertainty. Allow it to
lead you places. Be brave as it challenges you to exercise both your heart and your mind as
you create your own path towards happiness. Don't waste time with regret. Spin wildly
into your next action. Enjoy the present - each moment as it comes - because you'll never
get another one quite like it. And if you should ever look up and find yourself lost, simply
take a breath and start over. Retrace your steps and go back to the purest place in your
heart, where your hope lives. You'll find your way again.

Last night, I chatted online with my sister in Australia. She wanted to Skype but I wasn't
up for it.   She said, They do spinal taps all the time so don't worry, you WILL be fine! It
is silly to tell someone not to worry though because who wouldn't be worried at such a
time? So, worry, but believe that all will go well! A happy Thankgiving would be finding
out what is wrong and getting it fixed! That's great that you walked without the canes or
walker, Marty! Wooo hoooo! Keep up the good work! Did it hurt?

My cousin sent me an e, saying, DO YOU KNOW -IF YOU REALLY THINK THAT B6
WAS THE PROBLEM, then don't have the spinal tap, or go to the Glenrose, or anything.
The specialist said maybe you didn't have to be off work - so - why don't you wait.
Because you are getting better every day - maybe you are right. They can always
reschedule the spinal tap - and the Glenrose.
So I responded at six thirty in the morning:
        Yeah, I dunno. I think the chiropractor did some damage, even though the specialist
said "no". And B6 is a factor, apparently. Maybe it's those 2 things together. They say the
lumbar puncture, or spinal tap, isn't a big deal for most people. I'm going to go for it --
unless I chicken out at the last second! I've never chickened out at the last second. My
brother and sister told me horror stories, which didn't help!
        I sent our other brother a card with a note saying, I feel your pain.
        They say spinal fluid is liquid gold as it holds a lot of information for medical
researchers.
        ocs say I have a high threshold for pain. And I know how to meditate. So!
                PS Happy Thanksgiving!

She said, I could have told you horror stories about spinal taps, too, but I thought I'd wait
until after yours and you had recovered.

I thanked her.

This Is Not Spinal Tap!

I got a "Get out of jail FREE" card, a last minute stay of execution from the governor, my blindfold was taken off and the firing squad was told to go home. The doctor canceled my lumbar puncture, or spinal tap.

I didn't chicken out. I showed up early and waited a long time. The spinal tap was scheduled for nine and at ten thirty I was taken to the operating room, given a gown, took off my shirt and tee-shirt, and put it on, backwards, to sit for another ten minutes, waiting for the doctor.

If you are nervous or scared of something that is about to happen, the waiting and anticipation might be the worst part.

I meditated, cleared my mind, and remained calm so we could carry on.

The doctor came in an apologized, saying he should have done his homework, he thought the lab could handle it, but they weren't ready, didn't have everything they needed, and my spinal fluid would have to be stored for days, since it was Saturday, and the start of a long weekend, so the protein might not be in good shape for study by the time it got to Edmonton on Tuesday ....

What a relief! I said.

He looked a little surprised.

Well, I said, as you can see, I'm walking, now. And I'm walking more, and better, every day.

Yeah, he said. I noticed. And he said, You know what, I'm going to re-schedule the lumbar puncture for Wednesday.

I told him I had to go to Edmonton on Wednesday for an appointment for an EMG at the Glenrose.

Well, let's schedule it for the 17th, he said, and if you don't show up, I won't be surprised or upset.

Frankly, he said, I don't think you need it. The specialist is just trying to cover all the bases. If you're walking, your body is healing itself, and we don't have to worry about the cause or the cure.

Anyway, you don't have to worry about the procedure, he said. But you're a big guy and we might have to move the needle around a bit to find the right spot, and that can be painful ....

So, it's up to you, really, he said. I'll schedule it for the 17th, but I won't be mad if you don't show up.

I understand, I said. I got it.

I walked out feeling as though I was on death's row but got released.

I thought of stories and movies about prisoners who get a stay of execution at the last second, and how life-changing it has been for certain prisoners to have the blindfold removed while the firing squad is told to pack home and go home without firing a shot.

"Many years later, as he faced the firing squad, Colonel Aureliano Buendía was to remember that distant afternoon when his father took him to discover ice."

That's the opening line of One Hundred Years of Solitude by Gabriel García Márquez.

Dostoevsky was led before a firing squad and prepared for execution. He had been convicted and sentenced to death for allegedly taking part in antigovernment activities. However, at the last moment he was reprieved and sent into exile.

Dostoevsky's father was a doctor at Moscow's Hospital for the Poor, where he grew rich enough to by land and serfs. After his father's death, Dostoevsky, who suffered from epilepsy, studied military engineering and became a civil servant while secretly writing novels.

His first novel, called Poor People, and his second, The Double, were both published in 1846. The first was a hit, the second a failure.

In1849, Dostoevsky was led before the firing squad but received a last-minute reprieve and was sent to a Siberian labor camp, where he worked for four years.

He was released in 1854 to work as a soldier on the Mongolian frontier.

He married a widow and finally returned to Russia in 1859.

The following year, he founded a magazine, and two years after that he journeyed to Europe for the first time.

In 1864 and 1865, his wife and his brother died, the magazine folded, and Dostoevsky found himself deeply in debt, which he exacerbated by gambling.

In 1866, he published Crime and Punishment, one of his most popular works.

In 1867, he married a stenographer, and the couple fled to Europe to escape his creditors.

His novel The Possessed (1872) was successful, and the couple returned to St. Petersburg.

He published The Brothers Karamazov in 1880 to immediate success, but died a year later.

In the movie called Love and Death, Woody Allen in the role of a Russian soldier named Boris Grushenko is executed by a firing squad while waiting for a last-minute reprieve.

He appears as a ghost afterwards, before leaving with the Grim Reaper.

In the original Casino Royale, the Woody Allen character escapes from one firing squad, only to run headlong into another firing squad. He jumps over the firing squad wall shouting, "So long, suckers!" only to land in front of another firing squad.

Woody Allen plays Jimmy Bond, James Bond's neurotic nephew who informs the firing squad he has a "low threshold for death".

I walked out of the hospital, got into my car, drove up to the lake, admired the view of the big blue body of water surrounded by black spruce and aspen trees with leaves that have turned golden and yellow, and I felt like going to the provincial park to see how far I could walk on the trail that I hiked in the summer.

When I returned to reality, I drove over to the rec centre, instead, walked 200 steps from the handicapped parking to the reception area, and asked about the indoor walking track. I needed to know if they had an elevator, as the beautiful, cushioned, flat, oval track is on the third floor.

And, yes, of course they had an elevator to the third floor.

And while I was walking the long concrete sidewalk in front of the rec centre, didn't I hear two little kids calling my name, sounding excited, as they rode their bikes, beside their moms. It was two of my youngest students, one in Grade Three and one in Grade Four. The yelled my name and rode their bikes up to me to say, We miss you.

They were close to tears.

I told them I missed them, too.

When are you coming back to school? they said.

I'm working on it! I told them. The doctor says I won't be back for a while, but I'm fighting back.

They told me what they were up to, at school, and how they liked me more than the supply teachers they had for the past month.

Well then I'll work twice as hard to get back in there, I promised.

And they rode off to catch up with their moms, who were already in the parking lot.

After that, I drove over to a restaurant that had a Thanksgiving special, but it was packed, with no room in the parking lot, so I decided to go home.

But when I parked the car, I felt like doing something to celebrate, so I walked over to the bookstore and picked up an order for school plus a new book for myself. I bought the new, hardcover, novel by Paolo Coehlo, called The Winner Stands Alone.

Do I know how to celebrate?!

Woo-hoo!

The sun is shining, my nest is ready, by the window, I have a good book to lie flat with, and I am feeling relieved I didn't have to have the spinal tap -- relieved like a a guy who escaped from the firing squad to live his life as a free man, again.

Yee-haw!

Love Tree posted this quotable quote: "Be content with what you have; rejoice in the way things are. When you realize there is nothing lacking, the whole world belongs to you." ~ Lao Tzu

After the trips to the hospital and the bookstore, I slept in the sun for a few hours, after writing for a few minutes, and then I went to Baxter's Herbs, a store on the main street of the town formerly known as Grand Centre, and I got some St. John's wort to promote nerve healing.

Here's what I've been thinking: What did the damage to my back and legs and feet? Was it work, a chiropractor, B6, all three? And who cares. I can walk, again. I don't have to have a spinal tap. It's time to celebrate Thanksgiving. -- With St. John's wort.

I walked a little funny, and I got tired after 1,000 steps, but after a rest I could do another 1,000.

To celebrate, I took myself out to dinner. The Thanksgiving Special at Humpty's featured turkey, roasted potato, corn, stuffy, gravy, and apple pie.

They don't serve pumpkin pie until Monday, which is the actual day of Thanksgiving.

And I devoured that food like a hungry man digging in to his last supper before an execution.

It tasted fantastic, as food always does when you are feeling ecstatic.

And I wondered, when was the last time I felt ecstatic?

The last time I felt ecstatic was the day my comeback began, with that "Rocky" moment at the top of those Post Office steps.

I had a flashback to other ecstatic moments at Thanksgiving in other years. There were dozens of them. They made me think of my favourite Thanksgiving movie: Alice's Restaurant, starring Arlo Guthrie as himself, and featuring his song called Alice's Restaurant.

I went home and watched a short version of the movie on Youtube.

Thanksgiving Movies

That night I got a few e-mails from family and friends, asking how the spinal tap went, and I sent them all the same report. My friend in Edmonton said she prayed for me at the time I was supposed to be having the spinal tap. I thanked her sincerely.

She recommended a movie to watch while I was flat on my back and then I played Scrabble online while watching Tom Hanks in a movie called Charlie Wilson's War. It's a drama based on a Texas congressman, Charlie Wilson, and his covert dealings in Afghanistan, helping rebels in their war with the Soviets. It starred Tom Hanks and Julie Roberts, but they appeared to be mis-cast, to me.

Wilson partnered with CIA operative Gust Avrakotos to launch Operation Cyclone, a program to organize and support the Afghan mujahideen during the Soviet war in Afghanistan. The move was adapted George Crile III's 2003 book Charlie Wilson's War: The Extraordinary Story of the Largest Covert Operation in History.

Philip Seymour Hoffman was nominated for an Academy Award for Best Supporting Actor for his role as the CIA guy.

Rotten Tomatoes gave it 81%.

Not me!

I wanted to see Julie Roberts and Tom Hanks doing something else, not raising a billion dollars to buy weapons.

There was a Zen story I liked at the end of the movie about a Zen master who says "That depends" when he is told stories of good news and bad news.

Gust Avrakotos tells the story. He says, There's a little boy and on his 14th birthday he gets a horse... and everybody in the village says, "how wonderful. The boy got a horse" And the Zen master says, "we'll see." Two years later, the boy falls off the horse, breaks his leg, and everyone in the village says, "How terrible." And the Zen master says, "We'll see." Then, a war breaks out and all the young men have to go off and fight... except the boy can't cause his legs all messed up. and everybody in the village says, "How wonderful."

Charlie Wilson said, Now the Zen master says, "We'll see."

Day 16 Of The Comeback

I would like to say I got up early and worked out all day, beat my previous total of 2,000 steps, worked on balance, strength, stamina, and so on, but the truth is .... I got up at six, ten, and twelve, got my computer to read me this story, and slept a lot.

At noon, I made macaroni with beef, did some writing, and wondered why my energy was so low.

But then I rationalized it, saying resting my back was important, too.

My back was just a little sore and my legs felt asleep, below my shins.

There wasn't much happening on my Facebook page.

Susan Swan, the Canadian novelist, author of The Biggest Modern Woman in the World, kept reporting on rave reviews for her new novel, The Western Light, which uses Georgian Bay as a character. The Vancouver Sun praised The Western Light as a "spellbinding tale, a many-layered coming-of-age story".

Amazon.ca gave it 5.0 out of 5 stars.

One reviewer said, Mouse Bradford is back -- and that is wonderful news. In this prequel to her award winning The Wives of Bath, Susan Swan returns to Madoc's Landing, that small town on the shores of Georgian Bay that shaped Mouse (and her). In a splendid evocation of the 50s, The Western Light speaks of memory, searching and love -- both requited and unrequited. It is great read.

Another reviewer said, The Western Light, was for me hard to put down, Susan has a way of telling a story that is almost whispering the story into your ear, effortless enjoyment, a movie played in your senses. Interesting characters, a realisitc time period 1950's, weaving in a bit of comedy, with suspense and realizm. I highly recommend this novel.

Wally Lamb, author of She's Come Undone, posted pics of himself speaking at the University of Connecticut, some foreign editions of his books, and invitation to one of his workshops: Hey, writers: Come join me on Italy's spectacular Amalfi Coast next March for the first-ever Praiano writers conference. Workshops in poetry, creative non-fiction, and fiction facilitated by Suzanne Levine, Lary Bloom, and yours truly. Read all about it at www.praianowriters.com but hurry. Only 12 spots available. Ciao!

Also, he was celebrating the publication of a special 20th anniversary edition of She's Come Undone.

I'm also Facebook friends with some other famous and future authors, including my main mentors, Clark Blaise, W.D., Valgardson, and Dave Godfrey, but they rarely post anything on Facebook.

There's also Bharati Mukherjee, Penn Kemp, Margaret Atwood, Beth Harvor, Heather Haley, Barbara Gowdy, Sylvia Fraser, Joan Barfoot, Diane Schoemperlen, Bobbie Ann Mason, Susan Musgrave ....

And then there's Gary Barwin and Stuart Ross. And bill bisset.

And the authors and writers I've met through the novel marathons: Mel Malton, Richard Thomas, Christina Kilbourne, Bartha Batiz, Jacqeline

I went to grad school with Wally. We workshopped She's Come Undone, helped him write it, and watched his life change as Oprah picked it for her book club. It was a bestseller, he was a millionaire, and it appears as though he has lived a charmed existence ever since then.

I could not help thinking that if Susan Swan was an American author, writing about Long Island Sound, between New England and New York, instead of Georgian Bay, she would be on Oprah, given honourary degrees, and invited to do workshops in exotic places, like Wally Lamb.

Oh, Canada!

As for me, well, I'm happy I can walk, again, and I'm looking forward to beating my new record of 2,000 steps.

-- Maybe later today ....

Then again .... Maybe not.

Dr. Google may have the answer to my mystery. A website called Better Medicine at www.localhealth.com had some information about foot numbness. It said there were a numberof Orthopedic causes of foot numbness.

In short, it said, Foot numbness may also occur due to moderate to serious orthopedic conditions that injure or damage the nerves, including:

Back injury, Nerve entrapment or nerve pressure, such as from sitting too long, Degenerative disk disease, Herniated disk, and a few other things.

The x-ray report said I had DDD, an emerg doc said I had a back injury from a chiropractic adjustment, and the CT scan indicated a bulging disc.

What about sitting too long?

I've written over 100 books. -- Could that be the problem?

Or is it all of the above.

If that's the answer, how come nobody I've seen has a treatment for the problem, other than rest?

Well, I don't feel like doing much more than resting, today.

Maybe I can find Love And Death, the Woody Allen movie, online.

After watching Love And Death and then Charlie Wilson's War, again, I thought about my back, again. I was feeling frustrated, after six weeks of seeing doctors and having tests without getting much information or anything that would help, other than being told to "take it easy". Reflexology helped enormously. The specialist had scared me, talking about cancer, ALS, MS, AIDS, and long term disability.

I remembered hearing about a book by somebody known as The Back Doctor, so I Googled him.

Twenty-five years ago, Dr. Hamilton Hall wrote the book called The Back Doctor that changed the way a lot of people looked at back pain. He advocated activity instead of bed rest. In place of braces and girdles, he suggested that patients follow a program of specific exercise. He took the mystery out of back pain. And now he had a new book out, called A Consultation With the Back Doctor.

Well, it wasn't that new; it was published in 2004.

In it, he says, "Perhaps the final answer for our lack of progress in conquering neck and back pain is its extensive, unfortuante, and unnecessary medicalizaton. For over seventy years, the medical profession has worked to turn neck and back pain f rom an unfortunate human condition into an illness. It has created a disease for which it has no cure."

That rang a bell.

He went on to say that, "No drug, treatment, or surgery offers a panacea. Yet the pain of a sore neck or a sore back can be controlled and, in most cases, abolished, at least for a time."

How? you ask.

Exercise.

Day 17: Monday, October 8

What happened to the comeback?

On Day 16 I walked very little as I felt I needed rest. My back was sore, a bit, on and off, and I didn't have a lot of energy. Also, if rest was required to help my back and legs, then I thought yesterday was the perfect day for it. There wasn't any place I had to be or anything I had to do. So .... rest.

Today is Thanksgiving. I feel a little bit more energized than yesterday.

I got up at four and did some research and writing about The Back Doctor, then went back to sleep until nine. I started the exercises I had discovered the night before.

I started doing the exercises thinking they were too soft. First I did the shoulder shrug. -- No problem. Next I dropped my head to the left and then to the right and brought it up again. -- I could feel a little bit of pain in my back and my butt.

After that, I did some qigong exercises to get rid of stale energy and pump up new energy, and that went well.

All of a sudden, I felt hungry, after that.

On Facebook, my cousin posted a funny link with the comment, Martin, this is what you need. It was an article about an Underwater Treadmill at www.bookofjoe.com.

This was their claim: As the future of performance training, this contemporary exercise environment allows the user to train to a high level whilst at the same time minimizing body stress, it said, and it had several pictures of people in gym clothes and running shoes running on a treadmill encased in glass with water up to their knees or waist.

It was called a Hydro Physio Lifestyle Underwater Treadmill.

I Googled it, to see if it was a hoax, and found more pictures on the Runners World website. It said the Hydro Physio Underwater Treadmill cost £35,000 or $55,000, had 4 water heights, water temperature controls, resistance jets for increased intensity workout, and goes to at least 9 KPH or 5.5 MPH.

There was one for animals, too, called Aqua Paws.

Instead of feeling hungry, I had a wild desire to go swimming.

I tried a few swimming strokes, standing up and moving my arms, and decided that would have to wait for another day.

While I was up, I decided to get going, head out, go to the drug store for milk, but I felt good, while I was driving, so I drove over to the grocery store. Sobey's is a big store and the thought of walking up and down those aisles was intimidating, for the past month or so. But I knew I could handle it, now, so I parked in the preferred spot, walked in, and grabbed a cart.

I wandered around the store, instead of being methodical, because it was entertaining, in a way, and a little bit of exercise, and it was fun to see people and talk to a few of them. What I noticed was, as they say, everybody is fighting his or her own battle. Several people walked like me, or more awkwardly; one guy was in his pajamas and looked like he had some mental issues; a pregnant woman with a baby on her hip gave me a look that said "Omg this is a lot of painful work". A few other women smiled and said "hi". A guy in the check-out line told me where to find the pumpkin pie, hidden by the deli counter, on the racks with all the bread.

    I took my groceries out to the car, returned the cart, and walked through the store, again, to get a pumpkin pie.

    I laughed at myself because all of these little things seemed to be so big to me. Just being able to walk well enough to make it through a good-sized grocery story several times was something I had always taken for granted, but now it was like a great gift.

On the way home, I made a connection between the vision and dream I had and the advice given by The Back Doctor: My rehab and return to work was up to me. There was no doctor or physiotherapist who would identify the cause and solution to my mobility problems. I already had a lot of information and had a few ideas about how to get back into fighting shape. And, of course, I had my team, the group of incredible healers, the guys I saw in my huddle, when I called them in, just waiting for me to call the play so they could go into action and make my plan happen.

    On the way home, I also thought about my brother with the chronic, severe, back issues. -- I had written a piece about our Uncle John, who I often thought of but rarely wrote about, and now it was time, perhaps, to write something about my brother. I often thought about him but almost never wrote about him.

Oh, Brother

At my brother's second wedding, held in his house, I was asked, at the last second, to make a speech on behalf of the groom.

Sure, I said.

With a glass of wine in my hand, I stood and told the group a few things about my brother. To be funny, I told them that he was born feet first and we -- the people in my family -- thought he did a lot of things backwards, but we felt he was finally going in the right direction by getting married to the woman who was willing to say "I do" on that day.

He did something heroic, as a kid, a couple of times, and got his picture in the paper. The first time, he was playing around in an area that was off limits to the kids in our family, on a big granite bedrock outcropping with cliffs and caves and loose boulders of all sizes. One of the guys he was with fell down the big hill and hit his head on some rocks, so he was left bleeding and stunned, if not temporarily unconscious, part way down the side of the hill, stopped by big boulders.

My brother kept his head and got to the other kid quickly, without running, kept him calm by talking to him, and helped him get to the closest house, so they could call for help. The guy who fell could not walk very well and he was disoriented, so my brother half carried him around the hill and up the street.

A few years later, he was working on becoming a life guard, sitting in the shade at the little beach in the town where we grew up, reading a book about snakes, when somebody in the water screamed, Snake!

Everybody in the water got out and ran down a long dock to the shore and my brother ran in the opposite direction so he could jump in the water and grab the snake.

We never saw water snakes, where we had swimming lessons, so hardly anybody knew what to do, or what kind of snake it was, or if it was dangerous.

My brother jumped in the water without going underwater, kept his eye on the snake, swam up to it quickly, grabbed it right behind the head, and tossed it onto the dock.

A lifeguard followed him but panicked a bit when he saw the snake. He kicked it and it landed in the water again.

My brother grabbed the second time and tossed it on the dock but this time he caught it a little bit too far below the head and it twisted around to bit him on the hand.

In high school, at the end of Grade Thirteen, he was named Athlete of the Year and Best All Round Boy. He played basketball and he wrestled.

He won the Across The Lake Swim at the annual regatta on Gull Lake.

He was in air cadets and wanted to be a fighter pilot but found out he was colour-blind, a bit, and could not get into the pilots' program. So, he followed his older brother into education and became an elementary school teacher.

As an adult, he played rugger, or rugby, for two decades, and one year he was named captain of a mid-Ontario all-star team that toured Wales and went undefeated.

He got a pain in his back that was so severe he could not walk around his classroom. He had to stop teaching phys ed and switched to an elementary school classroom with chairs stationed at intervals around the room so he could lurch from one to the next to get around to see his students' work.

But then he had to go on long term disability because the pain was just too much.

He had a couple of operations, had a disk removed, then a spinal fusion operation, but the pain remained.

He could walk, again, but he had to take a lot of powerful painkillers.

Despite that, he became a regional bridge champion. He had always liked playing cards but after his time playing sports was over, he took up the card game called bridge, and he was almost impossible to beat.

And, as Forrest Gump used to say, That's all I have to say about that.

Well, it's not all I have to say. But that's all I'm going to say.

-- I hope it doesn't sound too much like an obituary or Lives Lived column in The Globe And Mail.

Writing a Lives Lived essay about a family member or a close friend who has died is an emotional experience, but it can be very satisfying, they say. You share the story of the person as you knew them, and create a lasting record of who they really were, beyond their CV.

A Lives Lived essay is a mini-biography that makes readers feel they too have gotten to know the person a little, perhaps over a drink or two. Unlike a formal obituary, which stresses external achievements, a Lives Lived essay paints an intimate portrait of a person, complete with light and shadows.

It doesn't shy away from foibles and weaknesses, but includes them in the context of strengths and loveable qualities. A couple of anecdotes that show us what the subject was like are better than a series of rose-tinted phrases.

The essay should contain a bit of basic information, such as education, marriages, children and career, but it's even more important to tell us what the person was like.

As Forrest Gump said, That's all I have to say about that.

-- He ain't dead yet!

Whenever I think about my brothers, these days, I think of a book by Thich Nhat Hanh that claimed Jesus and Buddha were brothers.

Metaphorically speaking, of course.

I like that idea.

And I like my little idea, that Jesus and Buddha play on the same football team with Moses, Isaiah, Confucius, John The Baptist, Mohammed, St. Francis of Assissi, Gandhi, The Dalai Lama, Lao-tzu, and maybe Mother Theresa, alongside some of my mentors, including W.O. Mitchell and Eli Mandel.

After writing about my brother, while thinking about Jesus playing football, I turned to the internet and found out that someone had posed this quesition on Yahoo: What football club would Jesus support?

Yahoo! Answers voters chose these responses:

Glasgow Celtic. They play in Paradise. I rest my case.

Southampton - The Saints.

Jerusalem FC

Liverpool and they STILL CANNOT WIN!!

Jesus plays golf bro. Sorry

Best Answer - Chosen by Voters: He invented windsurfing. He is not into football.

Day 18 Of The Comeback

Love Tree did their daily Facebook posting, and I liked it. This one said, At the centre of your being you have the answer; you know who you are and you know what you want."

My former girlfriend in Edmonton woke me up at 8:30, on her way to work, and said the same thing, basically. I told her I thought my condition was caused by hips out of alignment and B6 plus a hit by a chiropractor that hurt my sciatic nerve, and not one of the scary diseases like cancer, AIDs, MS, ALS, West Nile, lyme ticks, and on and on. That's good, she said. Well, it sounds like you've got a good handle on it.

I told her I didn't want to go to Edmonton on Wednesday because it seemed unnecessary at a time when rest and exercise were, in fact, necessary.

She said, Well, it's good to get those other things ruled out.

And I said, It's better to get what you need than run around chasing long-shots.

She had to agree with that.

But then she saw a snowshoe hare and had to get a picture of it, so we hung up and she called me back after a few minutes.

But I had to go to Jiffy Lube.

At Jiffy Lube, I had to get in and out of the car a few times, stand around for fifteen or twenty minutes, and it was no problem.

My leg muscles feel tight, or something, as though they need to be warmed up and stretched.

Day 19 Of The Recovery -- On Steroids

The Love Tree did a double posting in the morning: By choosing your thoughts, and by selecting which emotional currents you will release and which you will reinforce, you determine the quality of your light. You determine the effects that you will have upon others, and the nature of the experiences of your life. -- Gary Zukav

Even if we don't genuinely feel that we are able to 'let go' we are able to say 'May this seemingly negative connection be our link to waking up. ~Pema Chodron

The phone rang and woke me up but I was happy. My brother phoned last night and we talked for over an hour and I was happy. My symptoms are still the same, but I feel happy.
    Is it the result of feel-good philosophy? Or is it the fact that my new doctor gave me stronger drugs?
    Hint: My philosophy of life did not change overnight!

My former girlfriend phoned on her way to work to report that it was snowing but not sticking down in Edmonton, so I rolled over to look out my window and saw that Cold Lake had been transformed into a winter wonderland. The black spruce trees outside my window were white. So were the rooftops and the other trees and the grass and trucks that were parked outside overnight. Everything was covered in white stuff except the roads and sidewalks. And it looked beautiful.

She sounded worried about the stories she was working on, in Edmonton, about oil sand, the movement to refine it here, and Chinese investment here. That's what she was working on.
    "Don't worry," I said. "It's all good."
    That's not what I usually said about oil and Canada's place in the world. In the summertime, I spent a lot of time working on a thriller novel about terrorism in the tar sands. But this morning it looked to me as though it would all work out.
    Those drugs must be good, she said.
    She told me that lately she has been making the same wish for all the people she knows who are going through difficult times, and it was no longer "hugs", it was now "hugs and drugs".
    And I had to agree with her.
    The three drugs the doctor prescribed for me were definitely working.

My brother told me all about Thanksgiving, and how they had renamed it Misgiving.
    He said he took his new girlfriend to dinner at our other brother's place.
    Why? I said. I thought you liked her!
    He laughed and said, It wasn't a test or anything.

He had another motivation.

He said his girlfriend had kids and she thought her little family was dysfunctional because her kids were grown but still came home and acted like kids, instead of adults. So my brother said, If you think this is dysfunctional, you ain't seen nuthin' yet!

He described Thanksgiving dinner with our brother and his family and it was nothing new but this time it made me laugh instead of feel bad.

I had already taken the drugs the doctor prescribed.

Instead of driving from Cold Lake to Edmonton in the snow, at a time when nobody had snow tires on their vehicles, yet, since it was summer a few days ago, I slept in and felt fortunate I didn't have to drive anywhere. The only place I was supposed to go was a few blocks away. I had an appointment with the gifted reflexologist.

I was very happy I had kept that appointment and not the one at the Glenrose.

What would you rather do: Drive for six hours in the first snow to be poked and prodded, with electrodes stuck into your skin, at the big rehab hospital, great reputation or not, or sit in a comfy chair and get a great foot massage?

Ha!

I phoned around looking for a physiotherapy place that could get me in sooner rather than later than the 22nd. And then I noticed Love Tree had posted two more quotable quotes on Facebook. Usually, they just post one a day. I wondered what was going on with them.

The first one said, "There will come a time when you believe everything is finished. -- That will be the beginning."

The second one was from Lao-Tzu and it said, The key to growth is the introduction of higher dimensions of consciousness into our awareness."

In the time it took to write down those two quotes, I got a response from the physiotherapy clinic I had called just before that. It was a private clinic in the next town so it probably wasn't "covered".

Did I care?

Yes!

But I said, "Sign me up" anyway as they had an opening on Monday. That would be the 15th. Instead of waiting until the 22nd for an appointment in Lac La Biche or the 30th in Cold Lake, I could get into the place in Bonnyville on the 15th. "Sign me up!" I said.

Health First Physiotherapy had a website that said, Physiotherapists have advanced understanding of how the body moves, what keeps it from moving well and how to restore mobility.

The website had some more info on physiotherapy and what they had to offer. Here's what it said:

What is Physiotherapy?
Physiotherapists manage and prevent many physical problems caused by illness, disease, sport and work related injury, aging, and long periods of inactivity.

Physiotherapists are skilled in the assessment and management of a broad range of conditions that affect the musculoskeletal, circulatory, respiratory and nervous systems.

Here are some of the ways physiotherapy can help:

Address physical challenges associated with back pain, arthritis, repetitive strain injury etc.

Attend to sports injuries and provide advice on prevention and recurrence

Direct care for children with paediatric conditions such as developmental delay, fractures and cardiorespiratory conditions

Get you back on your feet after surgery

Help you manage the physical complications of cancer and its treatment

Manage incontinence

Maximize your mobility if you have a neurological disorder such as stroke, spinal cord injury or Parkinson's disease

Oversee rehabilitation in your home after you have been ill or injured

Provide pre- and post-natal care and attend to other women's health conditions

Treat neck and back pain and other joint injuries

Work with you to treat and manage respiratory and cardiac conditions

After finding physio, I felt like falling asleep again, but first I went to Youtube and played "Don't worry, be happy", the song by Bobby McFerrin.

Here is a little song I wrote
You might want to sing it note for note
Don't worry be happy
In every life we have some trouble
When you worry you make it double
Don't worry, be happy....

There were a few versions of the song on Youtube so I played them all. The Bob Marley version was great.

The phone rang a few minutes before my alarm went off. It was Nola, the nurse from the Edmonton Clinic, looking for results from the spinal tap. I told her it was postponed because the doctor and the lab weren't ready. I also told her I had found a doctor and thought he was good and I had booked physiotherapy for Monday.

Good, she said. It's important to have a doctor you think is good. And it's very good that you're going to physiotherapy sooner.

I told her I was still walking a bit more every day and she said that was excellent.

Mecell phoned while I was writing up my notes on the nurse's call. She said I was late for my appointment with her.

The last hour had flown by in seconds, somehow. I felt as though I had lost an hour.

"Don't worry," she said. "I have lots of time today, so just get over here as soon as you can."

It was another great Mecell visit. She said she could tell that I was taking drugs because my feet responded differently. She said it made it a bit harder for her to do her job, but she could still do it.

She told me she had a story she wanted to share with me as she had a client who was in the same shape as me but was a thin, young, beautiful woman from Northern Saskatchewan. She was a model but she developed a severe back pain and had been going to doctors for months without any relief. Her mother was talking to the reflexologist and Mecell said she gave her a gift certificate so she could try something different.

I specialize in taking pain away, she said, so what does she have to lose? But she was hesitant.

To make a long story short, the mother took the daughter in, Mecell worked her magic, and she walked out smiling and laughing.

Her mother looked shocked as she had not heard her daughter laugh for months or seen her smile. She said, You look beautiful, again, because you're no longer just grimacing in pain!

As I was leaving reflexology, my phone rang, as the Primary Care Network had an opening, they said, if I could get there in the next half hour.

A nurse fixed up my ears for me.

She asked me how I was doing and pushed a bit so it wasn't just small talk so I told her. She responded by looking up all my information on the computer right in front of me.

She said the drugs I was prescribed yesterday weren't noted in the system, yet.

While I was in the hospital, I grabbed a sandwhich and a salad plus a V8 and then headed home to make some phone calls and cancel a few appointments since I found alternatives that were earlier. And then I phoned to make sure I had a follow-up appointment with the doctor in Lac La Biche.

It was Wednesday and I was happy I did not have to see another doctor for a while or go to any appointments until Monday, when I would go to physiotherapy in the morning and then have reflexology in the afternoon. And that Friday, the 19th, at noon, I would see Dr. Birkill again.

It was close to four o'clock when I got home. I spent around an hour on the phone, making and cancelling appointments, so I forgot about going to the drug store to pick up the Dr. Ho belt they ordered for me, as I felt a need to lie flat for a while.

In the evening, I played Scrabble, chatted online, and exchanged a few e-mails, and then looked for a comedy to watch on Netflix. My choices for the night were the Ghostbusters movies, which I hadn't seen since they first came out. Ghostbusters, in case you missed it, is a supernatural comedy, multi-media franchise created in 1984. Its first product was the movie Ghostbusters, released in 1984 by Columbia Pictures. It centers around a group of eccentric New York City parapsychologists who investigate and capture ghosts for a living.

It was directed by Ivan Reitman and written by Dan Aykroyd and Harold Ramis. The film stars Bill Murray, Aykroyd, and Ramis as three eccentric parapsychologists in New York City, who start a ghost catching business. Sigourney Weaver and Rick Moranis co-star as a potential client and her neighbour.

The film was followed by a sequel, Ghostbusters II in 1989, and two animated television series, The Real Ghostbusters and Extreme Ghostbusters. As of October 2012, a third feature film still remains uncertain.

Day 20: Broke Back Martin

Pastor Malory flung himself off the bell tower and plummeted like a gigantic bird with broken wings, splattering his brains like so much bird shit when he hit the street below.
-- Mo Yan, Big Breasts & Wide Hips, 2012 Nobel Prize Winner

Yo, man! Woke up at ten to discover I had not won the Nobel Prize for Literature. Again.

Okay, I wasn't nominated, so I figured it was a bit of a long shot.

However, long shots often win the big prize.

This year, the Nobel Prize for Literature went to Mo Yan, from China.

You've heard of him, right?

And you can list all the Canadian authors who have won the Nobel for their books, too, right?

Yes you can! -- If you say nothing, you are right. -- Nobody in Canada has ever won the Nobel Prize for Literature.

Not Alice Munro, not Margaret Atwood, not Robertson Davies or Margaret Laurence. And not .... me!

This year the prize goes to Mo Yan, a Chinese author described as "one of the most famous, oft-banned and widely pirated of all Chinese writers".

"Mo Yan"—meaning "don't speak" in Chinese—is a pen name. The Chinese writer Ma Jian has deplored the lack of solidarity and commitment of Mo Yan vis-a-vis other Chinese writers and intellectuals who were punished and/or detained despite the freedom of expression recognized by China's Constitution.

The 57-year-old is the first Chinese resident to win the prize. Chinese-born Gao Xingjian was honoured in 2000, but is a French citizen.

Mo Yan has been censured for hand-copying Mao Zedong's Yan'an Talks on Literature and Art in commemoration of the 70th anniversary of the speech

The news about the prize came with an AP photo of the Chinese writer smoking a cigarette during an interview in Beijing.

The Swedish Academy, which selects the winners of the prestigious award, praised Mo's "hallucinatoric realism" saying it "merges folk tales, history and the contemporary."

The prize is worth about $1.2 million.

They praised Mo's "hallucinatoric realism" saying it "merges folk tales, history and the contemporary."

Mo was an unexpected choice by a prize committee that has favoured European authors.

What about Larry and Curly? many writers were asking. Why Mo?

The Chinese government disowned the only previous Chinese winner of the award, who had escaped from the totalitarian country to live in the west. A Chinese emigre to France, Gao Xingjian, won in 2000 for his inventive fiction, especially the novel Soul

Mountain. His works are laced with criticisms of China's communist government and have been banned in China.

I read Soul Mountain, translated into English, and I liked it a lot.

When Mr. Gao won, the communist leadership disowned the prize.

In China, national television broke into its newscast to announce the prize — and that is something almost never done on the tightly scripted broadcast that usually focuses on the doings of Chinese leaders.

Mo is best known for his 2004 novel called Big Breasts & Wide Hips.

He's written 11 novels and let's say a hundred short stories.

Jailed dissident Liu Xiaobo won the Nobel Peace Prize in 2010, which infuriated the Chinese leadership. The communist leadership also disowned the Nobel when Gao Xingjian won the literature award in 2000.

Gao's works are laced with criticisms of China's communist government and have been banned in China.

Mo chose his pen name while writing his first novel. Garrulous by nature, Mo has said the name, meaning "don't speak," was intended to remind him to hold his tongue.He had to mask his identity since he began writing while serving in the army.

His breakthrough book was the novel 'Red Sorghum, set in the anti-Japanese war. It was turned into a film that won the top prize at the Berlin International Film Festival in 1988.

Mo writes of visceral pleasures and existential quandaries and tends to create vivid, mouthy characters, the say.

I wouldn't know. I've never read him or heard of him.

Reports say he uses raunchy humour and has become more experimental, toying with different narrators and embracing a free-wheeling style often described as 'Chinese magical realism.'

The Nobel jury has been criticized for being too euro-centric, not to mention anti-Canadian.

Mo was among those getting the lowest odds on betting sites before the announcements.

Mo Yan, a pseudonym for Guan Moye, was born in 1955, the same years as me.

As a 12-year-old during the "Cultural Revolution", he left school to work in agriculture and later in a factory. In 1976 he joined the People's Liberation Army and began to study literature and write. In 1991, he got the Master degree in Literature from Beijing Normal University.

TIME described Mo as "one of the most famous, oft-banned and widely pirated of all Chinese writers". He has been referred to as the Chinese answer to Franz Kafka or Joseph Heller. His style is strongly influenced by the magical realism of Gabriel García Márquez. He uses dazzling, complex, and often graphically violent images.

His latest novel, called Life and Death Are Wearing Me Out, is about reincarnation.

It describes China's development during the second half of the 20th century through the eyes of a noble and generous landowner who is killed and reincarnated as various farm animals in rural China.

Mo's main protagonist in the novel, named Nao, is targeted during Mao Zedong's land reform movement in 1948 and executed so that his land could be redistributed.

Upon his death, Nao finds himself in the underworld, where Lord Yama tortures him in an attempt to elicit an admission of guilt. In subsequent reincarnations, he goes through life as a donkey, a pig, a dog, and a monkey, until finally being born again as a man.

Through the lens of various animals, the protagonist experiences the political movements that swept China under Communist Party rule, including the Great Chinese Famine and Cultural Revolution, all the way through to New Year's Eve in 2000. The author, Mo Yan, uses self-reference and by the end of the novel introduces himself as one of the main characters.

Big Breasts and Wide Hips is a novel that spans the 20th century, about a woman named Shangguan Lu, born during the Boxer Rebellion. Her father is murdered by German soldiers and her mother follows soon after, killing herself. She is brought up by an aunt, goes on to marry a husband "as useless as a gob of snot", and then has a series of liaisons that result in eight children, including the much-treasured son, "Golden Boy", who narrates the novel.

Shangguan Lu is described as an indestructable woman who outlives her eight daughters, the various men who fathered them, her husband, her ferocious mother-in-law, and her fellow villagers. They all, with a few exceptions, perish in the waves of war, famine, and Communist enforcement that bathe the region in suffering.

Mo tells the story of her son, Shangguan Jintong, the only child who outlives her. For his whole life, Mo writes, he basked in the "glorious tradition of Shangguan women, with big breasts and wide hips." He is obsessed with breasts and breast-feeding, which Mo describes with an extraordinary vividness

John Updike reviewed the book in The New Yorker, years ago, and he said, "So impressive and ardent are Jintong's evocations of nursing's primal pleasure that this reader was slow to realize that Mo Yan intended our hero to be not a healthily typical male but a case of arrested development." But it's comedy, he concludes, with a rough edge of accusation: evidence, Updike writes, that "bad societies offer no incentive to grow up."

The book caused controversy in China when it was published in 1995. Because of its sexual content and the depiction of class struggle contrary to the Chinese Communist Party line, the author was forced by the PLA to withdraw it from publication although it was pirated many times.

The People's Liberation Army (PLA) is the military arm of the Communist Party of China and the armed forces of the People's Republic of China. The PLA is the world's largest military force, with approximately 3 million members, and has the world's largest active standing army, with approximately 2.25 million members.

After Big Breasts & Wide Hips was translated into English a decade later, the book won him a nomination for the Man Asian Literary Prize.

His latest novel, Frog, about China's "one child" population control policy, won the Mao Dun Literature Prize - one of his country's most prestigious literature prizes - last year.

Life And Death Are Wearing Me Out

A couple of years ago, I asked a Zen master about reincarnation. For several years, I went to the Zen Forest, a retreat centre in the Far East of Ontario, every weekend, holiday, and summer, to do volunteer work, practice meditation, and learn from the Buddhist monk who was building the place.

I had many reasons for going, even though I'm not a Buddhist. I liked Zen and I felt spiritually exhausted. An old friend from Alberta had committed suicide after getting attacked by a grizzly bear and enduring countless surgeries that did not work. She had nightmares about grizzly bears performing surgery on her. An old girlfriend in Ontario left her husband, or was thrown out of his big house for fooling around, and then broke up another family to run off with the husband when his wife and daughters threw him out. I was working as a teacher and education had entered a period of decline, after the turn of the century, in Ontario.

Life and death was wearing me out.

Thich Thong Tri Thay, the monk and Zen master, showed me how to put together three powerful things: Zen meditation, solo massage like Reiki self-healing, and energy exercises like qigong.

Thay often talked about reincarnation. He said we might meet again in the Zen Forest, in the future, after this life. Sometimes he said, You could be reincarnated as a tree in the Zen Forest. Or a mosquito.

He also told me a story about reincarnation that made me "wake up".

One day he said, "I don't know where you learned this nonsense, but there is no such thing as reincarnation, soul, higher self, and the rest of that mambo jumbo rubbish!" he said. "You get one crack at this life and if you screw it up, that is it. There are no second chances. Either you get your act together and awaken, or you lie on your death bed and lament wasting this most precious of opportunities."

"But, Thay," I tried to argue. "Everybody knows that reincarnation is the religious or philosophical concept that the soul or spirit, after death, begins a new life in a new body that may be human, animal or spiritual depending on the moral quality of the previous life's actions. This doctrine is a central tenet of the Indian religions and is a belief that was held by such historic figures as Pythagoras, Plato and Socrates. It is also a common belief of pagan religions such as Druidism, Spiritism, Theosophy, and Eckankar and is found in many tribal societies around the world, in places such as Siberia, West Africa, North America, and Australia.

Although the majority of sects within Judaism, Christianity and Islam do not believe that individuals reincarnate, particular groups within these religions do refer to reincarnation; these groups include the mainstream historical and contemporary followers of Kabbalah. In recent decades, many Europeans and North Americans have developed an interest in reincarnation. Contemporary films, books, and popular songs frequently mention reincarnation.

"Shut up!" he answered. "It is all lies. There is no reincarnation and being lazy and wasting this life is the greatest sin you can commit."

And that was the end of that.

He told me the old Zen story about The Secrets Of The Universe. It's also known as the story of Totonaka and Master Blumise at the Nowind Monastery. Blumise gave Totonaka the same information and advice.

They were both transformed quickly, after that.

The transformation of Junjo was as remarkable as that of Totonaka. In a few short months, each of them got his body and mind in shape by ending their indulgent ways, and made enormous spiritual progress by meditation regularly, intensely, and seriously.

I left the Zen Forest after that and went back to work. I moved out west, to Alberta, grizzly bear territory, and took a job as a teacher in a place where they bragged about education being a very high priority.

I was happier as I no longer had a lot of questions about the universal secretes. Is reincarnation true? How about an afterlife? Karmic Laws? There is no such thing as reincarnation, soul, higher self, and the rest of that mambo jumbo rubbish, the monk and Zen master told me. -- Wake up!

Day 20 Of The Recovery, Part 2

Around noon, I tried taking a few steps. I walked from the couch to the kitchen. My legs hurt in a new way. They felt stiff and sore, as though I had just completed a marathon.

Part of me wanted to walk 3,000 steps and do a workout that combined qigong and yoga with the exercises recommended by The Back Doctor. -- But that part of me was not my legs.

My brain and my body had an argument. And my legs won.

"You're not going anywhere without me," they said.

So, I went back to bed and slept until almost five.

I had dreamy dreams about who knows what. In one of them, I met Mo Yan and said, Yo, Man, wussup? In another one, I met his indestructable woman and we hit it off so I wanted to ask her the questions on the Proust questionaire. In another one, I had won the Nobel Prize for Literature and I was described along the same lines as Mo Yan.

When he was twelve years old, he was the captain of his peewee hockey team, the Indians. Instead of going to the small high school in his hometown, where it was easy to win all the awards in sports and academics, he chose to go to Bracebridge and Muskoka Lakes Secondary School, which was four times as big, for more competition. His soccer team went to the All-Ontarios and his Jr. C. hockey team won the provincial championship. He started writing very young and was published while in high school as a regular columnist for the Bracebridge Herald-Gazette.

He attended the University of Victoria, Banff Centre School of Fine Arts, York University, New College Writers Workshop at the University of Toronto, for an Hons. B.A. in Indivualized Studies (Literature and Writing); Vermont College of Norwhich University, for an MFA in Writing; the University of Toronto, for a B.Ed.; Queen's University's Herstmonceux Castle for an Hons. Specialist in English; and the Stratford Campus of UT/OISE for A.Q. in Drama.

He had five books published by Oberon Press, Ottawa, by age 30.

He was the Arts Department Head of a small school in Muskoka, the Writer in Residence at the Bracebridge Public Library, the Pickering Public Library, and the Zen Forest. In his mid-fifties, he moved to Alberta and added an ATC to his OTC, and taught elementary as well as high school English.

A very prolific writer, he is the author of over 100 books and thousands of articles for Bleacher Report, Suite 101, and mainstream print newspapers. He was the founder of A Novel Marathon in Owen Sound, the Muskoka Novel Marathon at the Huntsville Festival of the Arts, The Great Canadian Winter Novel Marathon, and the Port Perry Poetry Marathon. He won the awards for Spirit and Most Prolific at the MNM.

He won the Balzac Award for poetry and four of his novels have been nominated for the Leacock Award for Humour.

He coached soccer for twenty-five years, basketball for a few years, swimming for a year, and drama for fifteen years. His senior drama group won the Festival Théâtre jeunesse de l'Alberta.

He is the first Canadian to win the Nobel Prize in Literature and was considered a longshot to win over Margaret Atwood and Alice Munro. He said, It's about time a Canadian was awarded this prize. Robertson Davies, Mordechai Richler, W.O. Mitchell, Clark Blaise, and both Atwood and Munro should have won the award before me."

When I regained consciousness, as Bobby Clobber used to say, I realized I had slept the day away, and I wondered what was in those drugs the doc gave me!

Day 21

A phone call from work woke me up around eight thirty. The principal said I should drop by for a visit because all the kids wanted to see me. But I said, I don't think they want to see me walking like a zombie. So she sang, You're so vain ....

She said there was a problem if I couldn't get the specialist to write a letter saying I should be off work and I should get Short Term Disability and Long Term Disability. I told her the new doc gave me a note but she pointed out there would be a gap.

You should push the specialist for that letter, she said, otherwise you will wind up going for weeks without pay.

My friend in Edmonton called right after that and it was hard to hear her but I think she said she was okay, her doctor said she looked pale, so she had to have some blood work done, and she said, Don't worry about the doctor's letter, just focus on getting better. Your doctor will give you some sort of note or letter later to cover the time you were off.

Instead of worrying about it, I checked e-mail and Facebook and then went to Yahoo.ca for the news of the day. Mo Yan was getting a lot of criticism the day after he won the Nobel Prize for Literature.

"Giving the award to a writer like this is an insult to humanity and to literature," noted Chinese artist and activist, Ai Weiwei told the British newspaper The Independent. "It's shameful for the committee to have made this selection which does not live up to the previous quality of literature in the award."

Ai's diatribe toward Mo appears to be rooted in part to his work on a book last year to celebrate the 70th anniversary of a speech given by Mao Zedong, according to TODAY.com.

Mao's speech, known as the "Speech at Yan'an Forum on Art and Literature" set the guidelines for appropriate subject matter for Chinese writers and artists of that period, calling upon them to focus on and espouse the merits of Communism and threatening punishment to those who did not bend to the will of the party.

Mo Yan and around 100 other Chinese writers and artists hand-copied paragraphs from the speech for the book.

That act, in conjunction with Mo's position as vice chairman of the government-backed Chinese Writer's Association, which has failed to voice support toward fellow writer Liu Xiaobo's Nobel Peace Prize victory, has raised the ire of artists like Ai who wonder just how committed Mo Yan is to free expression.

In China, state media gave the official stamp of approval Friday over the decision to award the Nobel Prize for literature to Mo Yan, giving him front-page coverage across the country.

The warm coverage of the award is unsurprising considering the prestige and recognition that China's ruling Communist Party will collectively bask in as a result.

But in another sense, the warm reception for the awarding is striking considering the anger and hysteria drummed up by Beijing following the 2010 awarding of the Nobel Peace Prize to imprisoned political dissident, Liu Xiaobo.

"This is the first Chinese writer who has won the Nobel Prize for Literature," gushed China's People's Daily newspaper. "Chinese writers have waited too long, the Chinese people have waited too long."

But critics of the Communist regime point out that Gao Xingjian, who won in 2000 in part for his critical writing of the government, was China's first winner of the Nobel for literature. He had been exiled to France by the time the prize was awarded.

Liu's name and the term "Nobel Peace Prize" remain blocked terms on China's twitter-like service, Weibo.

Earlier this week, a BBC report on Liu's imprisonment noted that the activist and his wife, who remains under illegal house arrest, have been facing extraordinary pressure to accept exile from China in exchange for their freedom.

(NBC News' Johanna Armstrong and Yanzhou Liu contributed to this report on stories in the BBC, Today.com, and The Independent.)

After that I had a shower and get ready to go out, but fell asleep for a few hours.
Deep sleep. Deep, restorative, sleep. -- That's what it felt like, anyway.

After two days of rest with very little walking, I ventured outside again. After stopping at the drug store, I went to work, and I hit a grocery store on the way home. Walking felt different. It was a little painful on both sides of my butt. Also, I didn't feel as though I could walk 2,000 steps. I had a strong desire to get home and go back to bed.

At work, the teacher replacing me wanted to talk, so after checking in with the head secretary and the principal, and getting a hug from the librarian, we walked into the staff room and yakked for an hour or so. -- She was a little anxious about doing a good job so I told her I thought she was doing great. And I told the principal that, too, when she came in. And hit her up for a computer to run the SmartBoard in the elementary English classroom.

Being in the school building reminded me of how much I love the place and the people, but standing, walking, and standing some more made me feel tired and reminded me my balance is terrible. I almost toppled over a few times.

When I got home, I found a long message on Facebook chat from a friend of mine in Vancouver who I call Zenda and she was reporting on the work she was doing in the world of healing. I've been encouraging her to do Reiki, to become a Reikimaster, because I have a feeling she would be good at it.

Zenda: I've been seeing a healer who is doing "theta healing" on me.

Me: I'll have to look up theta healing.

Zenda: Theta Healing is a powerful healing technique that addresses the limiting subconscious beliefs that hold us back from reaching our fullest potential, our most optimal health, and our deepest joy. Using ThetaHealing, limiting subconscious beliefs are instantly and permanently shifted into beliefs that are self-empowering and aligned with what we consciously desire. Because ThetaHealing taps into the unconditional love-energy of the Creator, countless people have experienced instant physical healings ranging from the reparation of a broken bones, to the disappearance of back pain, to instant and permanent remission from cancer. ThetaHealing is a phenomenal healing modality that is compatible with all other conventional and non-conventional treatments.

Me: so ... how do you do it!

Zenda: I think it would be best with a trained practitioner. They help you to uncover negative subconscious messages that you weren't aware of (in a meditation) and then undo them, with words and energy and replace them with something positive and in alignment. I've had 3 sessions. In one, I burst into uncontrollable tears when a particular memory came up (something I've never done before). She said it was very deep. At another one, I heard a Northern Flicker calling out in the background during the meditation, so we looked it up in a spirit guide book and its message was perfect for where I am right now! She is also using Reiki (when necessary) to move sluggish energy out during her theta healing. We just cleared out some draining ties to a few old relationships the other day, and the next day, someone associated to one of the relationships (over 30 years ago) popped up on my FB with a friend request! And BTW, she is only acting on what I say, and asks permission at every step along the way.

Zenda's report made me wonder. Were there negative subconscious messages I had to uncover and replace with positive messages for a dramatic healing?
    I thought I would stick with the mainstream medical program, plus reflexology, for now.

At the end of the day, I noticed this posting on Facebook:
"Every situation properly perceived, becomes an opportunity to heal." ~ A Course In Miracles

Day 22 Of The Comeback?

Be not angry that you cannot make others as you wish them to be, since you cannot make yourself as you wish to be. ~ Thomas a Kempis, posted by Love Tree

I put a question mark on this chapter title because things feel different, now. For eighteen days, I felt better and could walk better every day. I've been telling people, and myself, the same thing since then. But the truth is .... These past four days have been .... different.

Every day is different. The back pain came back, but it comes and goes. My legs have more feeling in them, but it's not a good feeling. They are stiff and sore.

"Have you overdone it?" my friend in Edmonton asks.

"Don't kick a guy while he's down," I told her.

After driving to LLB and back, meeting the doctor, getting stronger drugs, things have changed. I spent two days in bed. That was Wednesday and Thursday. On Friday, I dropped by the school where I work. People who hadn't seen me for two months looked horrified to see the way I walked. People who had seen me with a walker looked very happy to see that I could walk without assistance.

Last night, after yakking for over an hour at work, I drove home and went to bed, but then I watched a movie, traded some e-mails, chatted on Facebook, watched another movie, and realized it was going to be a long night.

Sleeping has not been a problem. Doctors assumed I wasn't sleeping and that I had a lot of headaches, but that wasn't true. However, last night was very long.

It's Saturday and I woke up at 9:30 ish with a novel in mind.

The last thing I read last night was my friend Zenda's notes on theta healing, whatever that is. She said "uncover negative subconscious messages that you weren't aware of (in a meditation) and then undo them, with words and energy and replace them with something positive and in alignment."

My first reaction was that I had already done a lot of work like that and I felt certain I no longer had any negative subconscious messages I had to undo and replace. I meditate, I write, I've uncovered my subconscious and discovered negative messages and I've erased them by re-scripting. I've written books like that.

But after staying up late and waking up a few times at night, I got up in the morning thinking about one of the schools where I worked for a time and did not have a good time.

Well, it was mixed. I had a good time working with students, coaching soccer and swimming, and drama, but I had a hard time working with the other teachers and the principal. I got along famously with the vice principal and with the school council and with the principal who hired me. But she got kicked upstairs and the guy who replaced her was a close talker with bad breath who reminded me of my father and a weasel.

I don't want to mention any names, so there's no lawsuit. Maybe I will write about it as fiction, change the names and dates, and deal with it that way.

As a teacher, I changed schools every six years, to stay fresh, and then, for six years, I changed schools every year. So, I've worked at ten different schools, including one in Alberta. -- Not including my first job as a teacher, in a prison in Manitoba, before I had training as a teacher, as it was not in a school, I was on my own, more or less, working with Frontier College, which is like Canada's answer to the Peace Corps.

Everybody in Canada has heard of the Peace Corps, which is American, but not everybody has heard of Frontier College, even though it has been around for a long time, putting people in place to work with adults who have problems with literacy.

My friend Jackie in Parry Sound said she had an idea for Nanowrimo, or national novel writing month, which takes place in November, and I've done that several times, but I wasn't planning to do it this year. I was planning to get back to work and I knew it would be a lot of work, taking over my classes from four supply teachers and getting them ready for exams.

If I was going to do Nanowrimo, and write 50,000 words in one month, I had a good idea for it. I would write about the high school from hell!

But I didn't want to do that. I wanted to be back at work by then. So, maybe I would have a weekend novel writing marathon and knock it off in two or three days.

It was not a big novel idea. It felt like a short novel idea.

It was just after the province-wide teachers' strike in Ontario, when the four teachers' unions came extremely close to overthrowing the government of a guy named Harris. And now the teachers of Ontario are preparing to go on strike again, even though the current premier has made it illegal for them to strike.

The government has frozen their wages and clawed back sick days, too..

They are unofficially working to rule, cancelling extra-curricular activities. In other words, they are still at work, doing their jobs, but they are not coaching teams or running clubs after school.

Back in the day, I joined in the strike. I was a picket captain and got right in to the thick of things. But when the strike was over, and there was an unofficial work to rule campaign, I didn't like it. And I didn't go along with it.

I thought the schools in Toronto and other cities should do it, because they would get media attention and their students had other options. I was working in a small school in a small town and if our students didn't play for a school team or join a school club, there were no alternatives for them in town. In the city, kids could walk down the street and join a community club or team. -- Many of them.

So, the teachers at my school shunned me. They wouldn't work with me or even talk to me. On the other hand, the principal, vice principal, secretaries, school council, and the students backed me more than one hundred percent. I was the Teacher Of The Year for three or four years. I was the Arts Department Head and we were working on a plan to turn our school into the designated Arts school of the region.

That job action by the teachers' union faded away, our principal got a promotion, and our school got a new principal, who was a jock. He wasn't just disinterested in the Arts, he was against the Arts. He was interested in sports.

We didn't hit it off.

To say the least.

Yes, maybe there is a novel in that!

Everyone who has gone through high school knows it can be hell.

Adolescence is one thing. But certain high schools are more horrible than others.

Imagine going to William McKinley High School, as seen on the TV show called Glee. Or North Shore High School, from the movie called Mean Girls. Harry Potter's Hogwarts was something else. Before that, there was Degrassi High, the Rosewood High of Pretty Little Liars, and Bates High School, for Carrie.

Twilight had Forks High School and Buffy The Vampire Slayer had Sunnydale High.

The Heathers had Westerburg High School and the Big Wolf On Campus had Pleasantville High.

Marty McFly had a hard time at Hill Valley High School.

I can think of half a hundred movies set in high schools, and I don't think I would have liked any of them: Dead Poet's Society, The Breakfast Club, Donnie Darko, Ferris Bueller's Day Off, Grease, Rebel Without A Cause, even Hoosiers and Fame.

Personally, I loved my high school. Bracebridge and Muskoka Lakes Secondary School. I think I was in love with it!

Instead of going to the small school in my hometown, I took the bus to the next town in order to go to a much bigger high school. So, I was the captain of the hockey team in one town, going to school with the guys in the town that was our arch-rival and nemesis in hockey. It could have been a horrible disaster. But I had a blast!

But that's another story.

It might be related, as I returned to my old hometown as a teacher, decades later.

Was I still angry about things that happened in that high school?

No. I was just angry with myself because I couldn't make myself walk!

I should heed the words of Thomas A Kempis, quoted by Love Tree on Facebook today: Be not angry that you cannot make others as you wish them to be, since you cannot make yourself as you wish to be.

Thomas A Kempis wrote The Imitation Of Christ, one of the best known books of Christian devotion.

One of his other famous saying is "In angello cum libello", which means, "In a little corner with a little book" and is a shortened form of a motto often ascribed to him. The complete saying is a mixture of Latin and Dutch and runs as follows: "In omnibus requiem quaesivi, sed non inveni, nisi in hoexkens ende boexkens", or "I have sought everywhere for peace, but I have found it not save in a little nook and in a little book.

I posted that as my Facebook status for the day.

Only one person commented on it. A friend of mine, from childhood, now living in Gulf Shores, Alabama, gave me an F for spelling.

I was at 66,666 words in this book I noticed. So I added these lines!

Day 23 Of The Recovery

This morning I woke up with a little spark of energy. Yesterday, Saturday, I took the day off to rest, rest, rest. The only walking I did was from the couch to the kitchen and bathroom and back. I was hoping all that rest would be followed by a sudden burst of energy.

At six thirty, I got up, onto my hands and knees, and did a few exercises, and that went well, but when I stood up I felt a sharp pain in my back. The only way to walk without pain was to bend over like a very old man with a cane. But I didn't use a cane. I walked to the kitchen like that and got an ice pack for my lower back. And then I could walk like homo erectus instead of like a gorilla.

My ankles feel better and my feet are better than they were, but not that great. The numbness and tingling or pins and needles is still there but it's not as intense. I have a lot more strength in my feet than when I went to the Edmonton Clinic and the doctor tested them.

He held the bottom of my feet, one at a time, and said, Press down the way you would if you were driving. And then he urged me to press harder and harder. I didn't have much strength at the time. But I do now.

He asked me to walk on the balls of my feet and the heels of my feet and I couldn't do it. But I can do that now.

Instead of crawling back into bed, I am trying to be optimistic and enthusiastic, as I know I need exercise but I don't really feel like it. The spirit is willing but the body sends different messages to the brain: It hurts, it's going to hurt, let's give in to fear.

In the battle between the body and the brain, guess who wins?

It's like the Leafs playing the Canadiens in 1967.

It's like the Oilers playing the Canucks in 2012.

-- That's a little joke. The Oilers should have a good team this year, finally, and the Canucks should be as strong as last year, at least. But who knows? The NHL season has not started. The players have been locked out by the owners. Lots of NHLers have defected to the KHL. That's the way I feel. My brain really wants my legs and feet to play, but it has locked them out, anyway.

On the internet this morning, I saw the news about a meningitis outbreak in the U.S.A. Two hundred people have died. A company that makes the steroid injections for people with back pain is getting the blame. Imagine: You have back pain, you try a lot of things to make it go away, the doctor sends you to the hospital for an injection of steroids, and you get meningitis from the serum. You go in with a pain in the back and you come out with headaches, slurry speech, and a lot of trouble walking. You have meningitis.

Nightmare!

Well, let's not focus on the negative, let's take the gorilla out for a walk!

In the movie called What About Bob, with Bill Murray and Richard Dreyfuss, they talk about a therapy called "baby steps", which is all about taking small steps instead of worrying about everything in the world. Instead of worrying about all the obstacles between you and a trip downtown, for instance, you focus on crossing the room, walking down the hall, getting into the elevator, walking down another hallway, crossing the parking lot, getting into your car, and so on.

Although I wasn't taking "baby steps", I was counting my steps.

First I went to the drug store, to get a few things, and I counted 500 steps. Next, I went to Walmart, for a few more things, and I counted around 1,000 steps. After that, I drove over to the provincial park and I counted another 1,000 steps.

One of the things I love about Cold Lake is the fact that it has a provincial park located so close to the little city. On my way there, I heard a CBC News report on the radio about provincial parks in Northern Ontario being closed to camping because of low numbers.

First they cut the budget for parks, reduced the number of people working in the parks, cut back on programs, and then, when fewer people used the parks, used that as an excuse to cut back even more. People in Northern Ontario were protesting that.

They were also protesting the sell-off of the railway, up north.

Toronto, the capital of Ontario, was so far from Northern Ontario, and southern Ontario was so unlike Northern Ontario, it is hard to imagine someone sitting in Toronto making decisions for Thunder Bay or Moose Factory. They are like different countries, not just separate provinces.

I'm not going to strain myself to make a metaphor here, saying Toronto is like your brain and Northern Ontario is like your legs and feet, and when you have trouble walking, Toronto is going to have about as much success fixing the problem as it does solving the problems in the NHL.

After walking through two stores, my legs needed a rest, so I sat in my car in the parking lot and listened to Micheal Enright interview the Archbishop of Toronto on the anniversary of Vatican II. They said it was getting to be time for Vatican III.

In Cold Lake Provincial Park, I walked from the parking lot to the beach and along the beach to a platform overlooking the lagoon and then up a little hill and back to the parking lot. Walking on grass, dirt paths, the beach, a gravel trail, was just a little more challenging than the flat, tile on concrete, floors of the stores in town. But it was more inspiring.

I love hiking and I haven't been able to get out there, into the natural world, for a couple of months, so it felt great to get a little taste of wilderness. The tall trees, the huge lake, the fresh air, and the autumn sun all gave me energy and lifted my spirits.

In the lake and on the pond, I saw ducks I had never seen before, but they were too far away for me to see clearly enough to identify.

It was only three degrees but I was warm wearing jeans, a tee-shirt, a shirt, and my barn jacket. This afternoon, it is supposed to be ten degrees warmer. I'm planning to go back and walk some more.

Is your glass half full, or half empty? It felt great to get outside and walk where I used to hike. But it made me feel sad to think that I could walk only a small fraction of the hike I used to do so easily.

One of the things you learn in Zen meditation is how to choose your thoughts. You practice letting go of your thoughts and that helps you develop the ability to pick and choose which thoughts you want to hold on to. Instead of feeling bad or sad about the fact I cannot walk anywhere close to the distance I used to be able to cover easily, I chose to feel glad that I can now walk a lot longer than I could a couple of months ago. -- A short hike in the woods sure beats crawling from your bedroom to the kitchen.

The Archbishop of Toronto had the same outlook on Vatican II. He said that old doctrines that were bad or sad were on their way out and they were going in the direction of celebrating those parts of the religion that made you feel glad.

On Facebook, Susan Swan reported from the WordFest down in Banff: An older, more frail looking Martin Amis spoke eloquently to a huge Banff WordFest audience last night about his new novel Lionel Asbo. Here are a few of his memorable quotes: "Mitt Romney looks like a porno star. He has one of those triumphant unlined faces that you only get from thinking you're on the way to a post millennial heaven." Two, "the novelist's gamble" is hoping their experience is universal. Three, all fiction since Samuel Beckett has come from the celebratory impulse and so does his new novel. Four, growing up with a controversial novelist father prepared him for the slings and arrows of the press. Four, (said afterwards over a cigarette in the hospitality suite): "Ah yes, the Booker. It glows in the dark. But remember, all prizes are consensual choices."

CBC Edmonton reported on Justin Bieber being in Alberta for concerts in Calgary and Edmonton. Apparently he said "Hello, Edmonton!" when he was on the stage in Calgary. Normally Calgary and Edmonton are very competitive, but the Biebe, they said, was forgiven by the crowds. And now Edmonton has Bieber Fever. His concert was sold out weeks ago and the only way to get a ticket now is to pay a scalper about four thousand dollars.

I have to say that I know this much is true: I've come undone but I'm wishing and hoping and thinking and praying for amazing grace.

Day 23, Part 2

After sleeping in the autumn sun, with the temperature around ten above zero, and the glass door open, for a couple of hours, I got in the car, drove to the park, and did the same circuit faster and with bigger strides so I had to walk around the parking lot, at the end to bring the total up to 1,000 steps. And then I did another 1,000 at Walmart.

My total for the day was well over 4,000 steps, and that was a new personal best since I lost my feet and legs, or they stopped working, or the time I could no longer walk.

That work-out made me tired and sweaty.

Each time out, I walked as far as I could, or until my legs were sore and my balance was gone, and I felt as though I was going to fall over.

Each time, it was my left knee that hurt the most.

Once again, I felt it was a short walk for a man but a giant step forward in terms of my recovery.

While I was walking, I wasn't thinking about much, as I was counting steps. It was like counting breaths, when you do walking meditation. I focused on my breaths, to some extent, but I walked faster and faster, which is not the way you do it when practicing walking meditation.

-- To walk meditatively, you co-ordinate your steps with your breathing, and you keep your focus on breathing in and breathing out.

Even so, I had a few thoughts.

First of all, I thought about the colour red and how much I miss it in the autumn out west and up north. Maple trees are missing from the local forest as they do not grow this far north, so the shades of red they add to the palette of autumn colours in the forest is missing.

Maple trees will grow up here, for a while, but they have not been able to survive the winters. When it is 40 or 50 below for a week or so, maple trees die, apparently.

I don't just miss them, I long for them.

The screen saver on my computer is a beautiful photograph of a forest of maple trees and all their leaves are a glorious shade of red.

Sometimes I Google images of maple trees in autumn.

And all those thoughts about red leaves, maple trees, and autumn led me to thoughts about Gabriel Garcia Marquez and his novel called The Autumn Of The Patriarch.

And that made me realize that this period of my life has taken me from the end of summer to the start of winter. It has been the worst fall ever. Autumn is my favourite season, usually, but this year it has been two months of missing work due to back pain and the inability to walk.

And what will winter be like? I'm wondering.

I cannot shovel snow or push stuck cars out of snowdrifts.

I want to skate. I want to ski.

I want to be strong enough to hike through the wilderness of the provincial park when the trail is covered with snow.

Last year, we had a mild winter, here, more like winter in southern Ontario than what I was told to expect up here. I've seen much more severe weather when I lived in central Ontario, Northern Ontario, and southern Manitoba.

It was unusual, the locals say. A fluke of nature. -- Nobody around here sees it as a trend caused by global warming.

Last year I wrote a book about it. -- Poetry. And the title was Celebrating Global Warming.

In this part of the world, we create and we celebrate global warming.

Canadians are concerned about global warming. But the vast majority of Canucks are happy to see our climate warm up by a few degrees.

Well, we would be, if it didn't mean wild weather and climate disasters in the rest of the world.

That's what I thought about while walking in the provincial park. My route takes me from a gravel parking lot to a hard packed dirt trail to a sandy beach and then up a little hill and across a grassy area, back to the parking lot.

I am looking forward to the day I can hike up the hill on the trail that is on the far side of the beach. It is long and steep, about one hundred yards at more than a 45 degree angle.

That will be the day I celebrate another "Rocky" moment.

The other thoughts I have while walking are about the novel idea I got a few days ago. Imagine a novel set in a high school after a big strike by the teachers with one teacher going against the union and the rest of the teachers going against him. Most high school novels and movies are about the students, or one student and their friends and enemies, or about one teacher who inspires his or her students. This one will be about a teacher who goes against the teachers' union to coach teams and run clubs to inspire students.

Robin Williams can play me, in the movie, and it will look like a sequel to The Dead Poets Society. What happened to his character after he got the boot from the private boys' school in the U.S.A. Maybe he moved to Canada and got a job in a public school in a small town a hundred miles north of Toronto.

However, instead of re-living that part of my life, and re-scripting it, maybe it would be smarter to write about something that doesn't take me back to hell and back.

The other thought I had while walking is that maybe whatever is going on with my feet and legs actually started a long time ago. Maybe my back pain and my walking issues are two issues that overlap.

In my twenties, I had burning feet. In my thirties, I was strong and healthy. In my forties, I had plantar's fasciitis. And when that pain ended, I felt an odd sort of numbness in my feet.

I remember going up north to play hockey with my brother, one winter, for the first time in decades, and my legs getting shaky while me feet felt somewhat numb. At the time, I put it down to being out of practice.

Cross-country skiing was a big part of my life in the wintertime until three years ago, when it just didn't feel right, any more. My legs felt okay but my feet felt different.

When I left Ontario for Alberta, I felt good, again, with no pain anywhere, and the strength and sensitivity returned to my feet. But I remember going to the ski hill, planning to try downhill skiing, again, when our school had a ski day, but feeling as though my feet were not responsive enough for me to do it.

I forgot about my feet and my legs in the spring, as they felt okay, but the balls of my feet felt funny in the summer and I had an odd little pain from ankle to knee in both legs.

Acupuncture took that pain away.

Also, I had a little bit of back pain. Decades earlier, I got orthotics for my shoes, and that solved the problem. But when my back hurt again, after I got over plantar fasciitis, I tried orthotics again, but they did not fix the problem that time.

There wasn't much pain.

It was nothing compared to the pain of plantar fasciitis.

And when that foot pain went away, I was so happy, I felt inspired to start a new life in China or somewhere else in the Far East, but thought I should go Out West, first.

The reflexologist says it is the result of old injuries, coming back for healing, again. It was an old skiing injury, an old hockey injury, and a car crash, causing my bones to be out of alignment, leading to back pain and problems walking.

-- I didn't know that old injuries came back to haunt you later in life.

Those were my thoughts while driving home from the provincial park, as I went past some houses that were already decorated for Hallowe'en, which is still more that two weeks away. A few houses in town have more Hallowe'en decorations than Christmas decorations.

That always makes me think of a guy in Collingwood, Ontario, who put up elaborate decorations for Hallowe'en, one autumn, and didn't take them down when the snow started to fall. He just added his Christmas decorations to the scene. So, he had the Baby Jesus surrounded by Frankenstein, The Mummy, a vampire, witches, and a lot of ghosts.

His neighbours didn't like that!

He got his picture in the local paper and a few visits from neighbours and the cops. He wasn't breaking the law, apparently, but he was offending a lot of people.

He said that wasn't his intention. -- He thought it was funny and fun and entertaining.

Your neighbours aren't laughing, the cops pointed out.

Day 24 Of The Big Comeback: Me And God

This morning, I woke up early, doing math. It was 49 days ago that I got a big zap of pain in my back and went to the chiropractor. Seven weeks. For 24 days, I went downhill. For eight days, I walked like a spastic. For eight days, I used two canes to get around. And for eight days, I used a walker. And this is day 24 of my comeback. For about a week, I used the canes again. And then I started walking like a zombie, or Frankenstein. Each day for the past three and a half weeks, I've walked a little further and better.

So, in the past seven weeks, I went downhill for three and a half weeks, and I've been making a comeback for three and a half weeks.

Math is not my thing. I'm an English teacher.

Yesterday, I walked approximately two miles. Four thousand steps equals roughly two miles.

My progress was steady until the doctor gave me drugs, and then I lost a couple of days. All I did was sleep, essentially, with a few trips to the bathroom and kitchen.

Can I go back to work? My balance is bad. My speed is low. My endurance is not good. When will I be back to normal? -- One week? Two weeks?

When I got the walker, I was shopping for a wheelchair, as it looked like I would need it for a long time, or forever.

My friend in Parry Sound, who has hip and knee replacements, says it's amazing how I battled back and never quit. I don't know if this is true or if it's just a matter of time. And rest.

Last night I saw a movie about a writer with a bad back. After enduring years of back pain, he got a big zap of pain that dropped him to the floor and he could not get up. He crawled to a chiropractor who gently readjusted his bones so he could walk again. He was amazed that he was pain-free for the first time in several years.

The movie isn't just about a writer with a bad back, of course. Here's the storyline: A reclusive author of spiritual books, is pursued for advice by a single mother and a man fresh out of rehab. Everyone wants to meet Arlen Faber, played by Jeff Daniels, the world famous author of the best-selling spiritual book "Me and God". The crotchety, disgruntled author simply wants to be left alone. For two decades, he's been successful in keeping his identity a secret. But all that changes.

A back injury leads the reclusive writer to begin dating his chiropractor. As Arlen's relationships with his newfound friends begin to grow, he must come to terms with his past and the realization that he doesn't hold all the answers.

They say he has cornered ten percent of the "God market". His book is a publishing phenomenon with huge sales, translated into a hundred languages, and great reviews around the world.

Eventually he goes public, on the 20th anniversary of his book's publication, and shocks a group of groupies by telling them he can't talk to God and never could. He says he had a lot of questions, felt frustrated because he couldn't talk to God, so he wrote down all his questions. The answers just popped into his head. And that's how he wrote the book.

"Are you saying you made it all up?" a reporter asks him.

"I don't know," he says. "Maybe. Maybe whoever's up there used my anger and pain as part of some divine plan. If He did, he sure as hell didn't let me in on it."

However, it is clear that the answers to his questions were profound and inspired. He says they came from him, not from some invisible super-being.

But it appears as though they did, really, no matter what he says.

Earlier in the movie, inspired by his love for a woman, he says some great things.

Arlen Faber: You know what He says. [points upward]
Elizabeth: No, I sure don't.
Arlen Faber: For you and you alone I have made this place, kaleidoscope of wonder to keep your eye upon as I turn turn the world.

Arlen Faber: The trick is to realize that you're always doing what you want to do... always. Nobody's making you do anything. Once you get that, you see that you're free and that life is really just a series of choices. Nothing happens to you. You choose.

He also says something great about kids:

Arlen Faber: I love kids. They're short, highly emotional people who don't know anything. They rely on their creativity and imagination to get by in the world. A world, I might add, filled with giants. Amazing feat.

The Huffington Post gave it a great review: "It's rare that a movie manages to be both graceful and biting at the same time, let alone smart, funny and sweet. But *The Answer Man* fills the bill," Marshall Fine wrote in the post.

"The book, a series of questions and answers in which Faber appears to be having an actual conversation with the Almighty, has brought comfort to millions. Most of them, it seems, assume Faber still has a direct pipeline to the Big Guy."

The Answer Man was a great movie, I thought, but Rotten Tomatoes gave it a low rating: just 30%.

That's two movies that reviewers gave low, or rotten, marks but I loved: A Thousand Words and The Answer Man.

Both of those movies spoke to me, as they say.

The first one said, "Make peace with the people in your life." The other one said, "It doesn't matter if you talk to God or your higher powers or whatever it is you believe in and it looks as though you have a direct line or you are channeling Spirit through your writing. Back pain could bring you to your knees anyway. Your spine has to be aligned."

I forgave and forgave and I sent my brother a card with a note saying, I feel your pain, bro. I had a vision of my healing team as a football team and I got better and better.

Was it stronger drugs plus rest in bed that did it? Was it just a matter of time? Or was it prayer, meditation, visualization, and exercise that brought me back.

I don't know. But I do know how I feel. Yes, I feel like shouting Hallelujah, it's a miracle, I can walk again!

I feel like listening to the Leonard Cohen song called Hallelujah, especially the version performed by k. d. laing at the Olympics.

Hallelujah, Hallelujah
Hallelujah, Hallelujah

Your faith was strong but you needed proof

I did my best, it wasn't much
I couldn't feel, so I tried to touch
I've told the truth, I didn't come to fool you
And even though it all went wrong
I'll stand before the Lord of Song
With nothing on my tongue but Hallelujah

Hallelujah, Hallelujah
Hallelujah, Hallelujah ....

The X Factor USA 2012 had a version of that song performed by Jeffery Gutt that brought tears to my eyes. It was right up there with Melanie Masson's version of the Janis Joplin song called Cry Baby on The X Factor UK 2012.

On Monday morning, I went to a physiotherapist for the first time in my life. I drove down to Bonnyville, the closest town, about 40 km away, and had a rather thorough examination and discussion with XXX at Health First Physiotherapy.

He said my back is not bad, and my knees are back, but my ankle and feet muscles have a way to go.

He said a little forward bulge of a disc is enough to affect nerves going to feet -- 3 of them, for 3 muscle groups -- so he's putting together an exercise program I'll get on Wednesday.

He thinks I could go back to work soon, rather than later, with a cane or two.

He got me to walk for him, he pressed on my spine, top to bottom, and tested my reflexes as well as my strength, in my knees and feet. He said my feet were weak but I told them they were stronger, now, than when I went to the Edmonton Clinic.

He asked me to stand on one foot and rise up on my toes, or the ball of my foot, and that was extremely difficult.

He took a Kleenex, got me to close my eyes, and touched my legs and feet in different places, to find out if I could feel it. I felt it most places, but not on the bottom of my feet.

He asked me to point my big toes up and try to keep them up while he pushed them down. -- I failed that test completely.

Even so, he said I'd likely be able to go back to work in the near future.

I'd be okay if everything went well, I thought, but I would be no good in a crisis or if some kid did something weird, so, as the Zen master said in the movie called Charlie Wilson's War .... "We'll see!"

But his prediction was good news, he said. And it was a lot better than shopping for wheelchairs and going on long term disability!

After physio, I drove back to Cold Lake and out to the provincial park. It was sunny and 15 and I felt pretty good so I thought I'd walk some more, to get some exercise. However, when I got out of the car, I noticed my legs felt stiff and sore, so I went home. -- I tried walking, but after 20 steps, or so, I felt as though my legs were going to collapse.

As soon as I got home, I went to bed and fell deep asleep.

Reflexology, that afternoon, was good. Mecell said she was surprised, even amazed, by the progress I was making, walking so much. She liked the physiotherapist's assessment.

At the end of our session, she asked me when I thought I'd be ready to go to work.

I told her the physiotherapist said I might go back with a cane or two before long.

She suggested going back part-time. And taking the walker.

I said, I'd love to get back to work as soon as possible -- but not too soon. -- I could do it now, if nothing went wrong, but I would be a handicap in the event of fire or even a fire drill. Never mind a shooter in the school.

She said, You want to be fully ready when you go back.

I nodded my head.

As always, she made my feet and legs feel better. When I left, I felt as though I was floating.

I went straight home and crawled back into bed.

In the evening, I traded e-mails and chatted on Facebook with the usual crew, then I watched The Answer Man, again, and fell asleep.

From the Ontario Teachers' Federation, I got the news that the premier of Ontario had surprised everybody by announcing his resignation. The OTF appeared to be happy about that. The teachers had supported McGuinty but he had frozen their salaries and taken away their right to strike.

The teachers took a strike vote, anyway, and about 98% said they were in favour of going on strike. They were already refusing to coach school teams and run after-school clubs, in a lot of schools in Ontario.

I searched the net for more information about the premier stepping down and saw there was some speculation about him leaving provincial politics for federal politics, and running against Justin Trudeau.

-- Good luck with that, I said.

The big news of the day was about a girl in B.C. who killed herself after being bullied online. "Cyber bullying", they called it.

My cousin told me about a heroin addict who was getting out of detox and what that was like. My friend in Edmonton told me she felt better after her blood transfusion, but was afraid her symptoms pointed toward cancer. My sister told me she had pains in her neck but hoped they would go away when she resigned from her job and moved on with her life.

I fell asleep thinking that most people are fighting big battles. Everyone you meet is fighting a hard battle, Plato said. And I was winning my battle, I was battling back, and my recovery was continuing relatively quickly.

I had been off work for six weeks but felt I could get back in the game in another two.

P.S. The full quote, from Plato, is "Be kind, for everyone you meet is fighting a hard battle."

The phone rang at 8:30 as my friend in Edmonton was on her way to the hospital for a test and wanted to check in. Reaching for the phone, I noticed my back was sore, again, so I decided to rest it, instead of going out and trying to set a new record or personal best by walking three miles. While resting my back, I did a little research into the book called Conversations With God.

The Answer Man, the movie about a guy who writes questions he wants God to answer, and then gets the answers by inspiration, reminded me of the books called Conversations with God by Neale Donald Walsch, written as a dialogue in which Walsch asks questions and God answers.

The first book series, An Uncommon Dialogue, appeared on bookshelves in 1995, and quickly became a publishing phenomenon, staying on the New York Times Best-Sellers List for 137 weeks.

The succeeding volumes in the nine book series also appeared prominently on the List.

His story is well-known: At a low period in his life, Walsch wrote an angry letter to God asking questions about why his life wasn't working. After writing down all of his questions, he heard a voice over his shoulder say: "Do you really want an answer to all these questions or are you just venting?"

When he turned around, he saw no one there, but he felt answers to his questions filling his mind and decided to write them down.

Important parts of Walsch's writings are mirrored within other well known spiritual writings and traditions. Wikipedia has this list:

Souls reincarnate to eventually experience God-realization (Sikhism / [Hinduism]/' 'Bhagavad-Gita).

Feelings are more important as a source of guidance than intellect (Rousseau).

We are not here to learn anything new but to remember what we already know (Hinduism/Plato).

Physical reality is an illusion (Sikhism / Hinduism/ Buddhism's concept of maya).

One cannot understand one thing unless he or she understands its opposite (Tao Te Ching).

God is everything. (Sikhism / Hinduism / Spinoza / Brahman)

God is self-experiential, in that it is the nature of the Universe to experience itself. (Hinduism/Hegel, and process theology as first outlined by Alfred North Whitehead)

God is not fear-inducing or vengeful, only our parental projections onto God are. Fear or love are the two basic alternative perspectives on life (Drewermann)

Good and evil do not exist (as absolutes, but can exist in a different context and for different reasons as Nietzsche).

Reality is a representation created by will. (Schopenhauer)

Nobody knowingly desires evil. (Socrates)

It's just a ride. (Bill Hicks)

Walsch wrote half a dozen books in the series and then a dozen or so guidebooks to go along with them, followed by related books, called What God Wants, Happier than God, When Everything Changes, Change Everything, and The Only Thing That Matters.

    A movie called Conversations with God, dramatizing the author's experience, came out in 2006. It chronicles the journey of a struggling man turned homeless, who inadvertently becomes a spiritual messenger and bestselling author.

    "Conversations with God" tells the true story of Neale Donald Walsch. The journey begins after he unexpectedly breaks his neck in a car accident and loses his job. He goes from being an ordinary guy to a homeless bum struggling just to stay alive. He writes when he is angry and bitter. The conversations that follow end up being read by over 7 million people in 36 languages. And counting.

    They say that the movie crew had foggy mornings just on the days they needed them, and the fog lingered only so long as it was needed to complete the scenes. And when they needed it, snow came down in huge flakes until the director said "Cut" on the last take, and then it stopped.

That movie is promoted online along with The Celestine Prophecy, The Secret, and One Week.

    The Celestine Prophecy was a huge hit, like Conversations With God. So was The Secret.

    One Week was a Canadian film that came out in 2009 about a guy who has been diagnosed with cancer and is told he requires immediate treatment but decides to take a motorcycle trip from Toronto across Canada to Vancouver Island.

    Along the way, he meets several people that help him with his dream of becoming a writer.

    The movie features the scenic backdrop of the Canadian landscape, including Toronto, Northern Ontario, the Prairies, Banff and the Rocky Mountains, and Vancouver Island, as well as an all-Canadian soundtrack.

    It opened at the Toronto International Film Festival and the Edmonton International Film Festival as well as the Kingston Canadian Film Festival.

I saw One Week, read The Celestine Prophecies, was aware of Conversations With God, but never saw the movie. I looked for it online but couldn't find it. But I did find interviews with Walsch, including one about his book called Happier Than God.

    He criticizes The Secret for being all about money or material things, instead of how to make the world a better place, and says his book tells the secret behind The Secret. And he says that secret is "do unto others as you would have them do unto you".

    Make sure that the other person always gets the better of the deal, he says. That's the secret.

He says happiness is not getting what you want, it's wanting what you get.
It's not about settling or resigning, he says, it's about celebrating what you get.
Move through life in a state of wonder, he advises.

So, here I am, lying on the floor with a sore back, struggling to walk, again, listening to Walsch talk about moving through life in a state of wonder, celebrating what I get, and wanting what I get.

Did I want crippling back pain? Did I

Did I want a month or three off work to watch movies, read books, and write books? Can I celebrate it?

Well .... If you put it that way .....

My back hurt until noon and I don't know what happened then but with that pain gone I rested comfortably and listened to my computer read me this story about a writer who loses the ability to walk and he has to be off work, as a teacher, but he writes this great book called Me And God .... No!

Conversations With God.

No!

Zen And The Art Of Healing Your Aching Back.

My friend in Edmonton phoned a couple of times to tell me about all the medical tests she has to undergo and how it has made her think about her life and death and wonder what she would do if she was told she had one year or two weeks to live.

She said that after her test today, she tried to get some info out of the technician who did her ultrasound, but all she said was, Well, you can buy green bananas.

What the heck does that mean? I said.

That's what I said, she told me.

She guessed that the cryptic comment meant she would be around long enough for the green bananas to ripen so she could eat them.

How long does it take a green banana to turn yellow, but not black, so you can eat it?

It depends how green they are. It could be a few days or a couple of weeks.

What would you do if you were told you could buy green bananas?

I'd get those organic green bananas. -- They never turn yellow!

Day 51: The Big C

Yesterday, I got a letter from the asebp, the Alberta School Employee Benefit Plan, saying that my school board told them I've been off work since September 11, and that I might get EDB, extended disability benefit, if I'm disabled for 90 days. That's 90 calendar days.

This would be Day 36.

Last night, after sleeping most of the day away, I went out to meet a few colleagues for dinner. The phone rang at 3:00 p.m. and Julie told me wing night at O.J.'s was on. So I told her, I'll see you there. It took me an hour to get ready and go. I had to circle the block because they don't have handicapped parking.

The last time they saw me, I was walking with two canes, so they looked impressed to see me walk in without a walker or anything. My walk was a little wobbly but, otherwise, I looked normal, they said.

In cross-country skiing, to go uphill, there's a move called the windshield wiper. You move your upper body back and forth, from side to side, as you herring bone your way up the hill. That's the way I walk.

I practice walking without moving my upper body.

My colleagues said they were tired because of work. Everybody has five or six different classes and not many prep periods. Plus supervision. One is planning a school trip, to the south of France, for next year. She is also the local teachers' federation rep. Another is running student council and planning the grad event. The other is new and just got back from the new teachers' conference in Edmonton.

Our little school has a big grad event. They plan for it all year and produce a nighttime school assembly for the graduating students, their families, and friends, that lasts three or four hours and includes speeches by each of them and for each of them, plus a slide-show or powerpoint presentation with pictures from all their years at the school.

The gym is decorated, including the ceiling as well as the walls and the stage, a caterer is hired to bring in dinner, there's a social time before and a dance afterward, and a photographer is hired even though everybody has a camera and thousands of photos are taken.

I've worked at big schools where the grad event is far less elaborate.

This year's grad group wants to hold their dinner and dance at a hotel, nearby, but it isn't working out as the timelines would be difficult to meet. They want you in and out fairly fast and our event requires a long set-up time with a tear-down time that is almost as long. So, it looks like the school gym will be decorated and used again.

Exams follow shortly after the grad event, so students write their finals in a gym that is still decorated for the grad event.

It's mid-October and they've already had several planning meetings.

At the restaurant, we talked about producing a skit, with music, poking fun at all the graduating students, highlighting a few of their foibles and memorable mistakes. It will be a music video with all the high school teachers imitating each of the grads, set to music.

The letter from ASEBP re: EDB has me wondering what happens if your disability period as 89 days or less. Does that mean you are on your own, financially, for three months or however long you are off work?

You pay into this insurance fund your whole career, hoping you will never need it, and when you do need it, is it really there?

After an hour with my colleagues, I headed home, again, for another quiet night of Scrabble and chatting online through Facebook while watching a movie. The Answer Man played in the background and I watched it while waiting for other people to make their moves or type their comments.

My friend in Edmonton phoned on her way home from work, just to check in, and she confessed to some of her fears about her health concerns. She didn't get a family doctor until recently, after moving to the city a year and a half ago, and the doctor has her booked for a few tests, after sending her to the hospital for a blood transfusion.

"I have a slow leek," she said. "I'm losing blood, for some reason, and they gave me three units at the hospital, which is quite a bit."

She has looked into the possible causes for losing blood and they all scared her. And her number one fear is The Big C: cancer.

The Canadian Cancer Society website has a lot of information, including a section on myths and controversies about cancer. It says they understand Canadians are concerned about cancer, but they recommend that you are cautious about any information or claims obtained from unmonitored sources, in particular the Internet. They say the Internet can be an empowering source of information, but a healthcare professional should be consulted before making medical decisions.

There is a long list of topics in the section on myths and controversies, including: Antiperspirants and breast cancer, Bras and breast cancer, Food additives, Microwaves and plastic containers, Oyster and soy sauce, Disposable water bottles, Sunless tanning products, Tampons and cancer, Lead in lipstick, Abortion and breast cancer, An alkaline diet and cancer, foods that prevent or cure cancer, Hemp and cannabis products, Sugar and cancer, Thermography, Fluoride in water, and Chemicals in sunscreen.

Inspired by my friend and her fears, I Googled "cancer industry" and got millions of hits for websites that call the cancer industry "evil" and complain that the cancer industry doesn't talk about cancer prevention, it only promotes expensive cures involving drugs and surgery. Apparently there is a war raging between alternative cancer therapies and allopathic medicine's standard of cancer care with drugs, chemo and surgery.

Linus Pauling, a 2-time Nobel Prize winner, was quoted, saying, "Everybody should know that the 'war on cancer' is largely a fraud."

I come from a place where an alternative cure for cancer is promoted, a little, because it was discovered there. A nurse named Renee Caisse created a product called Essiac -- that's her last name spelled backwards -- based on information given to her by local First Nations people. -- She called them Indians. There is a sculpture of Renee Caisse on the main street of Bracebridge and the community theatre attached to the high school -- my old high school in a new building -- is named after her. Bracebridge and Muskoka Lakes Secondary School has a new location right beside the Renee Caisse Theatre.

From the 1920s through the 1970s, ESSIAC® was promoted as a cancer treatment by Rene M. Caisse, a Canadian nurse, who claimed that it had been given to her by a patient and that the recipe derived from an Ojibwe people, from the Ojibwa tribe, in Bracebridge Ontario. In 1977, Caisse gave the ESSIAC® formula to a Canadian company called Resperin Corporation , which attempted to commercialize the product. The company became ESSIAC® Canada International Inc. in Ottawa.

The original formula is believed to have its roots from the native Canadian Ojibway Indians. The four main herbs that make up Essiac are Burdock Root, Slippery Elm Inner Bark, Sheep Sorrel and Indian Rhubarb Root.

There are those who say spinal surgery for DDD is a fraud, too. My brother had back surgery twice, and it didn't help at all. My other brother has been going to a chiropractor for fifty years. Doctors and chiropractors fight over this issue. My physiotherapist recommends decompression therapy, with an expensive machine. And then there's Dr. Ho's back decompression belt, regularly sold for $300.00, now on sale for half that price. My brother loves it and he talked me into trying it.

And guess what? -- I like it, too!

My favourite movie used to be "Whatever Works", the Woody Allen movie starring Larry David, set in New York City. It's classic Woody Allen: witty, ironic, and very funny. It's about fate and chaos and a sophisticated guy whose philosophy of life can be summed up in just two words: whatever works.

-- That's my philosophy in regard to back pain, these days.

So what if the Dr. Ho belt is advertised on TV with a lot of other products that look like gimmicks? -- Whatever works!

The other day I heard an interview on CBC Radio One with a writer who had a book and a performance piece based on her experiences in hospitals in Canada. Today I ordered the book online. My Leaky Body, by Julie Devaney, is sub-titled "Tales From The Gurney".

She takes on a journey through the health care system as she is diagnosed and treated for ulcerative colitis. In and out of emergency rooms in Vancouver and Toronto, she is poked, prodded, and abandoned to a closet at one point, bearing the helplessness and indignities of a system that seems hell-bent on victimizing the sick, the blurb says on Amazon.ca.

My Leaky Body argues for fixes to the system and better training for all medical personnel.

She is now on tour, going across the country, setting up a gurney on stage at workshops and conferences to teach Bedside Manners 101 and to advocate for repairs to the system.

The book came with a warning printed inside. It said, Experts have concluded that reading My Leaky Body may lead to: righteous anger, roiling bitterness, heart palpatations, mental alertness, unfocused indignation, rampant amusement, emotional distension, cantankerousness, ocean madness, slight depression, irrational sadness, irrational happiness, bouts of open weeping, lack of trust in health care professionals, heebie-jeebies, aching lung syndrome (from laughing), dry mouth, nausea, cabin fever, tennis elbow, writers' cramp. My Leaky Body is not to be taken internally. Use care when reading My Leaky Body while operating heavy machinery. If My Leaky Body leads to upset, please contact your local MP to make improvements to health care.

My experience with health care in Canada was quite different.

As a kid, growing up in Muskoka, Ontario, I thought we had first rate health care. We went to Fisher's Clinic in Gravenhurst and to the hospital in Bracebridge.

When I lived in Halifax, for a year, and in Victoria, another year, the doctors were excellent, I thought. When I lived in Toronto, I met a number of doctors, and thought they were almost good as the doctors in Muskoka. In Owen Sound, there was an excellent hospital. The doctors were okay, almost as good as in Toronto. When I moved back to Muskoka, there were excellent doctors in Bracebridge. Fisher's Clinic had added an acupuncturist. When I moved back into the Greater Toronto Area, I found excellent doctors in Pickering. And when I moved up to Port Perry, the birthplace of chiropracty, it took six months to get a doctor, but she was good. The hospital was okay. I met an acupuncturist in Port Perry who does "Western acupuncture" or medical acupuncture, hooking the pins that are used to electrodes and cranking up the electricity, and she did something nobody else could do: she healed my plantaar fasciitis.

Everything in Alberta is different, it seems to me. It's impossible to get a family doctor in Cold Lake, but I was able to find one, fast, once I figured out that you had look beyond your city limits. I met four doctors in the emergency department of the hospital in Cold Lake, and they were all good. There is a nurse practitioner working in the hospital and she was great. My doctor in Lac La Biche is excellent.

It seems odd that all the doctors in this part of Alberta come from South Africa. But nobody complains about that.

The Edmonton Clinic just opened and looks like something from the future transported to 2012. The Glenrose is legendary.

There is nothing like our Wellness Centre in Ontario, that I ever encountered. And, of course, by now I'm certain I've given you the message that I think my reflexologist is particularly gifted.

And my physiotherapist is excellent.

True, I had little to do with doctors in Ontario, aside from annual check-ups, a skiing accident, a bad hit in a hockey game, and a car crash. However, I've spent more time with doctors in Alberta in the past 50 days than in 50 years in Ontario.

So, I won't be ripping apart the medical system in this book or writing a one man show to take on the road in an effort to lampoon or satirize doctors or the medical system.

I think it would be great if the mainstream medical practitioners and the alternative practitioners could get together and all their services were covered by our insurance policies or our governments. And, of course, there is a doctor shortage, that is well-known.

Perhaps the new medical school in Northern Ontario will solve part of the problem and evolution will solve the other part as West meets East and alternative healing practices.

Physiotherapy was deceptively tough, again.

The first meeting with the physiotherapist included his examination, which didn't feel to rough, but left me feeling a little back pain the next day. The second one was an hour-long workout that took my legs out, for a moment.

Ten minutes on the treadmill, after some exercises for the stomach muscles, followed by leg work on the parallel bars, and I almost fell over. My leg muscles were stretched and then fatigued. After sitting down for a few minutes, there was a set of exercises using a big ball, and then I was put in the traction machine.

You lie on a therapy bed, on top of a two part harness, the physiotherapist wraps it around you and seals it with velcro, and then the machine pulls you apart, to decompress your spine, for ten minutes.

It looks like you are about to be drawn and quartered, but it actually felt good.

He gave me a print-out of the exercises, so I could do them for homework, and then he booked me for two more appointments for next week.

Will I be off work next week? I booked the appointments for times during the day, on Tuesday and Friday, assuming that my doctor would sign another note or write a letter saying I needed to be off for short term disability.

I left the physiotherapy place in Bonnyville feeling happy and looking forward to getting myself in shape, again, so I could do a lot of things in the near future, including work, coaching basketball after school, and skating at Christmastime.

On the way home, I stopped at Walmart and bought the Wally Lamb book, Hopin' And Wishin', and the short walk from the car to the store and back felt like it was about all I could do, for one day. I staggered from my car to the elevator and home, feeling exhausted but happy, and optimistic about the future.

    -- Knock on wood!

Day 27: The Way Of The Peaceful Warrior

This morning my friend in Edmonton phoned and woke me up at 8:45 and we talked about positive affirmations, provincial governments in Canada, and the book and show called My Leaky Body.

I told her my experience with healthcare in Canada was quite different than the way it's described in My Leaky Body. She said, Yeah, well, you're not a woman, your health concerns are quite different, and you're not in the big cities of Toronto and Vancouver.

Last night my brother phoned, so I told him about the decompression machine, and he said, Yeah, he's done that, but never told anybody about it.

I realized that he has been managing back pain for fifty years. It's a wonder it hasn't driven him insane.

This morning I noticed a couple of posts on Facebook.
1. Part of the success of Zen in the west has been the dropping of the more "religious" aspects, this article says.

It said that the popularity of Zen in the West began with Buddhist missionaries in France in the Sixties, when hippie culture was entrenched and millions of young people were protesting against the War In Vietnam.

Or, as they say in Vietnam, the War Against America.

The Zenmaster I studied with was not a missionary. He said that Zen is not a religion, it's a philosophy, and it can be attached to anything. So, you can be a Zen Catholic, a Zen Hindu, whatever.

A Zen Buddhist missionary showed up at the Zen Forest one day and argued with him about that. He told me Zen in the West was completely different than Zen in the East, and it was wrong to separate Zen from Buddhism.

Who was I to tell him that he was wrong? He had studied Zen in Bangladesh, India, Korea, Japan, and China. -- I told him he was wrong, anyway, because I had studied Zen in the West.

2. Love Tree posted a piece called 'Trusting in Spirit': Faith is our direct link to universal wisdom, reminding us that we know more than we have heard or read or studied- that we only have to look, listen, and trust the love and wisdom of the Universal Spirit working through us all. ~Dan Millman

I like Dan Millman and his work, especially his book, The Way Of The Peaceful Warrior, turned into a movie called Peaceful Warrior. It was about an American gymnast who got injured but made a huge comeback with the help of a mysterious, Zen-like, mentor. Millman broke his legs, badly, in a crash with a car while on his motorcycle. Doctors told him he would never compete again. But he made a great comeback. In the movie based on the book, Nick Nolte plays the part of the mysterious mentor. The movie was called "Rocky for the soul". I wrote a book about a Canadian gymnast and gave it the sub-title "The Way Of The Quiet Canadian Warrior".

The book was about Karen Cockburn, who overcame great odds to be the #1 trampolinist in the world. At the last Olympics, in London, last summer, her partner for synchro trampoline, who she mentored, was the only Canadian to win a gold medal.

Rosie MacLennan.

Cockburn finished in 4th place in Individual Trampoline, missing the bronze by the narrowest of margins. She won a Silver medal for the synchronized trampoline event with MacLennan and a Bronze medal for the Team competition.

Cock won a silver medal in the women's final at the 2008 Summer Olympics in Beijing. She won a bronze medal at the 2000 Summer Olympics in the individual event in Australia. She is the only trampoline athlete to have won a medal at every Olympic Games at which the event has been competed. She was selected as Canada's flag bearer for the closing ceremonies of the Beijing 2008 Olympics.

Both MacLennan and Cockburn are coached by Dave Ross at Skydivers Trampoline Place, just north of Toronto. Ross is the Canadian National Team coach. At Skyriders, his club, he coached four of the five Canadian Olympic trampolinists: Karen Cockburn, Mathieu Turgeon, Rosie MacLennan, and Jason Burnett.

I interviewed him at length while he was stringing one of the high performance trampolines he creates and sells around the world. His coaching style focuses on increasing the technical difficulty of the optional routine. Burnett, Cockburn and MacLennan, coached by Ross, hold the world records for highest level of difficulty in the Men's Individual and Women's Synchronized trampoline events.

He is a legend. Ross is like the mysterious mentor in Millman's book, The Way Of The Warrior. His elite athletes are legends.

How come you haven't heard of them? Well, we are Canadians. And that is the way of the quiet Canadian warrior.

We have a strong dislike for people who are all talk and no action, especially if they are Americans, as we are the opposite.

-- Oops! I shouldn't have said that! We're supposed to be quiet about it!

Yesterday, at physiotherapy, I thought about Millman, Ross, Cockburn, and company while I was working out on the parallel bars. It was the first time I did anything on the parallel bars since I was in high school. The physiotherapist said, We'll have you doing shoulder stands and handstands on the bars in no time. -- He was just joking.

I held on to the bars and tried to lift my legs, one at a time, to the front, to the side, and to the back. -- It hurt but I kept on going. After ten minutes, I walked about ten steps and then collapsed on the floor.

Well, the physiotherapist said, we've fatigued those leg muscles of yours.

He gave me a cane and suggested I sit down for a few minutes. After a brief rest, I was ready for more exercises.

Today I have to do them on my own.
-- No problem!

While I watched Peaceful Warrior, blew up my big Zon ball and started the exercises I got from First Health Physiotherapy.

First I worked on the transverse abdominal muscle. Contracting the transverse abdominals is difficult, they say, since most of us are not even aware we have them. The TAs are like a corset the wraps around your midsection. Think of it as a big hug you are giving yourself that is going to help your back, spine, nerves, legs, feet, everything.

My instructions were to hold that for ten seconds and repeat it ten times and do that a couple of times a day. I did it while blowing up the ball, watching the movie, lying down, sitting, standing, on all fours, and on the ball. I did it whenever I thought about it, whatever I was doing, for the rest of the day.

Next was the exercise I did holding onto the parallel bars. They were very hard to do in Bonnyville, at first. You tuck in your belly button, stand on one foot, and lift one leg, ten times, times three, twice a day.

Next you raise your hip and swing your leg out, away from the other leg. Three times ten, one or two times per day.

And then you lift you leg backwards and hold it up for a few seconds. Three times ten times two.

The first time I did it, it was a challenge. But the more I did it, the easier it got.

And then I tried walking across the room, and I fell, because the muscles in my legs were fatigued. Tired. Done.

The second time I did it, I was at home, and I doubled all of them.

I felt inspired by the movie and by the fact that I could do it and it was getting easier.

And I wanted to walk and run and skate and ski again.

In the movie, as in the book, Dan Millman crashes his bike into a car, is thrown over the car, and shatters his leg when he hits the ground. If you break your leg in a dozen or more places, they say it is not broken, it is shattered.

When I was a kid in Grade 7, I won a race, skiing on a toboggan hill, after a freezing rain, and at the finish line I wiped out and shattered my leg.

Dan was told he was lucky he would be able to walk again but he would never compete again. But, with the help of his mysterious mentor, he got back in the game. I don't want to spoil the ending of the movie for you.

I was told to forget skiing and skating, but I told the doctor to forget that. And I returned to skiing, skating, hockey, soccer, swimming, et cetera later that year.

Coincidentally, I lost something I had won, because I had a cast that went from my toes to my upper thigh. The year before, I won the MVP award, when I was the captain of my peewee hockey team, and we had a great season. The prize was a pair of tickets to see the Leafs play the Canadiens at Maple Leaf Gardens. I was thrilled, as I was a huge hockey fan, loved both teams, and had never been to The Gardens.

Because of my broken leg, that dream of a trip was shattered. They gave me a huge basket of fresh fruit, instead.

Fresh fruit in the winter in Muskoka, in those days, was expensive and hard to get, and no doubt helped with the healing of my leg.

Years later, I found out that it was a good time for a kid like me to avoid Maple Leaf Gardens. It was a dark period in the history of the team and the arena and there was some sort of sex ring operating in the big city, out of the famous hockey rink. Young boys were tricked, offered autographed hockey sticks, taken into empty team dressing rooms, and abused. There were court cases, abuse allegations, out of court settlements, and convictions, and the whole situation did not get as much attention as you would expect. It took years for the cases to go to trial and come to some kind of resolution. Some victims of abuse carry the scars forever, apparently.

I missed all that, thanks to my broken leg.

Instead, I went to a tournament in the city at the end of the season, my leg still in a cast, wearing my team hockey jacket, using crutches, and a newspaper photographer snapped a picture that appeared in the paper with the cutline "Fallen Warrior Watches Action From The Sidelines."

Five years later, I was captain of our midget all-star team, playing a game against our arch-rivals, and a kid stoned on angel dust tried to pick a fight with me. When I skated away, he went crazy, and on the next play, he hit me from behind as another kid dove at my feet and a third guy slid into me, baseball style, with his skate blade aimed at my face.

We didn't wear face shields, in those days.

I was knocked out for a few seconds and when I came to I could tell I had been cut by a skate around my right eye.

The doctor at the clinic stitched me up and ordered me to stay out of the game, saying I was very lucky I did not lose an eye.

I followed the doctor's orders and my coach was not impressed. He said I should have got back into the game and got even with the guys who ganged up on me.

He tore the C for captain off my hockey jersey and gave it to another guy on my team.

We only had a couple of meaningless games after that, at the end of the season, a tournament, and then our hockey banquet.

Every year, we went to the Little NHL tournament in Western Ontario, to play teams from that region, southern Ontario, and the nearby states in the U.S.A. We traveled to Wingham, which was Alice Munro's hometown, for the March Break, and we finally won the tournament with our midget team. The final game, for the championship, went into overtime. Our new captain broke up a fast break at the other team's blue line, slapped a shot that hit the goalie, and I skated in to flip the puck over the sprawling goalie, into the net, for the win. That was my last game.

The father and son hockey banquet was never my favourite event, since my father had left our family a decade earlier. My final hockey banquet was even more difficult, for me, because the guest speaker was the General Manager of the junior team closest to our town and he went on and on about the captain of our team and how much he wanted our captain to play on his team the next year.

I thought he was talking about our new captain. Years later, I realized he had been talking about me.

My teammates and coaches all told me I shouldn't play Jr. C, anyway, as it was a hackers' league and I was a skater, so I should play Jr. B.

There was a great Junior B team twenty five miles away, in Orillia, called The Terriers, who were provincial champions. They phoned to say they wanted me.

But I said, No thanks. I was finished with hockey.

Canada was still the best in the world, in hockey, but the Soviets had shown they could beat us, even though they didn't win the infamous Summit Series. The NHL was fighting with the WHL. Both leagues were filling up with fighters and the stars were turned into warriors in the Cold War when they went up against the soldiers masquerading as amateurs in the CCCP. And our hockey players were risking life and limb for far less money than they would be making in a few years. But we had no way of knowing that.

Bobby Orr, the superstar from just up the highway, in Parry Sound, could not play against the Soviets because he had injured his knee playing hockey. And his team, the Boston Bruins, abandoned him, so he finished his career with the Chicago Blackhawks.

I was finished with hockey.

Our Junior C team, the Bracebridge Bears, without me as captain, went on to win the provincial championship. A few of the guys got hockey scholarships to play in the U.S.A.

But I was finished with hockey and I never missed it for a second.

Okay, that's a lie. -- I missed it a lot.

When I was 33, I played hockey again, in a two team league, with teams made up of cops and teachers, and everybody except me drank beer in the dressing room after the games. We put together a team of teachers to play against our students in a game during our winter carnival at school, and I joined in the fun.

It was frustrating, playing the game again, as I remembered what to do in every situation but my body could not do it the way I did it decades before. Even so, I had some fun, scored some goals, and got some exercise.

The guys I play with and against, and the students who watched the students versus teachers game all said the same thing: How come you do everything so fast? It's just a game. It looks like you are trying out for a spot on the Maple Leafs.

Carl Brewer, one of the old Leafs, was trying to make a comeback with his former team.

I had no interest in doing that. And I felt as though I was going in slow motion, when I played hockey, again. Even so, I did a lot while I was on the ice.

Fast forward another decade and a half or so. When I moved back to Muskoka to take the job of Arts Department Head in my old hometown, I joined the teachers for a hockey game. They had two teams that played every Monday and Friday night. I showed up a little late one night, after the start of the season, and hit the ice with about half of the hockey equipment most guys had. My team was down by five goals.

By the end of the game, I had a few goals, and more assists, and my team won by one goal.

It wasn't much to brag about. The other team's goalie was our Phys Ed Head, and he was pretty good in net, but the number on the back of his jersey was not #1, it was 911. That's the number you call when you need help! In other words, he didn't take it too seriously.

Unfortunately, some of the guys did take the competition seriously. One of those guys got promoted to principal, at our school, a few years later, after spending some time at another school. And, apparently, he remembered that game. He told me he felt I had tried too hard and had embarrassed him and his team. -- Maybe that was why we didn't get along when he returned to our school as principal.

That was the only hockey game I played that year, in my old hometown, or anywhere, or ever again. -- I was finished with hockey. Again.

Hockey on TV never interested me. When Wayne Gretzky and the Edmonton Oilers took over from the New York Islanders dynasty, I followed their team. And when Michael Jordan played for the Chicago Bulls, coached by Phil Jackson, I saw quite a few of their games, too.

Instead of hockey, I coached soccer for twenty-five years, basketball for three years, and swimming for two years.

My high school soccer team had one great season, winning our league, district, and region, and going to the All-Ontarios.

The first soccer team I coached went from worst to first. They finished last, the first year I coached soccer, but I kept the same group of guys together, more or less, and in their final year of high school, we had a perfect season. We went undefeated to finish first in the league, swept the play-offs, won the district championship, and went to the regional finals, but finished one step away from the provincial finals.

Coaching soccer in my old hometown was a lot of fun. We got to travel around Muskoka and Parry Sound all fall, and autumn is the most beautiful time of year in that beautiful part of the world. I had a good time with my teams and it was fun to re-visit the soccer fields of my glory years. We did not win any championships but one year we did what no other soccer team in living memory had done at our school: We beat Bracebridge, my old school, and that victory, for us, felt like winning the world championship or the Stanley Cup.

Gravenhurst High School was a small school, with a student population of 400, and Bracebridge And Muskoka Lakes Secondary School had three times as many students.

I coached against an old friend of mine, who was on our team when we went all the way to the All-Ontarios. The referee was another guy on our team that year.

Whenever I think about those days, I hear Bruce Springsteen singing Glory Days somewhere in the background.

Glory days well they'll pass you by
Glory days in the wink of a young girl's eye
Glory days, glory days

That's the song I listened to while I did my workouts, trying to get my back and legs and feet working again.

Feeling inspired, I did all the exercises assigned as homework by the physiotherapist, twice, and then I did the other exercises on the sheets he gave me, too. The physiotherapist said to do sets of ten a couple of times, so I did sets of twenty several times.

He said to do TAs, leg lifts in three directions, the lumbar spine rotation, single leg lifts, and bridging.

Bridging was the hardest. With knees bent, lift your seat up, off the floor, until hips are neutral, without tightening your seat muscles.

So, I did TAs, leg lifts in every direction, the lumbar spin rotation, singlee and double leg bends, passive spinal curls, the curl and rock, cat stretches, back flexion, back extension wiht prone elbows and with extended arms, seated flexion stretches using the big ball, flexion diagonal stretches, seated side bend stretches, standing side bends, back rotation twists, back extensions, and the clamshell, or abductors.

I did not try the double leg extension, lying down and lifting both legs up, straight, as it said it was very advanced and you should not try it unless TAs are well developed as it can crete severe strain to the low back. I'd save that for another day.

I did all the suggested exercises three times as often, and the other ones just as many times. And I added a few others.

And then I did an exercise I couldn't do for the past fifty days or so: housework.

The exercise called housework involves bending, lifting, stretching, and also, the way I do it, throwing. I had fun tossing socks and shoes and bags of garbage around, since I couldn't throw a sock across the room a little while ago.

One of my favourite American writers, the poet and novelist Richard Brautigan, once said, "You have to count all the small victories. I do, anyways."

When I was finished, I looked at the time. My workout had lasted two hours and I had done a lot in that time. -- I was back in the zone, or the Zen zone, as Gretzky and Michael Jordan and Phil Jackson and my Zen master used to say: You focus on each little movement and it feels like you are going in slow motion but an observer would get the impression you are moving very quickly.

I wished I could click a button, the way Adam Sandler does in the movie called Click, and fast forward to the day I could go back to work, take a long hike, and start skating again.

But then I remembered one of the lessons from the Peacefull Warrior: Enjoy the journey. Don't think you will be happy when you reach this goal or that goal. Be happy now, in the moment, and enjoy the entire journey.

-- Great advice.

After all that exercise and inspiration, I felt like doing a little more, so I did a half hour of qigong or energy exercises, then hit the showers. By the time I was dressed again, my legs were shaky and so was just about every other part of me. I was shaking all over, so I played the song, "Shakin' All Over". -- I loved that song.

"Shakin' All Over" was a rhythm and blues song originally performed by Johnny Kidd and the Pirates. It was written by frontman Johnny Kidd and reached #1 in the United Kingdom back in 1960 when I was five years old. In 1965, The Guess Who covered the song on a single and later the same year released their debut album, Shakin' All Over, which reached #1 in Canada, when I was ten.

That's when I get the shakes all over me
Quivers down my back bone
I've got the shakes down the kneebone. Yeah!
The tremors in the thighbone
Shakin' all over

"Don't overdo it," everybody told me, "or you could have a relapse."

Day 28/53: Strange Dreams

Last night I had a hard time falling asleep and then I woke up every two hours or so and I woke up feeling tired at 8:00. When I got up, I discovered my legs were sore and stiff and my back was a little sore, so I grabbed an ice pack from the freezer and reached for a box of cereal.

For some reason, I read the ingredients listed on the cereal box and discovered it has B6, so I put it back in the cupboard and grabbed a banana and an apple.

For two nights in a row I had unusual dreams.

The night before, I dreamt I was at a poetry reading or perhaps a prose reading with Margaret Atwood, Clark Blaise, Dave Godfrey, and perhaps a few other people in a back room at a Canadian bookstore that used to exist on Yonge Street in Toronto. Margaret Atwood's face kept on changing so she looked like famous photographs of herself from different decades. I thought she was beautiful but somebody said the earliest picture of her looked like she just got out of Auschwitz and enjoyed it. Dave Godfrey said that to me decades ago. He was a friend of hers and introduced me to her, in real life, not in the dream.

There was a second part to that dream. For some reason I was in a back room filled with old paperback books and they fell on top of me. I made my way through them to a window that didn't open so I banged on it to get help from whoever was next door but I couldn't get their attention. Instead of calling for help, I climbed up, through the books, so I was on top of the little mountain of paperbacks, so I could breathe again.

Last night I had a dream about a big indoor complex with restaurants and escalators, including an art installation made of moving escalators that covered the walls and ceiling as well as the floor and went around a corner to the way out. When you were in the main room it looked as though you were surrounded by escalators. It was a relief to get out of the art installation and the mall or the complex. Down the street there was an arena for NHL hockey that looked something like the new one planned for Edmonton but when I got inside it turned out the game had been changed to a different venue which was an outdoor rink within the walls of the building and I saw Sidney Crosby walking to the game that other hockey players were already playing so I said to whoever I was with, Hey, there's the greatest hockey player whoever lived, and he smiled at me, then invited me to join them on the ice. He helped me get some hockey equipment, including a new style of skates that had extension on the blades so you were about six inches taller with your skates on.

We played five aside hockey with no subs and just a handful of spectators even though the teams had some of the greatest hockey players in history from the present and the past, all playing as though they were in their twenties.

Before I went to bed, I was worried about seeing the doctor and getting another note to be off work or a letter saying I should be off for a Short Term Disability. After midnight, I checked Facebook and noticed my friend in Edmonton was online, because she couldn't sleep, either, so we chatted for a while. She told me not to worry and I told her not to worry. She said she had a lot on her mind. She was trying to buy a car for one of her sons. We said goodnight and she said something about it being a good thing we were friends.

It's Friday, 2 degrees, there's rain in the forecast, and the sun is up but you can't see it because it's overcast.

My back and legs and even my feet felt fine all night long, for a change. But I sure was stiff when I got up! I staggered to the kitchen.

The more I moved, the better I moved, but my legs were still a little stiff and sore from the physiotherapist's exercises I did the day before.

What do I say to the doctor? He wanted to see me in ten days, after physio, to see how I was doing. So, how am I doing?

-- A little bit better, I'd say. I'm walking a bit more. Physio made me fall, after twenty minutes of leg exercises, on Wednesday. The decompression machine took my back pain away, again. Doing the exercises for homework yesterday left me shaking all over. I sleep a lot. I don't think I could handle work just yet. If there was a fire drill or anything like that, I would be a handicap.

It has been 40 days since my last day at work, 53 days since my back hurt and I lost my legs, and 28 days since I stopped going downhill and started to walk better.

Okay, so .... The doctor surprised me completely. Just when I thought I would have to argue about going back to work, he turned the tables on me. The specialist said, Activate your short and long term disability. But the doctor said, Let's get you back to work.

He said, You have an appointment for physiotherapy on Tuesday? Ask him to write up his assessment and bring it with you on Wednesday. -- Maybe we will ease you back to work, going half days.

My jaw dropped, as they used to say.

I jumped in the car and drove to school. I stopped, on the way, to tell the principal and our head secretary what was happening, and told the principal I was thinking of dropping in.

She said, By all means.

So I drove to the school, parked in the blue parking area, and interrupted the high school classes just before the bell.

All the students came pouring out of their classes to say hi and they all looked so happy. We shot the breeze for a bit and then it was time for them to head outside to catch their buses.

The elementary students were even more fun. But they also demonstrated some maturity. A lot of them ran up to me, arms opened, to give me a big hug, or to get one, but they stopped two feet in front of me and said, Oh yeah, we have to be careful of your back.

Somebody had coached them thoroughly.

Back issues are well known, apparently. Everybody knows somebody who has had a bad back, a slipped disc, a back that goes out, whatever you want to call it, even though those expressions are far from accurate, and everybody knows about pinched nerves, losing the ability to walk, losing feeling or going number not just in the feet and legs but beyond that, too.

And, as a bonus, the kids in Grades 3, 4, 5, 6, 7, and 8 all made cards or wrote letters, and they were collected in big piles, for me, all saying Get well, We miss you, You're the best teacher in the world, We miss your spirit, smile, and gummy bears.

Last year, I had a Grade 5 class that would do anything for gummy bears, and they haven't forgotten about it.

After the kids left, I talked with teachers, got caught up, a bit, and then some of the high school teachers decided we should all go out to dinner together.

Boston Pizza. -- Yum.

The elementary teachers gave me a hard time for not calling on them when I needed help. And then they said, You've lost weight and you look great. You can tell just by looking at your shoulders, your belly, your face, and your butt.

I said, Stop looking at my butt!

And then they gave me hell for not calling the school and asking for help. They said they were worried about me and wished I'd asked them to do something, like get groceries or whatever.

So. There ya go. -- Things sure can change a lot in a single day.

The high school teachers said going back to work for a half day, to begin with, is a good idea, because it takes a lot more energy than you are aware of.

My plan is to work out a lot this weekend and next week, show the physiotherapist that I'm good to go, check in with the doc, and be back in the classroom -- full-time -- by the end of the week.

When I got home, I told family and friends what was happening and my sister sent back a quick response: So glad to hear about the support and letters/cards etc from your school. Niiice eh?

I was horrified to hear of you going down at physio! Ouch! That must have been scary. Luckily you were with other people and professionals at that! What happened? Aren't you concerned that it might happen again, like when you are teaching or whatever??

You can't live your life worrying about what if you fall again, but it would be difficult not to be concerned I think. Right? I keep putting myself in your shoes and thinking about how I'd feel in your situation and I'd be scared of never being 100% again but I'd feel grateful that I was walking again and improving step by step. I'd be concerned about returning to teaching too soon and certainly returning full time after such a short period of recovery and physio etc.

Teaching can be exhausting, especially if you are favouring your back or worrying if you might go down any time... I know you must miss teaching and the kids and that you must be anxious to get back in the classroom, but...

Why not try it for half a day as advised and see how you go and work up to full time again? Just a thought. You were stiff after physio and you've been sleeping a lot as part of your recovery and you will likely need to get more sleep and rest more often while you continue to recover Marty. Why rush things?

I am concerned for you and don't want you to have a relapse, if it can be helped. I want you well and healthy and happy again, but in a way that you can sustain without exhausting yourself or doing too much prematurely.

There, that's my sisterly love and concern speech, for what it's worth. I want you back teaching again too, but not too soon if it will be detrimental to your progress!

Why not see how your physio etc goes and how you are feeling, and then decide whether part-time or full-time is in your best interests? I love you and your total recovery is my dearest wish for you.

Take care and I'll write more soon. Maybe we could Skype for a bit on your Sunday night/my Monday. What do you reckon? It would be good to see your handsome face!

Good luck with the physio etc. Keep on truckin'!

Healing hugs,

xoxoxoxoxoxoxox

So I said, thanks for the wishes and hugs.

Falling in physio was no big deal. I worked out for 20 or 30 minutes, my leg muscles got fatigued, so when I walked across the room, they gave out, and down i went. it was more like sitting down than falling.

my return to work depends on what the physiotherapist has to say on tuesday afternoon. he gave me homework. i did everything on his list twice, as directed, and then i did everything a few more times. and then i did all the stuff on the page that was not on his list a few times. and then i did a half hour of qigung. so, i"m coming back fast.

nerves regrow. i have strength in my feet, now, even my toes, and they were completely numb, just days ago.

the physiotherapist pressed on my spine where i've had the most pain. really pressed. and it was no problem. so, i quickly got over my fear of re-injuring the area.

i thought the doc was nuts when he said, back to work. mentally i was prepared for long term disability. but i'm "into" it now!

That night, I went to sleep early, watching The Answer Man, woke up, played the movie again, fell asleep until the ending, played it again, fell asleep until the end ....

At three in the morning, I was awake, again, so I did some writing and research. Just for fun, I thought I'd look up the meaning of my dreams on Dream Moods at dreammoods.com.

Here's what it said: To see books in your dream indicate calmness. You are moving toward your goals at a slow and steady pace. Books also symbolize knowledge, intellect, information and wisdom. In particular, to see an open book in your dream means that you are able to grasp new ideas with ease. If the book is closed, then it represents your allure and mysteriousness. Consider the type of book for additional clues. The dream may represent your calling into a specific field of work or an area that you need to devote more study to. Alternatively, the dream could be telling you not to judge a book by its cover.

The authors in that dream were Margaret Atwood, Clark Blaise, and Dave Godfrey.

I'm Facebook friends with all three. Blaise and Godfrey were mentors as they were my writing instructors when I was an undergrad.

Puns?

Atwood, Blaise, Godfrey ....

Well, I'll let you play with that!

Although I didn't take them very seriously, I liked the interpretations of some of the elements of my dreams.

Before I went back to sleep, I checked the weather forecast for the weekend. It was supposed to snow on Saturday, so my plan to go for a walk in the woods might not work out. But weather forecasts were often wrong, so .... Who knows? Anyway, I like walking in the snow.

Also, I sent an e to a friend of mine who leads trips to Spain and on The Camino, the spiritual hiking trail, asking her if she had a trip planned for the summer. And I explained that I had lost my legs and feet but found them again after making a vow that I would walk the famous trail.

The Camino de Santiago (the Way of St. James) is a large network of ancient pilgrim routes stretching across Europe and coming together at the tomb of St. James (Santiago in Spanish) in Santiago de Compostela in north-west Spain. The most popular route (which gets very crowded in mid-summer) is the Camino Francés which stretches 780 km. (nearly 500 miles) from St. Jean-Pied-du-Port near Biarritz in France to Santiago.

Some people set out on the Camino for spiritual reasons; many others find spiritual reasons along the Way as they meet other pilgrims, attend pilgrim masses in churches and monasteries and cathedrals, and see the large infrastructure of buildings provided for pilgrims over many centuries.

Walking the Camino is not difficult, they say. Most of the stages are fairly flat and on good paths. The main difficulty is that few of us have walked continuously for 10, 20 or 30 days. You learn more about your feet than you would ever have thought possible! And you also learn a lot about life.

After sending my friend a note, I chatted online with a kindred spirit who is the daughter of a cousin of mine, now teaching in Colombia, after teaching in England, Egypt, and Thailand. And she said, I've been thinking about that for years, and I've thought about it a lot in the past few days, so .... Let's go!

But then she confessed she has "wonky" knees and might need surgery. She said walking The Camino with me would be a goal she could set to inspire her recovery.

Day 29/54: The Nerve!

Let yourself be drawn by the strange pull of what you love. It will not lead you astray." -- Rumi

Saturday at seven thirty, I got up with a great idea: With my friend the reporter in Edmonton, I could write a Newcomer's Guide To Alberta, with all the info we wished we had when we moved here a year and a half ago, for the many people who will be moving to Alberta this year and in the near future as this province continues to be the economic engine that drives the rest of the nation.

Right after that, I started research on how to regrow nerves.

I found out that the peripheral nervous system has an intrinsic ability for repair and regeneration. Nerves are cord-like fibers that work within your nervous system to relay impulses, stimuli and messages from your brain to various parts of your body, such as your cells, skin, muscles and organs. If conditions underlying nerve damage, such as a herniated disc pinching nerve roots, for example, are treated, nerves have the opportunity to heal over time.

According to LIVESTRONG.com, a variety of vitamins may help regenerate damaged nerves.

They recommend Vitamins B6, B9, B12, and C.

However, the specialist at the Edmonton Clinic told me to avoid B6 because it causes neuropathy.

Foods rich in vitamin B-9 include liver, sunflower seeds, dry roasted soybeans, turnip greens, collard greens, asparagus, peanuts and pinto beans.

Foods rich in vitamin B-12 include beef liver, clams, yogurt, milk, pork, eggs, American cheese, tuna and chicken.

Foods rich in vitamin C include oranges, tomatoes, pineapples, raspberries, blackberries, spinach, cabbage, potatoes and green peppers.

St. John's wort, lavender and tumeric reduce nerve irritation and aid regeneration of nerves damaged through traumatic injuries, according to Medline Plus. St. John's wort may cause skin irritation and should only be taken orally as directed, and it can interact with many prescription medications.

According to eHow.com, good herbal combination's for nerve pain and regeneration are Nerve Formula by Dr. Richard Schulze and Dr. Christopher's Nerve Extract. Both use the same herbs for the most part, black cohosh root, blue cohosh root, blue vervain, lobelia, valerian root and skullcap herbs, wood betony.

Peterson's Field Guide to Medicinal Plants and Herbs says "St. John's Wort oils are listed in the official drug compendiums of Czechoslovakia, Poland, Romania, and the former Soviet Union and are widely used in western Europe. Externally, this oil is applied to bruises, sprains, burns, skin irritations or any laceration accompanied by severed nerve tissue."

Tribe, at tribe.net, says Cow Parsnip is another one known to heal nerve wounds. It seems as if this would be hard to find though unless you live in the Rocky Mountains and are able to gather it yourself.

What about deer antler tips? I've heard a tonic made from the tips of deer antlers are popular in Chinese medicine and with athletes. And no deer are harmed in the making of the tonic. Deer antlers are harvested in the summer, in China and New Zealand. -- What about Canada?

What therapies does Dr.Weil recommends for neuropathy?

Take one B-100 B-complex vitamin daily. The B vitamins are necessary for normal nerve function, and supplementing is a good preventive measure. Do not take more than 200 mg of B-6, as higher daily doses can actually cause symptoms of neuropathy. Take 100 mg of alpha-lipoic acid daily. This antioxidant protects microcirculation to the nerves. You can gradually increase the dose to 300 mg twice a day over the next month.

Dr. Weil's website also says acupuncture can help relieve the pain of peripheral neuropathy. Additionally, a practitioner of Chinese medicine can provide you with herbs that may speed recovery. You might also try reflexology for neuropathy of the legs, feet and toes. If a toxic exposure is the cause, time is your greatest ally - injured nerves will slowly recover, as long as the exposure has stopped.

Dr. Weil's website says reflexology is a therapy based on the principle that there are small and specific areas of innervation in the hands and feet that correspond to specific muscle groups or organs of the body. In this system, the nerve endings in the extremities provide a "map" of the rest of the body. Examples are the base of the little toe representing the ear, or the ball of the foot representing the lung. Through the application of pressure on particular areas of the hands or feet, reflexology is said to promote benefits such as the relaxation of tension, improvement of circulation, and support of normalized function in the related area in the body.

Dr. Weil thinks reflexology is of great value for relieving generalized foot pain caused by cramped or chronically tight muscles. He also believes that, like other forms of massage, it may release endorphins, leading to pleasurable, relaxed states.

However, he remains skeptical of claims that by massaging or applying pressure to specific points on the hands or feet, a reflexologist can alleviate problems in corresponding organs or other systems throughout the body.

His website also says Dr. Weil is a best-selling author, speaker and integrative medicine thought-leader. "Dr. Weil is a world-renowned leader and pioneer in the field of integrative medicine, a healing oriented approach to health care which encompasses body, mind, and spirit."

After doing that research, I made up a shopping list. And I had this odd thought: I've never been to Venice, I've always wanted to go to Venice, and I wonder if you can go kayaking in Venice.

It turns out there is a company called Kayak Venice and you can rent a kayak and get a guide who will show you where to go.

I'm adding that to my bucket list.

First, hike the Camino. Next, kayak Venice.

Before that, I want to go to Martyr's Shrine, in Midland, Ontario, again, and revisit some of my other favourite places in Central Ontario, too. -- Maybe during the Christmas Break ....

I told my brother about that last fantasy. He said, You've got a room waiting for you in Penetang.

He lives just outside of Penetan, in the country near Awenda Provincial Park.

The Martyrs' Shrine is a Roman Catholic church in Midland, Ontario, Canada, which is consecrated to the memory of the Canadian Martyrs, six Jesuit Martyrs and two lay persons from the mission of Sainte-Marie among the Hurons, as well as Kateri Tekakwitha, the first aboriginal woman from North America to become a saint.

It is one of nine National Shrines in Canada. The Canadian Martyrs were Jesuit missionairies who brought Christianity to Canada.

The Shrine honours the eight Jesuit saints who lived, worked and died here some 380 years ago. The martyrs were canonized by Pope Pius XI in 1930. Pope John Paul II visited the Martyrs' Shrine in September, 1984. There has been a longstanding history of devotion to Blessed Kateri at the Shrine, with four existing statues on the grounds, erected by Catholics committed to "The Lily of the Mohawks", as well as her relics.

Kateri Tekakwitha, baptised as Catherine Tekakwitha, known as Lily of the Mohawks (1656 – April 17, 1680), was an Algonquin-Mohawk Catholic and religious laywoman. She was beatified by Blessed Pope John Paul II in 1980 and Pope Benedict XVI announced at Saint Peter's Basilica that Tekakwitha was scheduled to be formally canonized on October 21, 2012. -- That's tomorrow!

Various miracles and supernatural events are attributed to her name after her death. First, her face, scarred by smallpox, cleared up completely. Next, her mentor, looked up to see Catherine kneeling at the foot" of her mattress, "holding a wooden cross that shone like the sun" A friend reported that she was awakened at night by a knocking on her wall, and a voice asked if she were awake, adding, "I've come to say good-bye; I'm on my way to heaven." She went outside, saw no one, but heard a voice murmur, "Adieu, Adieu, go tell the father that I'm going to heaven."

A chapel was built near her gravesite and pilgrimages honoured her there. Her physical remains were used as relics for healing. For some time after her death, Tekakwitha was considered an honorary yet unofficial patroness of Montreal, Canada, and Indigenous peoples of the Americas.

Fifty years after her death, a convent for Native American nuns opened in Mexico. They have prayed for her and supported her canonization.

The process for Tekakwitha's canonization was initiated by United States Catholics in 1884, followed by Canadian Catholics. In 1943, Pope Pius XII declared her venerable. She was beatified as Catherine Tekakwitha in 1980 by Pope John Paul II. In 2011, the Congregation for the Causes of Saints certified a second miracle through her intercession, signed by Pope Benedict XVI, which paved the way for canonization. In February of 2012, Pope Benedict XVI decreed that Tekakwitha be canonized.

Her miracles?

Joseph Kellogg, a Protestant child, caught smallpox, and was cured by a piece of wood from Tekakwitha's coffin. Father Rémy recovered his hearing and a nun in Montreal was cured by using items formerly belonging to Catherine.

As people believed in her healing powers, some collected earth from her gravesite and wore it in bags as a relic. One woman said she was saved from pneumonia ("grande maladie du rhume"), and gave the pendant to her husband, who was healed from his disease.

Tradition holds that Tekakwitha's smallpox scars vanished at the time of her death in 1680. Pope Pius XII in 1943 declared this an authentic miracle.

Pilgrims to her funeral reported healings.

In 2011, Pope Benedict XVI approved the second miracle needed for Blessed Kateri's canonization. In 2006, a young boy in Washington state survived a severe flesh-eating bacterium but doctors told his parents he was likely to die. The boy received the sacrament of Anointing of the Sick from a Catholic priest. His parents prayed through Tekakwitha for divine intercession, as did their family and friends, and an extended network contacted through their son's classmates. A Catholic nun, Sister Kateri Mitchell visited the boy's bedside and placed a relic of Tekakwitha, a bone fragment, against his body and prayed together with his parents. The next day, the miracle occurred as the disease stopped.

In the morning, I chatted online with Cheryl Cooper, author of Come Looking For Me. She is a writer I met at the Muskoka Novel Marathon. She asked me, a couple of times, "how are you". So I told her: Cheryl, the strangest thing happened. Back pain sent me to a chiropractor for the first time in my life. He fixed that, but as a result I couldn't walk! I worked for a week after that, lurching around, but then I was off for over a month. Canes, a walker, shopping for a wheelchair, then it turned around and I went back to canes and now i'm walking again and heading back to work in a week or so. Quite a journey to docs, specialists, acupuncture, reflexology, etc. So, now that I've been to hell and back again .... i feel like i'm in heaven! lol

But the truth was, that morning, even after a shower and breakfast, my back was sore and my legs were stiff and sore, and I was walking like Frankenstein, again. Or Lurch.

When I told my students that I would be back in a week or so, I was thinking, This is crazy! While I was talking to them, I felt weak at the knees, and had to sit down.

How am I possibly going to get ready to go back to work in a week.

Maybe I'll get my brother to go to Martyr's Shrine tomorrow for the canonization of Tekakwitha and say a little prayer for me.

Oh well. You know what they say: May you let go of your worries and hold on to the present moment. Ahhhh, yes. Right here and now is where joy can be found.

Day 29/54 Part 2: Snow!

After writing and researching in the morning, feeling distressed because walking was very difficult, again, I went back to bed and slept until one in the afternoon. When I got up again, the ground, cars, and rooftops outside were covered with snow. -- Just a little bit of snow.

Determined to get a few things accomplished, I ate breakfast and got ready to go outside. I wanted to take my shopping list with me and go to the grocery store.

Shopping for groceries, the laundry, e-mail, a movie, and some Scrabble was about all I did all day.

I got a great e-mail from Shirley, saying she felt like doing cartwheels, she was so happy I was getting ready to go back to work. And I chatted online with her sister, Lyn. She told me there was nothing more frustrating, to her, than getting contradictory advice from different doctors. Also, she warned me that when nerves start to regrow, they are very itchy.

But I hadn't felt anything itchy where my nerves were supposed to be returning.

The movie I watched was called The Candidate and it was funny but very coarse. Americans handle satire and irony differently than Canadians, I'd say. We have more experience and have developed more subtlety. There is nothing subtle about The Candidate. Calling the Koch brothers the Motch brothers isn't witty or even humourous, really. Will Farrell in the role of a governor from North Carolina, on his way to be vice president, possibly, was fairly funny, but nothing compared to his best roles.

I loved him in Stranger Than Fiction.

Maybe it was just me.

Nothing struck me as funny, today.

Maybe it was because my legs hurt and I had trouble walking, again.

My friend in Parry Sound told me to take the day off, forget about exercising, because I had a big week, physically and emotionally, so I deserved a break today, and it's useful to rest between periods of exercise.

Even so, I felt like I wasted most of the day. -- I didn't get going until noon and I didn't accomplish a lot in the afternoon.

At night I watched part of Forrest Gump, an old movie I like, but the internet service in the area went out when Gump was sent home from Vietnam.

Forrest's trouble walking, at the start of the movie, and his lieutenant losing his legs, in Vietnam, later in the movie, and being miserable while in the hospital, both reminded me of me.

-- Maybe it was a good thing there was a problem with the internet in the region. -- Not even Forrest Gump was making me laugh.

The snow melted. It rained, on and off, in the afternoon and evening. My legs hurt whenever I got up to walk, so I didn't want to walk very much. I hoped Sunday would be a much better day.

Day 30/55: Tekakwitha Day

A night of worries and wild dreams but not nightmares. At seven thirty in the morning, it was still dark as night because it is night as the sun rises later and later up here. It changes quickly from week to week but this is my second year on the 54th parallel, instead of the 45th, so I'm getting used to it. But I am not getting used to the idea of going back to work.

Don't get me wrong. Going back to work is my number one goal. But my recovery stopped the day after physiotherapy and entered a period of decline. I did a much better workout the day after physiotherapy than while I was at the physiotherapist's place, the next day I saw the doc, drove to Lac La Biche and back, visited my school, went to dinner with colleagues, and yesterday, Saturday, my back hurt a bit and my legs hurt quite a bit, so I did laundry and got groceries but did no exercises because I was beat, and this morning, so far, I'm not feeling any better than yesterday.

Last night, I went to bed early, watching Forrest Gump, until internet service was interrupted, and even though I felt grumpy, I had some entertaining dreams. They say we have a lot of dreams. I remember two. In the middle of the night, I got up and made a couple of notes on the first one. It was about some guys trying to put a pipeline across Lake Muskoka. The one I had this morning was about the future.

It started out looking something like Lost In Space and The Jetsons, two TV shows from the past that were about the future, and it morphed into something different about a city of lights built around a lake that looked like Gull Lake, in my old hometown, and spaceships that could fly out to a silver rainbow to be instantly transported to another dimension. It turned novelistic, with two main characters and a dozen minor characters, with a couple that wanted to get married, a black woman and a white man, and a ritual battle they had to go through, which also determined your caste, and the couple wound up in the highest level, which was very special, not just luxurious in familiar terms but also in transformative ways. The couple got to sample the highest caste, which included altering your body so you could look the way you always wanted, and the two people came out changed so much that they didn't recognize each other.

The black woman, who was slightly pneumatic, came out thin and white, with her curly hair straightened, and the man came out multi-coloured, but suntanned, and a lot more athletic looking. They walked right by each other.

Did they fail the test? Was it all about inner beauty?

During the fight scene, they were turned into animals, including some wild and ferocious beasts, but the winners were brilliant and beautiful dogs.

Do I need the online dream interpreter to decipher these dreams? I don't think so!

The morning dream was about a desire for transformation. I want my life back. I want my back, legs, and feet to work again, so I can get in shape and go back to work. I want my nerves to regrow and get strength back into my knees and feet. And I don't want to lose that woman who lives in Edmonton.

The dream about a pipeline under Lake Muskoka was about old friends in the place where I grew up, doing new things, in conflict with each other. I think that all it means is I am a little homesick and would like to revisit that part of the world.

The internet is down, or out, locally, but my iPhone still works, of course, and I got a note from my sister in Australia, saying, It's good the Dr. suggested going back to work when possible.

But that's not what happened.

He said, You should go back to work.

He didn't examine me or ask me anything.

When I balked, because I was in shock, and could barely walk, he said, Well, what's stopping you?

I told him I could take a thousand steps, then had to rest, had no balance, my feet were numb, I felt I was getting better, but wasn't quite ready for a return to the classroom. I could, possibly, handle a perfect day, but would be a liability if anything went wrong. If I had to break up a fight, chase after a student, go outside for a fire drill, respond to a shooter in the school, I'd be a liability instead of an asset.

He said, You're a teacher? You can sit down a lot of the time. We can't control unlikely situations. Go to your physiotherapist and get him to write up his assessment and bring it to me on Wednesday.

He didn't give me a 'doctor's note' to be off work.

And I was not happy when I left his office or that town.

Although I got excited about going back to work, I really could not imagine doing the job in the condition my condition was in.

Yesterday I walked stiffly with sore legs on numb feet and got shots of back pain so I was exhausted after a trip to the grocery store. Picking up a laundry basket and walking was a challenge.

And so I dream about going home in a future when science can transform me into a fully functioning version of me.

These days, I have more faith in religion than science.

Yesterday, on Facebook, there was a funny note posted as a little letter from Science to Religion, commenting allusively to recent events making international news. It said:
Dear Religion,
Yesterday I dropped a man from space to the Earth, safely, while you had a little girl shot because she wanted to go to school
Sincerely, Science.

Today is the day an aboriginal woman from North America, whose relics are kept close to my old hometown, becomes a saint, officially. Miracles of healing have been attributed to her, things science and medicine could not fix or cure or help, including the horrible flesh-eating disease.

I've seen several doctors and all they've done is examine me and send me for more tests. A physiotherapist gave me exercises which seemed to help, for a while. The person who has helped enormously calls herself a reflexologist but she is not recognized by the mainstream medical community and I think she is a gifted healer, possibly a saint! She's got something going on, and it's not just a foot rub with knowledge of which part of the foot represents which part of the body, which is what reflexology is all about. She is using Reiki, energy healing, and intentions, taking away pain, soothing nerves, releasing endorphins, making me laugh, and much more. She also helped me with some practical things, like getting a walker when I could no longer walk with canes.

After yogurt and an apple, I did one set of the exercises assigned by the physiotherapist.
    No problem.
    Not much energy. A little bit of pain in the back and legs. But no problem.
    Doing the bridge was difficult. I couldn't get much height. The rest of the exercises were fairly easy.
    For breakfast, a new cereal with no B6.
    After breakfast, I did a second set of all the exercises. -- No problem.
    Also, I tried a double leg lift, while lying on my back. -- That was difficult. -- With great effort, I got my feet and legs off the ground, but my feet were only about two inches above the carpet.

It's 9:15 and the internet is still out. Facebook, on my iPhone, has a quote from Love Tree that appears to be perfectly timed for me: "When sadness comes to us, remember it arrives only as a visitor, not as a permanent guest."

For the rest of the morning, I slept, and had a warm dream, then got up at 11 to do the third set of my exercises.
    In the dream, I was in my old hometown, or close to it, and I was working as an alternative practitioner, doing reflexology and using other healing modalities including Reiki, Quantum Touch, and energy healing. My client was an old friend of mine, from childhood, who was in a lot of pain and said he had not been able to find relief or anybody to help him.
    Tom Montgomery.
    While he got comfortable on my healing table, I looked in a big old phone book for a reflexologist but could not find one.
    Why was I looking for a reflexologist when I was a reflexologist?

After making my client feel better, in the dream, a very warm feeling came over me, even though it was cold outside and cool in the room where I was sleeping. My lower back, my legs, and my feet were all bathed in heat that felt quite healing.

As I got up, I had the feeling that I was healed. But my feet felt the same: that odd combination of numbness and pins and needles.

The internet was still not working. I checked e-mail on my iPhone. On Facebook there was a message that said Canadian author Sylvia Fraser posted a picture of me and it got several comments, but it turned out to be spam from Costco. On Youtube, I watched a new rap video with actors as Mitt Romney and Barak Obama, debating and insulting each other with lots of rhymes and a steady beat. -- Funny.

After the third set of my exercises, I made lunch. Toasted tomato sandwich, orange juice, a banana, and a p/b/j sandwich. -- Everything tasted great.

After lunch, I felt like doing something, but I didn't know what. Going outside for a walk? More exercises? Listening to some music? Something!

Read a book, fell asleep for a few hours, got up, still no internet, phoned, and East Link said there was a big car crash somewhere in or around Cold Lake and it took out a couple of telephone poles plus internet service. At first they thought service would be restored by morning but now they are saying it won't be back for another four hours or so, or around 7 p.m.

That must have been a horrible crash!

My sister in Australia wanted to Skype today.

As the Zenmaster says, "We'll see."

Dinner: Grilled balsamic chicken with angel hair pasta, tomatoes, roasted garlic, and balsamic sauce.

Dessert: A book!

I feel as though I could devour a novel. Maybe the Wally Lamb story called Wishin' And Hopin'. -- It's not even 300 pages long.

Wally's novel is about Felix Funicello, who is related to Annette Funnicello, supposedly, and she's in the story, too .... In the end, we are told the famous singer and movie star had MS and was blind and in a wheelchair, and there was an address at the back of the book for donations to a charity that supports MS.

Wishin' and Hopin' was a departure for Lamb: a short, comically nostalgic novel about a parochial school fifth grader, set in 1964. It got mixed reviews.

Halfway through the book, I took a break from reading to try the internet, to see if Skype would work, but it was still down. When I phoned East Link, the service provider, they said the car crash was quite horrible and they wouldn't be able to get Cold Lake back online until 2:00 in the morning. -- I tried it at 2:00, and it wasn't working then, either.

Around the time my sister wanted to Skype with me, we tried chatting online, instead. At the same time, I heard from friends in Edmonton, Ontario, and Alabama. Trying to communicate with all four of them, typing on my iPhone, was fun.

-- Not!

After half an hour or so, I said goodnight to all four of them: "GTG TTYL".

Day 31/56: Nerve Formulas

The internet was reconnected, locally, like nerves regrowing in limbs with neuropathy, and first thing Monday morning I went online to research rehab for my feet and toes. The first thing I found was a disturbing abstract from the Department of Rehabilitation Medicine, University of Washington, School of Medicine, in Seattle.

It said, The goals of rehabilitation for patients with peripheral neuropathy are to maximize and prolong independent and safe locomotion and function, inhibit physical deformity, and provide access to full integration into society. The services of multiple disciplines are required to fully achieve these goals and include physicians, nurses, therapists, orthotists, social and vocational counsellors, and psychologists. Treatment is goal-oriented, using various modalities including exercise, bracing, adaptive equipment, medication, and surgery.

How's that for a jumpstart to a Monday morning.

My recovery was not going the way I had hoped. Initially, my ability to walk improved dramatically from day to day. That continued up until the day after my first day at physiotherapy doing exercises. At home, I did all the exercises and then some, and then I did them again and again.

The next day, I drove for two hours, for a doctor's appointment, and two hours back home, dropping in on my students and colleagues at school, saying I would be back in a week or so.

And I spent the weekend struggling to get through the exercises because my legs felt stiff and sore and I was dog-tired.

After getting lots of rest and eating well on the weekend, I started the week feeling tired and doubtful about the doctor's decision about going back to work. Frankly, I was feeling anxious and the opposite of inspired.

Facebook, this morning, says: You will find that it is Necessary to let things go; simply for the reason that they are HEAVY.

Ignore the semi-colon and the capitalization.

That little gem comes from a Facebook group called Rebels for Consciousness. Their home page has a lot of inspiration and a lot of it is humorous. Some of it is profound and some of it is just silly.

How about this one: Always be yoursellf. Unless you can be a unicorn. Then always be a unicorn.

They also quoted Eckhart Tolle: "The primary cause of unhappiness is never the situation but thought about it. Be aware of the things you are thinking. Separate them from the situation, which is always neutral. It is as it is."

Eckhart Tolle is a Canadian, living in Vancouver, born in Germany who has become the most popular spiritual author in the U.S.A. and the most spiritually influential person in the world, according to the New York Times and the Watkins Review.

His first book, The Power of Now, published in 1997, reached the New York Times Best Seller lists in 2000. The Power of Now and A New Earth sold an estimated three million and five million copies respectively in North America by 2009. Tens of millions of people sign up for the webinars he has done with Oprah Winfrey.

Tolle is not identified with any particular religion, but he has been influenced by a wide range of spiritual works.

Personally, I've never liked that expression "It is as it is" or "It is what it is" or any of the popular phrases that sound like quotes from Buddha and even claim to be quotes from the Buddha. There are several websites devoted to Fake Buddha Quotes.

Some of these quotes fly around the internet but who knows where they come from. Here's a few examples:

"You cannot travel the path until you have become the path itself'

"Do not dwell in the past, do not dream of the future, concentrate the mind on the present moment."

"He is able who thinks he is able."

Don't get me wrong. I'm not saying Eckhart Tolle uses fake Buddha quotes!

Day 54, Part 2: The Rise Of The Guardians

My brother phoned first thing in the morning to tell me he could not get into Martyr's Shrine for the ceremony to mark the moment Takakwitha became a saint because there was a huge crowd and the parking lot was filled.

The Globe And Mail called it "an an act of atonement" when the Vatican made Kateri Tekakwitha the first native Canadian saint, saying "all canonizations are political to some degree, but the canonization of Kateri Tekakwitha, the first indigenous Canadian saint, was more political than most. The First Nations considered it a key step in the Vatican's long and haphazard campaign to repair relations with a people it had mistreated for centuries."

She was canonized Sunday in St. Peter's Square by Pope Benedict XVI, along with six other saints. An estimated 80,000 pilgrims gathered in the square, according to Canada's national newspapers, including thousands from American and Canadian indigenous communities who claim Kateri as their own.

To the delight and amusement of the Italians in the crowd, the Globe And Mail reporter added, many of them wore colourful traditional costumes, such as feathered headdresses and leather-fringed tunics.

To some, she represents unconditional commitment to Jesus, they claimed, and to others she represents a bridge between Catholic and native spirituality and culture.

My brother asked me if I was going back to work sooner, rather than later, and pointed out that teaching is a good profession as a supply teacher can take my place and insurance can cover my absence, to some degree. He said he knew a lot of teachers who went on LTD, over the years he worked as a teacher.

My morning exercises were interrupted by a phone call from work with an invitation to join the high school teachers and admin for lunch. The principal made spaghetti with meat sauce and a cake for dessert and the elementary teachers were out of town for a professional development day.

It was great to see the gang again and get a hug from the principal. Everybody teased me at lunch and our head secretary said, We have to give you a hard time now, all at once, because you've been gone for a while and we haven't been able to tease you a little bit every day.

What a group. Honestly. I love this gang of teachers and administrators.

I took in my copy of Wishin' And Hopin' and gave it to the elementary English teacher as it is about a short kid named Felix in Grade Five and we have a student who is just like him. Well, something like him.

My supply teacher and I discussed our classes and students and I attended a meeting of the high school teachers, briefly, and then said, "So long!"

It was sunny but below zero so I felt inspired to try going for a walk at the provincial park. The wind off the lake made it feel like ten below, but it still felt good to get outside and breathe the air in the forest and get some exercise. I took around one thousand steps and then just leaned against my car in the sun for a while.

At the drug store, after the short hike, I asked the druggist for a recommendation for neuropathy, but he said painkillers and time were the best medicine, and I already had painkillers.

After walking through the drug store, my legs were shaky and sore, so I headed home.

While I was there I found a novelization of the new movie called Rise Of The Guardians, about Jack Frost, Santa Claus, The Sandman, The Easter Bunny, and The Tooth Fairy, getting together, so I got a copy, thinking it might be a good book form my Grade Three class.

Rise of the Guardians was an upcoming 2012 American 3D computer-animated fantasy-adventure film based on William Joyce's series, The Guardians of Childhood, and The Man in the Moon short film by Reel FX and Joyce, produced by DreamWorks Animation and set to be released on November 21, 2012.

Jack, Santa, Sandy, the Bunny, and Tooth get together to take on The Nightmare King and protect the children of the world.

So, it was not my best day, but it was better than the past two or three days, so I felt optimistic again. And that felt good!

Just in case the physiotherapist had to plan something different for an assessment, I phoned to tell him that was what the doctor wanted. His assistant told me it takes more than 24 hours for an assessment to be written up. He wasn't available on Thursday, so I moved my appointment with the doctor from Wednesday to Friday.

So, now I won't be going back to work on Friday. And that makes me feel a lot better.

I felt as though I had to buy some time, to give my legs and feet a little bit more of an opportunity to heal.

But who knows what the physiotherapist will say? Maybe he will say I need another month to get going again.

Day 55: A Plea For Integrative Medicine

Although I went to bed early, fell asleep fast, and slept soundly, I woke up around three with this idea in my head: The title of this book should be "A Plea For Integrative Medicine".

Where did that come from?

I had no idea.

In the evening, and as I fell asleep, I felt worried, even anxious, about getting the physiotherapist's report and a doctor's note, going back to work too soon, trying to do a demanding job when I was far from one hundred percent, and falling down on the job, both metaphorically and literally, so I got sick and had to be off even longer.

My friend in Ontario with the double hip and knee replacements and a lot of experience with doctors teased me, when we chatted online, about being worried and anxious. "That's not like you," she said. "Don't develop my bad habits!"

Worry and anxiety are side-effects of neuropathy, I told her.

But when I Googled that, I discovered the opposite is true: worry and anxiety sometimes cause neuropathy.

Anyway, I fell asleep and had a dream about traveling in a foreign country, possibly China, and running away, or defecting, and that was a very strange and disturbing dream, for me. When I awoke from that dream, I decided to take charge of my dreams, like the director of a movie, as I have learned to do, and stop that dream, or dream about something else. And I did.

Instead of just rolling over and going back to sleep, I got my computer to read part of this book to me, as I often do. And a few hours later I woke up with this idea: the title of this books should be "A Plea For Integrative Medicine".

It wasn't me who was directing this dream and I had a funny feeling about where that idea was coming from.

So, I got up and Googled it, but "A Plea For Integrative Medicine" did not turn up anything useful.

Next, I Googled "integrative medicine" and found exactly what I was looking for.

Wikipedia had an entry like a little essay that said what I wanted to say:

Integrative medicine or integrative health is a neologism used by health practitioners and organizations to describe the combination of practices and methods of alternative medicine with conventional (or orthodox) medicine (or biomedicine).

Next, I Googled "Deepak Chopra neuropathy" and that led to some interesting hits.

Day 55: Part 2

The San Francisco Giants were down two games in the National League finals but battled back to win. The Detroit Tigers beat the New York Yankees. So it was The Giants versus the Tigers in the World Championships. Underdogs were becoming champions.
Meanwhile, I was coming from behind in my battle against neuropathy, but at eight in the morning, things were pretty much the same as yesterday.
This little book of mine is now 250 pages long.
Today I'll go to physiotherapy for the third time, get an assessment, and I might have to drive up to Lac La Biche to get a doctor's note to take into work.

It's six below zero and I'm up and at 'em like Atom Ant. In other words, I feel more awake than usual and more energized. And that's a very good thing!
Last night, I had no back pain, for a change. And my morning shower went a lot more smoothly. That makes me feel a lot better! -- Maybe it's the EFT.
Physiotherapy was fascinating and the physiotherapist was fantastic. Everybody at Health First Physiotherapy in Bonnyville is first rate, I'd say. Quick, kind, and good communicators. And knowledgeable, too.
This time, my third appointment and second work, they took me in early, again. They started me off with a muscle stimulator, a battery operated device attached to two little black pads that were taped to the small of my back, and a hand-held controller that let me increase the amount of electricity on a scale of zero to one hundred. It felt good, not painful, so I cranked it up as high as it would go and I wore it for the hour-long workout.
Ten minutes on the treadmill, increasing the speed every minute, was the warm-up, and that was followed by leg lifts on the parallel bars, using two hands for balance, then one hand, and then no hands. -- Except I could not balance on one foot, so I couldn't do it with no hands.
But I tried.
Hard.
Next came side stretches while sitting on the balance ball. -- No problem. And leg lifts on the balance ball while raising the opposite hand and arm. -- Not much of a problem.
But the final exercise was a problem. Kneeling squarely on a mat, without the ball and then with the ball, lifting one leg and the opposite arm was not a problem on one side but it was a problem on the other side. My left leg just didn't go in that direction, no matter how hard I tried. -- I worked up a sweat, just trying to lift my leg, even without lifting the opposite arm, but I barely got it off the ground.
At that point, the physiotherapist told me he was going to recommend, in the assessment he was writing for the doctor, that I took another two or three weeks off work to develop core strength and balance.
Your balance is zero, so it's a health risk, or a hazard, as you are likely to fall down.

And that sounded about right to me.

The final ten or fifteen minutes of the workout was without the muscle stimulator and on the decompression machine.

You are supposed to relax while on the decompression machine and let it pull you apart, to open up your spine a bit, and doesn't hurt at all. In fact, it feels good.

While I was lying there, I overheard the physiotherapist talking with a new patient and getting her story in chronological order. She hit a moose, was taken to hospital on a spinal board, x-rayed, and released. At home she got worse, couldn't move, was in great pain, so she went to her family doctor, who sent her to Edmonton for a CT scan, and she was told she had severe whip-lash. After a short period of rest, with pain-killers and other medication, she was supposed to go to physiotherapy.

She clipped the front legs of the moose with her car, it swung around so the body smashed against the driver's door and the head smashed the windshield. The moose was dead and the three passengers in the car were not injured.

The physiotherapist remained calm as the woman told her story. I got the impression he heard stories about people running into moose all the time.

I grew up in moose country but rarely saw any in the bush, never mind on the road. Traveling from Muskoka to Parry Sound or Algonquin Park, the odds of seeing a moose are not bad, but I rarely saw one and never ran into one.

When a moose is spotted near the highway in Algonquin Park, cars pull over and park by the dozens, and a big crowd of people with cameras take hundreds of pictures.

The only moose I've seen in Alberta, so far, have been dead, at the side of the road, after being hit by a car or truck. In the past year, I've seen three of them.

My new exercises, to be added to my old exercises, are the alternate arm and leg extensions that were so hard to do, arm and leg lifts while sitting on an exercise ball, double leg bends while lying on my back, seated flexion stretch, and seated side bend stretches.

For the flexion stretch, you sit down, spread your legs, bend over and put your hands on the floor, for fifteen seconds, sit up for five, and then do it again and again.

For all of these exercises, you pull in the TAs. And don't forget to breathe.

The drive to Bonnyville and back was uneventful. It's an overcast day, the temperature is still below zero, the ponds by the highway are frozen, and the river looks like it is freezing over, too. I saw a few magpies, crows or ravens, three dogs running wild, and a dead red fox in the middle of the road.

When I got to Cold Lake I went to the Energy Centre to ask about joining the fitness club and found out their elevator is broken. The physiotherapist recommended walking on the track at fitness centre and using the treadmill, rather than walking in the woods, with the bears. "You can't tell when your leg muscles will get fatigued and you're going to fall over. So, if you're on the treadmill, you can stop it, get off, and sit down. If you're on the walking track, there are stations with mats where you can rest. -- And there are no bears."

Will the doctor agree with the physiotherapist's assessment? He asked for it.

I'm worried about getting a doctor's note. I phoned the doctor's office this morning and they said they would get the doctor to write a note for me to be off work and then phone me.

Some of my colleagues decided to cook for me. On Tuesdays, a group of us gets together for wing night, and after wings they presented me with a bag of homemade goodies: ham with pineapples, sour something jelly .... And they promised there was more to come.

They were not surprised when I told them the physiotherapist said, Two or three more weeks.

Although the food was not necessary, I appreciated the effort, the gesture, and the love.

My sister sent me a longer e-mail. She said, I was horrified to hear of you going down at physio! Ouch! That must have been scary. Luckily you were with other people and professionals at that! What happened? Aren't you concerned that it might happen again, like when you are teaching or whatever?? You can't live your life worrying about what if you fall again, but it would be difficult not to be concerned I think. Right? I keep putting myself in your shoes and thinking about how I'd feel in your situation and I'd be scared of never being 100% again but I'd feel grateful that I was walking again and improving step by step. I'd be concerned about returning to teaching too soon and certainly returning full time after such a short period of recovery and physio etc. Teaching can be exhausting, especially if you are favouring your back or worrying if you might go down any time... I know you must miss teaching and the kids and that you must be anxious to get back in the classroom, but...

Why not try it for half a day as advised and see how you go and work up to full time again? Just a thought. You were stiff after physio and you've been sleeping a lot as part of your recovery and you will likely need to get more sleep and rest more often while you continue to recover Marty. Why rush things? I am concerned for you and don't want you to have a relapse, if it can be helped. I want you well and healthy and happy again, but in a way that you can sustain without exhausting yourself or doing too much prematurely. There, that's my sisterly love and concern speech, for what it's worth.

Before wing night, I was so tired I almost skipped the weekly event. They called to let me know it was "on". At the time, I was lying down, watching big flakes of white snow fall in front of a row of black spruce trees. While we were at the restaurant, I noticed the snow continued to fall. After I got home, I hit the sack again but got up around midnight and saw that a lot of snow had fallen. It covered everything in sight, even the sidewalks and streets. On The Weather Network, I found a cool picture of Cold Lake in the snow. There was a very artistic picture of the waterfront with snow on the sidewalk arcing along the shoreline with a row of lights in the background. It looked like Paris in the wintertimes. I posted a link to it on Facebook.

It's early for snow in this part of the world. Usually, there is no snow until November, or sometime after Hallowe'en. At the restaurant I pointed out that it's starting to look a lot like Christmas before Hallowe'en.

At home, I watched the trailer for The Nightmare Before Christmas and thought about writing a story or a novel about that guy who decorated his house elaborately for Hallowe'en and never took that display down when it snowed but just added a lot of Christmas decorations so that The Grim Reaper stood beside Santa Claus and ghosts flew with angels and so on. But I didn't want to dream about that, so I got my computer to read this book to me, again.

Late at night I noticed a posting from an old friend of mine who was now living down south in the U.S.A. She said she was crying, thinking about all those who have passed away. "Crying. Life. Missing those who passed." Later, she added, "I miss too, someone to hold me in my grief....but this too shall pass."

She was a friend from my childhood -- my girlfriend in Grade Six -- and we had re-connected in our mid-to-late fifties.

As a response, I posted a link to Bob Dylan singing, "Death Is Not The End".

Before I turned in, I listened to the song a few times.

When you're sad and when you're lonely
And you haven't got a friend
Just remember that death is not the end
And all that you held sacred
Falls down and dowe not bend
Just remember that death is not the end.

Not the end, not the end
Just remember that death is not the end.

When you're standing on the cross-roads
That you cannot comprehend
Just remember that death is not the end ....

Day 56: Yoga

The day after physiotherapy, my legs were stiff, first thing in the morning, but after a warm-up, I tried the leg lifts that were so difficult the day before and discovered I could do them. Maybe it depended upon how fatigued my leg muscles were.

The things the physiotherapist got me to do would have been so easy a couple of months ago.

Losing my balance was never a problem, before, either.

I've noticed my leg muscles are quite different, now. The definition is quite different.

A couple of years ago, I did a lot of yoga. Skiing was out, so I looked around for another form of exercise. There was no gym or rec centre in the little town where I was living, but Port Perry had a great little yoga studio called Port Perry Flowyoga. After I got to know the gang at the yoga studio, they asked me to teach Zen meditation and qigong.

For years, I'd been both attracted and repelled by yoga. I'd tried it here and there but had never found a place where I felt comfortable. I had a great introduction to yoga one summer at the Omega Holistic Centre, down by Woodstock, in New York State, with the yoga instructor who worked with Deepak Chopra. He presented a three day lecture series based on his new book and that meant his audience spent a lot of time sitting in those legless meditation chairs while listening to him, so he encouraged everybody in his audience to join in the yoga classes in the evening, so we could stretch our limbs and be in better shape for sitting still several hours a day.

That yoga group was big and there were lots of yoga practitioners at every level, including newcomers, like me, and some very advanced people who did some complicated and impressive poses, balancing on one foot or on their hands and demonstrating incredible flexibility.

Our instructor assured us we were not expected to be able to do gymnastic stunts. "Your yoga is just for you," she said numerous times.

Maria Carr, the owner and lead instructor at Flowyoga in Port Perry, repeated the same advice. She coached me through the first level and got me into a more advanced class. I got quite attached to the studio and the group, enjoyed the feel of the place, even the aroma and the music. A lot of guys complain about yoga music, but that studio used a good deal of Deva Premal, and there is nothing like listening to her singing and chanting The Moola Mantra as you rest in savasanah or the corpse or cross position after an hour-long workout.

Yoga isn't like any other kind of a workout. It feels gentle, compared to an hour at the gym, or a basketball or hockey game, but you sweat more. Maria said that was because of all the twisting movements. "You wring yourself out like a rag and get a lot of moisture out," she said.

Other yoga studios in the region got bigger crowds with trendy approaches, especially hot yoga and candle lit yoga. But our group did not get into the trendy stuff and everybody dressed relatively conservatively. They joked about the expensive and revealing outfits people wore to hot yoga.

Was there a yoga studio in Cold Lake? I didn't think so.

There was one in the Harbour View Community Centre, I was told, but it had closed. Google informed me there was yoga in the Seniors Centre in Cold Lake North, called Cold Lake Yoga.

I had never been to the Seniors Centre and I didn't think my schedule would allow me to make it to any class on a regular basis. My recovery had turned into full-time work. Every day I had an appointment with a doctor, specialist, or physiotherapist, and that meant traveling for half an hour, an hour and a half, or three hours, and back. Aside from those appointments, my time was taken up with errands, physiotherapy homework exercises, and sleep. Appointments, exercise, sleep: that was my life.

Yoga is not all about balance, in my experience, but balance is a big part of it, and mine was at zero. Any of the many standing poses that required standing on one foot were out, for me, while I was working on my recovery. I kept trying the basic poses. I could do downward facing dog but the tree pose was now impossible.

Tree Pose, called Vrksasana in Sanskrit, helps strengthen your thighs, calves, ankles and back. It can also increase the flexibility of your hips and groin. Your balance and concentration can also be improved with practice. This Yoga Pose is recommended for people who have Sciatica and flat feet, they say.

STEP 1: Start with the Mountain Pose.

STEP 2:As you exhale, place your left foot on the inside part of your right leg, close to the groin area, with the toes pointing downward.

STEP 3: As you inhale, stretch your arms sideways to form a T, palms facing down.

STEP 4: As you exhale, bring your palms together in prayer position.

STEP 5: Raise your arms overhead, keeping your palms in prayer position. To maintain balance, it helps to focus your eyes on one point in front of you and keep on breathing through the belly.

I looked forward to the day when I could do yoga again.

Until then, I enjoyed the cartoons and comments of a Facebook group called Yoga Dawg, that made fun of yoga, especially the music, and made me laugh.

Right after I wrote that, I found Yoga For Beginners on YouTube, and I did it. There was a ten minute Part One that started with a warm-up on hands and knees, lifting and lengthening your back, rounding your spine, and then lifting your shoulders, repeatedly, coordinated with your breathing. After the warm-up, you start in mountain pose, which means standing up straight, belly button in, shoulders back a bit. Breathing in, you raise your arms over your head; breathing out, you bring your arms back down again. After doing that a few times, with your hands over your head, interlace your fingers and turn your palms up, then stretch to one side and the other. Next, you bring your arms down and bend forward, put your hands on the ground, then step back one leg, bring it back, stretch the other leg back, and then bring it forward again. You stay in those positions for a minute or two and breathe. Next comes the downward dog position, one of the most famous yoga poses. From there, step into a forward bend, come up with a straight back, exhaling, to mountain pose again.

Next, step your left foot back and stand with the right foot pointing forward with the left foot at a 90 degree angle. Move your hips to face forward and then bend forward so your right knee is over your front toe. Bring your arms up, over your head, beside your ears, and take a few breaths. Bring your arms down, turn so your left foot is in the lead, bend into that so your right leg is straight and your left knee is over your right toe, then lift your arms and breathe.

That pose is called Warrior 1.

End in the mountain pose and breathe.

Warrior 2 is next.

You end in mountain pose, and that's it.

And I did it.

-- I was shaky and I could really feel it at the back of my legs. But I did it.

Part Two was just five minutes, with similair moves, and I was able to do that, sort of, too.

Yogatic with Esther Ekhart was do-able, I discovered, so I tried a few more, starting with Moon Salutation. Triangle Pose, Pyramid Pose, with squats and lunges, was more challenging and I couldn't do much of that.

Esther Ekhart had a ten minute yoga for abs and core strength. That was a real challenge but I did most of it. And I was sweating by the end of it.

Start by sitting on the floor with your legs straight out in front of you and you back straight at a 90 degree angle from the floor. -- Not everybody can do that.

BodyRock.Tv had a three minute video demonstration of the roll-over commander push-up. It's a push-up with knee tucks while you are up. When you go down, you roll over and then do it again. It was called How To Lose Belly Fat In 1 Week. The "beginners" version was fun to do.

Those three videos gave me a pretty good thirty minute workout, first thing in the morning.

Esther Ekhart's Yoga For Balance was impossible, for me, as it featured standing on one foot at a time.

After yoga, starting at six in the morning, I found Deva Premal doing the Moola Mantra on YouTube and let it play while I did the physiotherapist's exercises, old and new.
    -- Phew!

These exercises, at this stage, are both depressing and inspiring, as I keep comparing what I can do now to what I could do before. If I compare it the rest of my life, until two dozen days ago, it's not impressive. But if I compare it to what I could do for the past fifty days or so, well, that's another story!
    For instance, the physiotherapist asked me to stand on one foot and lift the other leg. Yesterday, after ten minutes on the treadmill, I couldn't do it. Today, after half an hour of yoga, I did it. Two months ago, I'd thinking nothing of doing that. Now I think it's a significant accomplishment and I got excited and inspired when I finally did it.

On Facebook, Love Tree posted a quote from Positive Thinking that says, "Never be afraid to fall apart because it is an opportunity to rebuild yourself the way you wish you had been all along."
    "Would you be willing to know, that everything is working out perfectly? Even when things look to be falling apart... what if, they're actually falling together?"

Day 56, Part Two: The Cure For Death By Lightning

After two and a half hours of physiotherapy exercises, yoga, and the commander roll-over push-ups, I felt good, but I did not feel like walking or working out any more. The reflexologist at the Wellness Centre phoned to say, "Don't come in for your appointment because the roads are too slippery since there was a lot of snow last night and they haven't sanded the roads, yet."

Instead of venturing out into the winter wonderland, I stayed home and read a book: The Cure For Death By Lightning, by Gail Anderson-Dargatz, a Canadian author from B.C. who went to UVic, like me, and became the star of the Creative Writing Department as she got a huge advance for her first book while she was still an undergrad. Her novel was nominated for the Giller Prize, was awarded the Ethel Wilson Fiction Prize, and became a bestseller in Canada, selling over 100,000 copies, and in Great Britain, where it won a Betty Trask Award. -- That's a great big number for a book by a Canuck.

By the way, there is no cure for death by lightning, of course, because, well, there is no cure for death!

After reading, I did some research, online, and found a cure for neuropathy.

The McVitamins website, at http://www.mcvitamins.com/neuropathy.htm, claimed to have a remedy for this problem: "You might have heard of the new type of vitamin B1 being produced, called Benfotiamine," their website said. And Methyl B12.

It sounded too good to be true and most things that sound too good to be true turn out to be un-true.

Maybe I would order it later.

It was the day of the first game in the World Series, but I was too tired to follow a long baseball game. Besides, I don't watch TV. Instead of watching a long game, I sat on my exercise ball and saw Moneyball, the movie starring Brad Pitt, about the Oakland Athletics and the revolutionary use of statistics to evaluate baseball players. That sounds boring but it was a very good movie. Brad Pitt's character took his small market team all the way to the World Series. But I won't ruin the ending of the story for you. If you are a baseball fan, you already know.

Game 1 of the 2012 World Series between the San Francisco Giants and the Detroit Tigers was an upset of record-setting proportions. The Tigers were expected to win as they have a pitcher who is considered the best in baseball. The Giants pitcher had a big night and the Tigers pitcher looked unusually ordinary. And the big story of the night was the three home runs by one player called The Kung Fu Panda.

This sounds like fiction, but it's true.

The Giants' "Panda", Pablo Sandoval, joined Reggie Jackson, Babe Ruth and Albert Pujols as the exclusive club of baseball players who hit three home runs in one game in the World Series. And he did it faster, at his first three at-bats.

The San Francisco Giants beat Justin Verlander and the Detroit Tigers 8-3 to take Game 1 of the championship. No doubt, there will be a movie made about that game or this series.

While I was playing online Scrabble, my brother phoned that night for an up-date on the saga of my idiopathic neuropathy. Was my sore back, weak knees, numb feet, and nerve damage causing muscle failure caused by a chiropractic adjustment, or was it something else? Doctors disagreed. The physiotherapist said it was a bulging disc affecting two nerves controlling muscles in the knees and the muscle groups of the foot.

My Scrabble score went up to 1640, with 250 "bingos". But, frankly, I had played just about all the Scrabble I could stand.

Day 57: Changes

Do not dwell on "what is wrong", but instead think about all that is right!... Positive Thoughts

Trust that your soul has a planand even if you cannot see it all know that everything will unfold as it is meant to. -- Deepak Chopra

The morning of the 57th day of my neuropathy saga featured one change: My feet still tingled, had that odd combination of numbness and pins and needles, but there was a new feeling, too. My feet hurt.
    That had to be a good sign. Pain would be better than numbness.
    My back was a little sore and my legs were stiff. I slept until six in the morning but felt too tired to do another two and a half hour of yoga, physiotherapy exercises, and commander roll-overs.
    -- Maybe later.

That afternoon I watched Wimbledon, the movie, a romantic comedy about a washed-up tennis pro named played by Paul Bettany and an up-and-coming tennis star played by Kirsten Dunst. Bettany is an English professional Tennis Player in his thirties whose ranking has slipped from 11th to 119th in the world. He never really had to fight for anything as his wealthy but not close family easily put him through studies and allowed him to pursue his Tennis ambitions. He earns a wildcard spot to Wimbledon. He bumps into Dunst, the American rising star of female tennis and falls in love with her.

    Her interest in him changes his perception and gives him the strength to win. As their love grows, Peter's game becomes better and better, but her game goes downhill. It's fairly predictable, but it was still good enough to get me cheering and I might have had a tear or two in my eyes at a few points.
    Bettany was great, as usual.
    He was great as Chaucer in A Knight's Tale and he was another kind of great in A Beautiful Mind. I never saw Master and Commander. He was something else as an Opus Dei monk named Silas in and The Da Vinci Code.

Watching Wimbledon gave me the shot of energy I needed to head outside in the cold and ice to mail a letter at the Post Office. It was a little like a re-make of that "Rocky" moment I had when I climbed those stairs with canes instead of using a walker for the first time and Rocky II when I did it without canes, so this was Rocky III. Ice created an additional obstacle but it really wasn't much of a problem, and that made me feel singing James Brown's most famous song: "I feel good!"

Day 58: Forecasts Change

... We sit up there in the blues
bored and sleepy and suddenly three men
break down the ice in roaring feverish speed and
we stand up in our seats with such a rapid pouring
of delight exploding out of self to join them why
theirs and our orgasm is the rocket stipend
for skating thru the smoky end boards out
of sight and climbing up the appalachian highlands
and racing breast to breast across laurentian barrens
over hudson's diamond bay and down the treeless tundra where
auroras are tubercular and awesome and
stopping isn't feasible or possible or lawful
but we have to and we have to
          laugh because we must ...
("Hockey Players," The Cariboo Horses, 1965)

First thing in the morning, I found an e-mail from Jean Baird, a cultural activist in Vancouver, married to the famous Canadian poet George Bowering, saying, Thanks to everyone who contributed to help us complete stage one. Now we move to fundraising for upgrades and to build an endowment.

She sent me a press release with the headline: A NEW LIFE FOR THE AL PURDY A-FRAME and the sub-head: Work now turns to RAISING FUNDS TO UPGRADE AND INSTALL a writer-in-residence.

The press release said the A-frame home built in 1957 by the late Al Purdy, one of Canada's greatest poets, has been assured of preservation and a continued vocation as a place for writers to gather and work.

Thanks to the generosity of his wife, Eurithe Purdy, who dramatically reduced the asking price for the property, and donors from across Canada, the A-frame was acquired on October 9 by the Al Purdy A-frame Association, a newly incorporated national non-profit organization with a mandate to promote Canadian literature and Canadian writers.

"Now we can turn our attention to the next phase of this effort," said Jean Baird, president of the association. "It's not only a celebration of Al Purdy's legacy, but a mission to educate today's students on the value and worth of Canadian literature, and to preserve the Purdy home as a retreat for future generations of Canadian writers."

The A-frame, a lakeside cottage in Prince Edward County, was the centre of Purdy's writing universe and one of the most important crossroads on Canada's literary map. In their 43 years residing there, the Purdys hosted a who's who of Canadian authors: Margaret Laurence, Milton Acorn, H.R. Percy, Michael Ondaatje and hundreds of others.

The association plans to begin work on upgrading the property immediately, and hopes to have its first writer-in-residence installed next summer and working in local schools by fall 2013.

The association gratefully acknowledges the generosity of all donors to the project to date, including writers, poets, publishers, academics, students, booksellers, librarians, lovers of literature and, especially, Eurithe Purdy, who was crucial to the success of this effort.

Special thanks are extended to major donors ($5,000 to $40,000): The Good Foundation, Avie Bennett, George Galt, The Chawkers Foundation, The Glasswaters Foundation, The Metcalf Foundation, Michael Audain, Jeff Mooney and Suzanne Bolton, Leonard Cohen, Rosemary Tannock, Tom and Helen Galt, and Josef Wosk.

Fundraising efforts continue and are critical to the next stage of this project—upgrades on the property are required and the association will be building an endowment. Online donations are being accepted through PayPal at www.alpurdy.ca.

Purdy's A-Frame was not far from the Zen Forest. It was on the other side of Belleville, about a 45 minute drive away. I had helped out with the fundraising campaign a bit by writing an article about Purdy and the A-Frame campaign for the magazine from Prince Edward County and participating in a reading at the Belleville Public Library with George Bowering and some local poets. We all read Purd's poems, instead of our own, and I got to read The Hockey Players, so I wore my Sean Avery hockey jersey and it was a great night for poetry in Purdy's old hometown.

Also, I wrote a book of poetry called Al Purdy's Ghost, using his famous voice, and a lot of people said it sounded like I was channeling "The Voice Of The Land".

Last night I had another strange dream and it seemed to go on for a very long time. I was in the dream and so was my sister and the dog we used to have and I had sore legs but they were healing because I spent a lot of time meditating. It was quite vivid but that's all I recall. There was something about this book in the dream, too.

The dream faded away as my alarm clock went off. My iPhone plays a piano riff. That's my alarm clock. It was time to get up for a long day with physiotherapy in Bonnyville at 9:15 and a doctor's appointment in Lac La Biche at 2:30 in the morning with a lot of driving from Cold Lake and back.

My back and legs felt the way they used to feel the morning after a hockey game, I noticed, when I was a teenager. But that feeling never lasted long. Now it lasts a long time and I haven't been playing hockey. My feet are full of pins and needles, but it's fading away, with the numbness, and that familiar feeling is being replaced by the sort of pain you feel after a sprain.

It was minus 5, going up to minus 3, then down to minus 13, for my journey around central north eastern Alberta and The Lakelands.

That was the forecast but it never got that cold. It was only about 5 below, at the coldest. And outside of Cold Lake, there was no snow anywhere, except a little dusting here and there.

The weather wasn't the only thing that changed. My physiotherapist told me earlier in the week that he was going to recommend two or three weeks off work. He suggested four to six. The doctor gave me a note for four weeks and recommended to me that I re-book the Glenrose for an EMG.

Physiotherapy was good, but hard, as usual. The battery operated electrodes taped to your hairy back. Ten minutes on the treadmill, going faster every minute. Ten pound weights, up from five pounds, feel like one hundred, strapped to your ankles, for a hip exercise on a step between the parallel bars. Toe raises on a board. That bridge exercise, with an exercise ball, that I can barely do. Then something new: kneel on the floor and pull an elastic resistance band with two hands.

Pulling the band was no problem but getting into the right position was so painful I could barely focus on the exercise.

And the battery died on the electrical nerve stimulator.

After that, being drawn on the decompression machine was ten minutes in heaven.

After physio, driving from Bonnyville to Lac La Biche, I seriously doubed my ability to go back to work. I could barely walk. I was stiff, sore, and it was a lot of effort to move. I felt as though I was about to fall.

Close to Lac La Biche, a deer tried to commit suicide by running in front of my car. However, I saw him in the forest a long way back and watched him sneak along the tree line and then bound over a ditch, onto the road. Three other deer were on the other side of a fence, in the forest.

Why did the deer want to die?

I didn't want to die. On the positive side, I did more in physio than before, felt my calf muscles for the first time in months, and demonstrated some strength in my feet, pulling on a black rubber band wrapped around the ball of my foot.

When I went to Edmonton to see the specialist, I had zero strength in my foot and could not move a little elastic band.

The physiotherapist gave me a copy of the letter he sent to the doctor. Here's what it said: Left greater than right multi-factorial L4, L5, nerve root compression.

-- Whatever that means.

Mr. Avery did not complain of high levels of pain, although he had soreness in his back following a back injury 6 weeks ago. Following a chiropractic adjustment, Mr. Avery felt he deteriorated and developed difficulty walking (not due to pain but to poor balance).

His reflexes were absent bilaterally.

With excessive standing and walking, the compression of the L4, L5, and S1 nerve roots increases, which causes weakness of ankle and lower leg muscles, including the more proximal gluteous muscles. This puts him at high risk of falls.

An active strengthening program may take 4 to 6 weeks.

His presentation is somewhat unusual.

That's not exactly the way I would describe it. I'd say I had a huge back pain when I went back to work after the summer, on August 29th, I went to the chiropractor on August 30th, a Saturday, and could not walk after that. I went to "emergency" three times in September as my ability to walk disappeared. I staggered for a week, walked with canes for a week, and then could only get around with a walker. The reason for falling was not a lack of balance, it was weakness in the knees. My feet are numb, unresponsive, extremely weak, or hurt or have that pins and needles feeling, especially in the balls of the feet.

The physiotherapist also mentioned GBS syndrome.

George Bernard Shaw? I said to myself.

At lunch, in Lac La Biche, I looked up GBS and found out it stands for Guillain-Barré syndrome. It is rare, as the physiotherapist said, but some famous people have had it, including Joseph Heller, Andy Griffith, William "The Refridgerator" Perry, and Scott McKenzie, the guy who wrote the hippie song about going to San Francisco and wearing flowers in your hair.

Franklin D. Roosevelt had it. So did Serge Payer, the hockey player, who played in the NHL with the Panthers and the Senators. He started the Serge Payer Foundation, which is dedicated to raising money for research into new treatments and cures for GBS.

Heller, the author of Catch 22, wrote a book about it called No Laughing Matter.

His best known work is Catch-22, the famous novel turned into a movie about American servicemen during World War II. A man tries desperately to be certified insane so he can stop flying missions. The movie starred Alan Arkin.

The title of his book entered the English lexicon to refer to absurd, no-win choices, particularly in situations in which the desired outcome of a choice is an impossibility, and regardless of choice, the same negative outcome is a certainty. A catch-22 is a paradoxical situation in which an individual cannot avoid a problem because of contradictory rules. It's a no-win situation. You are damned if you do and damned if you don't. No matter what you choose, something bad will happen.

It's funny, in a darkly satirical way.

Heller's big influences were Kafka, Beckett, Celine, Faulkner, and Jaroslav Hašek. He was a big influence on Kurt Vonnegut.

In 1981, Heller was diagnosed with Guillain-Barré syndrome, a debilitating syndrome that was to leave him temporarily paralyzed. He was admitted to the Intensive Care Unit of Mount Sinai Medical Hospital the same day and remained there, bedridden, until his condition had improved enough to permit his transfer to the Rusk Institute of Rehabilitation Medicine, in 1982.

The book reveals the assistance and companionship Heller got during this period from a long list of famous friends, including Mel Brooks, Mario Puzo, and Dustin Hoffman.

Heller eventually made a substantial recovery.

In 1984, he divorced his wife of 35 years, Shirley, to marry Valerie Humphries, the nurse who had helped him to recover.

Speed Vogel writes about helping Heller, his friend for twenty years, through his rehabilitation. The pair wrote alternating chapters. They chronicle Vogel's rise through society as he stands in for Heller, even traveling to the Cannes Film Festival, while at the same time Heller is becoming more and more helpless.

Although Heller's disease is debilitating, the book is full of humour and never self-pitying. Heller's only lament is letting an insurance policy lapse, resulting in his out-of-pocket expenses of $120,000 in medical costs. -- That's the American medical system for you. What happens if you don't have an extra one hundred thousand or so sitting in a bank or stuffed in a mattress at home? It bankrupts your and your family and leaves your poor for generations.

It's no laughing matter!

While I was driving around Alberta, I got a few updates on the big news story from the region. The bizarre crash into the school in St. Paul turned into a real tragedy as one of the students hit by the vehicle that crashed into the school died in hospital.

One of the three Grade 6 girls in Edmonton hospital after being hit by a minivan Thursday morning, died, according to CBC Edmonton.

The girl was one of eight people hurt when a minivan slammed through a classroom window at Racette Junior High School in St. Paul, AB.

Richard Edward Benson, 46, is charged with dangerous driving, resisting arrest and possession of a controlled substance. There's no word whether more charges would be added.

At the end of the day, there was another update. Apparently the driver was the father of kids at the school and was dropping them off when the accident happened.

His brother was contacted and he said the guy should be in hospital, not in jail, as he was beaten badly a few years ago and has not been "quite right" since then.

My doctor's appointment made it too late for me to drive to Cold Lake in time for the end of the school day but I got there in time to see the end of a volleyball game.

On my way in, I had ten great conversations with kids and teachers. And when I stepped into the gym, the Grade 5 and 6 girls on the volleyball team interrupted their game to scream out my name. A couple of minutes later, after they won the final point of the game, they all ran over for a careful hug.

And to ask me if I remembered to bring the gummi bears.

Ten more conversations that were just great and I was out the door again, missing the place more than ever.

PS: My Scrabble score went up to 1655.

PPS: The Positive Thoughts posting of the day: You must believe you can. You must find the place inside yourself where anything is possible. It starts with a dream. Add confidence, and it becomes a belief. Add commitment, and it becomes a goal in sight. Add action, and it becomes a part of your life. Add determination and time, and your dream becomes a reality.

Day 59: Almost 60

After driving from Cold Lake to Bonnyville to Lac La Biche and back, going to physiotherapy, seeing the doctor, dropping in at work, watching a minute or two of the volleyball game, having two dozen quick conversations, I was tired by the time I got home. After dinner, a little writing, e-mail, and Scrabble, I hit the sack around eight o'clock, and I was out like a moose hit by a Sunfire.

My new Scrabble score: 1662.

My computer read to me as I fell asleep, but I didn't hear much of the story before I was in unusual dreamland. At four in the morning, I was up again, feeling sore all over, but laughing at myself, as I lurched from bedroom to kitchen.

Why was I laughing?

Was it because I had a little flashback to the time I was in the specialist's office in Edmonton, with my cousin, the doc, his two students, being told to stop laughing, be serious, because I couldn't walk, being reminded that I was a grown-up and I didn't have to follow his advice but he had a lot of expertise and saw a lot of cases like mine every day.

None of the other doctors or healers or anyone else had agreed with his assessment. And what was his big conclusion, anyway. "There's something going on down there and I don't know what it is, so I'm ordering more tests."

Idiopathic neuropathy. That means 'there's something going on with your nerves and we don't know where it came from or where it's going, but we do know that in a lot of cases it clears up on its own as the body heals itself."

And I wanted to say, "This is what you're an expert in? This is your specialty?"

You work in the the fanciest new building and you test me with a safety pin, give me a hard time for nervous laughter, and tell me you don't know what it is or where it came from, you order more tests but you say in a lot of cases it just goes away by itself.

All my test results from bloodwork, x-rays, and a CT scan are online but you can't access them because you don't know how to work the computer?

And you wonder why I'm laughing?

An alternative practitioner I talked to said, Well, doctors can only check your vitals, really, and prescribe drugs, or set you up for surgery.

The guy at the Wellness Centre says they are good at dealing with certain medical emergencies, but they are interested in disease, not health care.

A cynical friend of mine, whose father was a doctor, with a Ph.D., not an M.D., with a cross appointment in Biology and Chemistry, and worked in hospital labs, said it's like a factory system or assembly line where they take you in, drug you, and schedule you for surgery, no matter what's wrong with you. They prescribe pills and pull out the knife. All they know is disease and they have a narrow focus, looking at one thing at a time, instead of thinking holistically, looking at your whole life and your hole body and considering all kinds of options from other healing traditions that have been around for thousands of years longer than their own assembly line funded by drug companies.

-- You have to laugh.

Where I live, there are doctors and physiotherapists, but you can't get in to see them. It's Kafkaesque. You have to drive around all day to towns an hour or an hour and a half away to meet with a doctor for a few minutes.

-- That's laughable.

Why are you laughing? the specialist says. You can't walk. This is serious.

We laugh to stop ourselves from going crazy.

Obviously.

-- That's how I'm feeling at four o'clock this morning. -- I'm ranting and raving after braving a long drive at the end of autumn in Northern Alberta, traveling through the oil patch at the edge of the prairie and the Boreal forest, where moose and deer head for the highways during hunting season to commit suicide.

And I am certain that my own assessment and prescription is far superior. I prescribe reflexology with a gifted alternative practitioner, lots of movies, especially comedies and sports come-back stories, including A Thousand Words and The Answer Man, plus music, rest and exercise, good nutrition, vitamins, and I don't know about that nerve formula. How about a TENS Unit, the battery-operated nerve stimulator used in physiotherapy, for use on the feet, at home?

Google told me that Walmart sold them.

Google told me Walmart also sells resistance bands.

Cold Lake has a Walmart super store.

It's a good place for a long walk!

Maybe it was inspired by a piece I heard on CBC Radio about Conrad Black saying he was not a convicted criminal, he was just one more victim of the American justice system that operated like an assembly line in a factory designed to deliver guilty verdicts.

Black was back in Canada after serving some jail time in the U.S.A. and he was ranting at some Brit who called him a convicted felon. He called the Brit an ignorant prig, or words to that effect.

I wasn't a big fan of Black's or anything. I used to work with his brother and his wife, in Toronto, when I worked as a public relations writer, right after I got my first university degree. But Black always makes you think.

If the American justice system was designed to deliver guilty verdicts, what was the American medical system designed to do? And how was Canada different than the U.S.A.

Black appeared to be happy to be back in Canada.

A bit before nine o'clock I was chatting online with friends far away. My friend in Alabama said she was going away for a few days to go scuba diving in the Caribbean. My friend in Ontario told me her kid took her car and ran into a curb. The licence plate fell off. But the car was fixable, she said. And my friend in Edmonton told me her doctor said she might have cancer.

"No way," I said. "He's only guessing. -- I bet he wants you to have a bunch of tests. -- Right?"

She said that was right and she was freaked it but wasn't planning to tell anybody else because she didn't want to freak them out.

I tried to convince her she didn't have any of the usual symptoms, except one, and that could be caused by a number of other things.

And I think she believed me because she said she felt a bit better after talking about it. "Thanks for listening," she said.

Amazing, I thought. This was a woman who would move heaven and earth if one of her boys got sick or needed something to save his life. And she didn't want to tell them or her sister, who was a nurse, or anyone else. -- What a woman!

Instead of going to Walmart again, I tried Canadian Tire, and I found the resistance cord and one of my Grade 9 students. First I went to a nearby drugstore. Actually, I went to a herbalist, to see if they had a nerve tonic, but they didn't so I just bought some alpha-lipoic, one of the ingredients in the nerve formula that was being promoted on the internet. Next stop: Sobey's for groceries. Total steps: 1,500.

Alpha lipoic acid was something I had never heard of. So, of course, I looked it up online, and found out about it from About.com's alternative medicine pages. They said it is a fatty acid found naturally inside every cell in the body. It's needed by the body to produce the energy for our body's normal functions. Alpha lipoic acid converts glucose (blood sugar) into energy. Other names for it include lipoic acid, thioctic acid, and ALA.

Alpha lipoic acid is also an antioxidant, a substance that neutralizes potentially harmful chemicals called free radicals. What makes alpha lipoic acid unique is that it functions in water and fat, unlike the more common antioxidants vitamins C and E, and it appears to be able to recycle antioxidants such as vitamin C and glutathione after they have been used up. Glutathione is an important antioxidant that helps the body eliminate potentially harmful substances. Alpha lipoic acid increases the formation of glutathione.

Alpha lipoic acid is made by the body and can be found in very small amounts in foods such as spinach, broccoli, peas, Brewer's yeast, brussel sprouts, rice bran, and organ meats. Alpha lipoic acid supplements are available in capsule form at health food stores, some drugstores, and online. For maximum absorption, the supplements should be taken on an empty stomach.

They had a sub-section titled Why People Use Alpha Lipoic Acid and the number one use they listed was Peripheral Neuropathy.

They said peripheral neuropathy can be caused by injury, nutritional deficiencies, chemotherapy or by conditions such as diabetes, Lyme disease, alcoholism, shingles, thyroid disease, and kidney failure. Symptoms can include pain, burning, numbness, tingling, weakness, and itching.

Alpha lipoic acid is thought to work as an antioxidant in both water and fatty tissue, enabling it to enter all parts of the nerve cell and protect it from damage.

Preliminary studies suggest that alpha lipoic acid may help. In one of the largest studies on the use of alpha lipoic acid, 181 people took 600 mg, 1200 mg or 1800 mg of alpha lipoic acid a day or a placebo. After 5 weeks, alpha lipoic acid improved symptoms. The dose that was best tolerated while still providing benefit was 600 mg once daily.

The sun came out, so I got inspired to go to Cold Lake Provincial Park for a picnic lunch. It was five below but that's not too cold for a quick bite on the picnic table by the big lake. It's 200 steps from the parking lot to the lake. While walking, I counted steps; after snacking by the lake, I counted breaths for twenty minutes.

The lake looked so beautiful, I took its picture.

Leaving the park, I saw a red fox. In my lifetime, I've seen a few red foxes, briefly, at a distance, but this one walked right in front of me and then, after crossing the road, it stopped and looked at me. From a distance, I thought it was a dog or a coyote, but when I got closer and saw its true colours, and that big, bushy, red tail, there was no question what kind of a wild animal it was.

The big lake and the wild fox reminded me that "Life is not measured by the number of breaths we take, but by the moments that take our breath away."

That quote comes from Carleton cards, a Canadian greeting card company, but it is often attributed to George Carlin or somebody else. -- But I like it.

On my way out, I took down one of the winter tires for my car, instead of two, as I usually do. Bringing in groceries took two trips, so I took down a second snow tire when I went outside again.

It was challenging, walking anywhere this morning, as my legs were stiff and sore, and muscles hurt when I moved, but I kept going anyway. After taking about 2,500 steps in the morning, I felt like going out again in the afternoon.

It was the day of Game 3 in the World Series and the day of a play-off game for the Cold Lake football team. My school does not have a football team so a couple of our guys play for the Cold Lake High School team. One of them got munched at the start of the season and was out of action with a broken ankle. The other guy was still in the game and I thought it would be good to see him play a game.

They play in the Northern division of Alberta's high school football league, in what is called the Wheatland Football League. There are a dozen teams: Bonnyville Voyageurs, Central High Rams, Cold Lake Royals, Edwin Parr Pacers, Holy Rosary Raiders, Lac La Biche Huskies, Lloydminster Barons, St. Paul Regional Lions, Vegreville Vortex, Vermilion Marauders, Wainwright Commando, and Westlock Thunderbirds.

The Royals were in the play-offs with the Barons, but it turned out the game was in Lloyd, so I didn't go.

I got a note from the cousin I call Saint Shirley. She said, i am glad yu have a bit more time off work. i know you were getting better, but you want to make absolutely sure you are %100 before you go back to work. once you get off the disability, it is hard to get back on again.

nice weather we are having eh? personally i hate it...ughh its going to be a long winter..

The Cold Lake Ice had a game, but they were on the road, too. Our winning Jr. B hockey team played in the North Eastern Alberta Junior B Hockey League with the Killam Wheat Kings, Lloydminister Bandits, Virmillion Tigers, Wainwright Bisons, St. Paul Canadiens, Saddle Lake Warriors, and Vegreville Rangers.

The Rangers, Warriors, and Canadiens all had logos like NHL teams on their jerseys. The Ice had the coolest name, I thought. And they were the league champs the last two years in a row.

Cold Lake was in first place in the North division and Wainwright was in first place in the South Division, so the winner of Saturday's game would move into first place overall.

Instead of heading out to a football or hockey game, I headed back to bed for a three hour nap. Suddenly, after lunch, I felt very sleepy. So .... I slept.

When I woke up, I looked for No Laughing Matter on Amazon.com and got a different idea about the book. It said that Heller was jogging four miles a day, working on his novel called God Knows, and getting divorce. He was feeling perfectly fine one day -- but within twenty-four hours he would be in an intensive care unit and remain hospitalized for nearly six months.

God Knows was a satirical, irreverent book about King David.

Heller described Guillain-Barré syndrome as a debilitating, sometimes fatal condition that can leave its victims paralyzed from head to toe.

Amazon described the book as an inspiring, hilarious memoir of a calamitous illness and the rocky road to recuperation as only the author of Catch-22 could tell it. And it said, "No Laughing Matter is as wacky, terrifying, and great-hearted as any fiction Joseph Heller ever wrote."

In the afternoon, I went to Walmart and walked another thousand or so steps.

My favourite old couch is worn out, I must admit, as I liked sleeping on it, and it has got a lot of use in the past couple of months. My colleagues, at work, were joking about it the other day, for some reason. -- None of them has ever seen it, so I don't know what they were going on about. Anyway, I have a great bed with a very firm mattress and a comfy mattress topper, but I like sleeping on the couch even more. Go figure, as Joseph Heller used to say. But it was time to replace the old thing.

For the past two months I've been sleeping on the floor, a lot, on a Thinsulate camping mattress, plus some blankets for extra padding, right in front of the big, double, glass doors, leading to a balconey, as I was spending a lot of time in bed and the view from there is much better than in my bedroom. Also, the internet connection is nearby, so I can watch a movie, if I feel like it, while resting my back.

Walmart was full of people, as usual, and it was the usual mix of Canadian air force, Cold Lake First Nations, metis from a nearby settlement, anglophones and francophones from the city of Cold Lake, and shoppers from nearby towns in Alberta and Saskatchewan. And the parking lot, as usual, we decorated with a dozen big black birds that could be huge crows or small ravens, plus a few white seagulls.

And it hit me as I walked out of the store and into the fresh air .... I love this place.

After a month and a half of living week by week, always hoping, thinking, and believing I'd be back at work the next week, all healed, but then getting another week, and another week, everything has changed. Now I know I will be off for a month. So, I made some changes.

My couch has been replaced. It looks heavy but it is very light, so I was able to move it to a spot where it looks good and nobody will be tempted to sit on it. In it's place there is a self-inflating camping mattress, by the window doors.

It will be better for my back.

Instead of sitting on the couch while I work at the computer, I will sit on the exercise ball, and that will be good for my back, too. That was something I used to do all the time, and then sometimes, and then forgot about, and have started to do again, and now I will do it all the time.

Tramadol has been replaced by vitamins. The painkiller I was prescribed has run out. A multivitamin has replaced the B complex vitamins I used to take.

I feel good about these changes.

This coming week, I have fewer appointments to see doctors and so on. The focus is on physiotherapy.

Today I did none of my exercises, but I walked a good deal, but only around 3,000 steps. And I bought the big elastic bands called resistance bands, for my homework from physiotherapy.

What else can I do?

I'm still thinking about ordering a TENS Unit and those new neurology drugs online.
-- I'll sleep on it!

Day 60: Ice Cold Feet

Sunday was day 60 and it felt like a landmark day, but when I got up at 7 in the morning, there was nothing different, except it was a little darker outside than it was on Saturday morning. There was still snow on the ground as it had been for a full week before Hallowe'en.

In the U.S., Hurricane Sandy was in the news, as it looked like it was going to hit New York City. Hard. There were a lot of hurricanes but nobody was blaming global warming.

The big election in the U.S. featured presidential debates but nobody was debating global warming.

On CBC, I heard some singer say the end of the world, caused by global warming, was a sure thing, so it wasn't like some Bible story or strange prophecy about global destruction, it was just something we all had to wrap our heads around.

I couldn't believe it.

The Giants had won Game 3 of the World Series, beating the Tigers 2 to 0, so it looked like they had a lock on the baseball championship as they were leading three games to zero. The NHL was still locked out and nobody liked the players much more than the owners in this dispute. The Oilers young guns were playing together in the AHL and turning the Oklahoma City Barons into the hottest team in the league. The Nuge was a year older, ten pounds heavier, playing stronger, and winning a lot more face-offs. It looked like the NHL lock-out was just what he needed.

"We'll see," as the Zenmaster likes to say.

The Barons weren't in first place, but they were getting more attention from the media than any other team in the league. They had three of the NHL's first round draft picks and a couple of other guys who looked like they weren't too far behind.

A Facebook friend posted the new, 9-billion-pixel image of the Milky Way galaxy showing more than 84 million stars. That was ten times more than previous studies.

Somehow, that led me to the website for Power 4 Patriots, which was started by a guy in the U.S.A. who wanted to tell everybody how to make their own power plant, with solar and wind energy for your own house. He said there was a conspiracy by big oil, big electricity, and big government to make big bucks out of energy.

I didn't want to worry about that or anything else other than my feet, legs, and back.

Yesterday my friend in Edmonton told me her doctor said she might have cancer.

I was worried about that, too.

My sister sent me an e saying she was going to see Jersey Boys at the theatre. I sent her an e saying I heard an interesting interview on CBC Radio about a guy who quit his job at Goldman Sachs by publishing his letter of resignation in the New York Times. And he got a book deal out of it. I suggested she could do the same sort of thing: write a letter of resignation, publish it in the biggest newspaper in Australia, get a book and movie deal out of it, and use the publicity to launch her private practice.

By 8:30 the sun was rising, so I could see it was another cloudy day and the snow was still all over the ground and rooftops, if not on the streets, anymore.

For me, it was going to be a day of physiotherapy exercises and I was looking forward to it.

On Facebook, I got a chat message from Ontario. My friend and former business manager said she was taking her daughter and her granddaughter to my old hometown, Gravenhurst, which she called G-hole, so her daughter's newborn could be taken to the prison on the edge of town, to meet her father. Or so the father could see his daughter for the first time.

My sister sent me a note, too. She said, So, 4 more weeks off work and maybe 6 weeks eh? Sorry Marty, I know you miss it and want to get back there but you need to take this time to really recover and get strong and balanced so you can do all you need to do in your teaching position. It is pretty exhausting work remember (it took you all summer to recuperate) so don't go back until you are fighting fit! Do the exercises etc and practice lots of self care so you can get back in the game!

My thoughts and prayers are with you always Marty.

We are having a lovely, sunny day here in Adelaide. Niiiice!

I'm just getting ready to leave for the theatre to see Jersey Boys, the story of Frankie Valli and the Four Seasons. It should be great! I hope I am seated with some fun people who will be singing along and clapping along etc and not just sitting there like statues! lol

Frankie Valli and the Four Seasons had a huge hit with a song I liked a long time ago called Sherry, which was better known as Sherry, Baby. -- I wrote a book with that title a few years ago, dedicated to the first girl I kissed, back in Grade Six, who was named Sherry. We were girlfriend and boyfriend when that song was the number one hit in Canada and the U.S.A.

Sherry sent me an e saying she would miss playing Scrabble with me when she went scuba diving in the Caribbean for a few days. She said she hoped the water was calm and the area wasn't disturbed by hurricanes.

Hurricane Sandy could be the biggest storm to hit the United States mainland when it comes ashore on Monday night, bringing strong winds and dangerous flooding to the East Coast from the mid-Atlantic states to New England, forecasters said on Sunday.

Sandy could have a brutal impact on major cities in the target zone like Boston, New York, Washington, Baltimore, Philadelphia, one of the most densely populated regions of the country.

New York City's subway, bus and train service will be suspended on Sunday evening, which could bring the country's financial nerve center to a standstill.

The major Wall Street exchanges said they planned to open as usual on Monday because they have alternate facilities they can use.

The Obama administration estimated it could affect 50 million people, and the storm was already disrupting transportation systems.

More than 700 flights, including international ones, were canceled on Sunday and nearly 2,500 more were canceled for Monday, FlightAware.com said.

Forecasters said Sandy was a rare, hybrid "super storm" created by an Arctic jet stream wrapping itself around a tropical storm, possibly causing up to 12 inches of rain in some areas, as well as heavy snowfall inland.

Ontario may see the worst Sandy when it hits early next week as the so-called Frankenstorm continues to grow, the Canadian Hurricane Centre said Saturday.

Spokesman Bob Robichaud said while rainfall amounts are still hard to predict, southern and eastern Ontario could see between 50 and 100 millimetres late Monday and early Tuesday.

"That's certainly in the realm of possibility for that part of Ontario," said Robichaud in an interview on Saturday. "It looks like southeastern and eastern Ontario might be getting the most rainfall out of this."

Robichaud said those areas will also see high winds, although they will likely not hit hurricane strength. He said 80 km/h winds are a possibility.

Facebook gave me the news about the local teams. It was another big road win for the Cold Lake Ice with a 2-1 win in Wainwright versus the Bisons. Niko Bourget got both goals for the Ice who improved to 6 wins and 1 loss on the season. Next game: Friday night at the Energy Centre vs Wainwright.

The Cold Lake Royals football team won their game in Lloyd by a score of 40 to 14.

Cold Lake can't lose for winning.

It's like Obama.

I wanted to get aboard that bus again!

It was cold outside and I didn't feel energized, or rested, so I went back to bed a bit, after getting up at 7, and when I opened my eyes again, around ten, I noticed two things:

1. It was snowing outside, again.
2. My feet were cold.

The fact that it was snowing again, a couple of days before Hallowe'en, was remarkable because it doesn't usually snow until after Hallowe'en, in this part of the country.

The fact that my feet were cold was more remarkable as they had not felt hot or cold for the past sixty days.

After I wrapped my feet in blankets, they warmed up again. -- That's normal, for most people, but it was new for me.

My feet felt warm and cold again! Sensation was returning to my feet! The nerves in my feet were coming to life again!

Woo-hoo!

Happy? I was as happy as I felt when my team won a hockey, soccer, or basketball game. It was like scoring extra points or a touchdown and winning a football play-off game. It was like winning a face-off, scoring a goal in overtime, and winning a hockey game. It was like going to the provincial drama festival and winning the award for best production. It was like getting a book published and getting a rave review.

I gave my feet a rave review. "Way to go, guys!" I said to them.

"Now, let's get to work and keep this comeback rolling!"

My morning workout started with toe lifts. I was supposed to get up on my toes and stay up there for two or three seconds, ten times in a row, then rest, and do that three times, twice a day. I started with 100.

Next I did a little dance move to the music I had playing: James Brown was wailing, "I feel good!" My dance move was simple, but I couldn't do it for the past couple of months. I put one foot behind the other and brought it back again and then moved the other foot behind it and back again, for the length of the song, about three minutes.

Feeling inspired, I tried a balance test I had failed for the past sixty days. You stand like a surfer with one foot behind the other. It was shaky, for the first ten seconds, but then it smoothed out, so I stood like that, surfing in the kitchen, for a full minute.

Let the bells ring out and the banners fly, I said to myself. Poppa's got a brand new bag!

What was different? My balance was better, my feet felt hot and cold, and I could do a lot more toe lifts than before.

Two days before, I had started taking a multivitamin. One day before, I started taking alpah-lipoic. My mattress was softer. Sixty days had gone by, since this adventure began. Four physiotherapy visits plus the exercises given for homework and a few walks in the park.

Maybe it was knowing I did not have to go back to work in a week.

Maybe it was because I was highly motivated to get my legs back under me and on my feet again.

Maybe it was because of the prayers everybody I knew, no matter what religion they were, told me they were saying for me. -- I had Catholics, other Christians, Buddhists, Jews, New Agers and atheists praying for me!

Wally Lamb posted an odd note on Facebook. He said, Hurricane's a-comin'! Sure hope I get our spiffy new generator figured out before the power goes off. Wishing all my FB friends in the storm's path a safe and sane next few days. Take care.

If I lived in New York or New England and was worried about the Frankenstorm called Hurricane Sandy, I'd head over to Wally's place. That guy has been touched by angels, starting with the time Oprah picked his first novel for her Book Club, I would say.

In Cold Lake, Alberta, we don't have to worry about Frankenstorm hitting on Hallowe'en. True, it's unseasonably cool, and we've had snow for over a week, already, which is odd, but Hurricane Sandy will not hit us or effect weather this far north.

In the U.S., they are likely blaming us for their bad weather, as usual. "Cold air from Canada is causing blah, blah, blah." They say this hurricane started with Arctic winds meeting warm water where the Caribbean Sea joins the Atlantic Ocean.

How was Sherry doing, Scuba diving in the Caribbean? I wondered.

Next in my personal exercise program -- my PEP -- was something for my feet.

You sit on the floor or on a chair, anchor some tubing to a solid objet, attach the other end around your foot, your you foot inwards as if to look at the sole of your foot. Repeat 3 X 10, once or twice a day.

After that, you turn your foot outwards, the same number of times.

You do that with both feet, of course -- one after the other.

Sitting on the exercise ball, I did it 30 times, for starters.

Try this: stand, keeping both legs straight, with one leg overhanging a stair and the other on a stair. Lower the pelvis of the side overhanging the stair and then level it, ten times on each side, three times each.

If you don't have a stair, use a big, thick, book.

Which book? you ask. How about The Winner Stands Alone, by Paolo Coehlo.

On top of a big dictionary.

-- Done.

-- Next?

The reverse bridge, using the exercise ball. Try holding it for 20 seconds, three times, and then 30 seconds, three times.

Why do I find that one so hard?

Susan Swan posted a link on Facebook that led to her blog, so I read it, for the first time. It was called Sunday Morning Writers Blog: A Letter To My Younger Self.

She wrote a bit about one of her fictional characters, called Mouse, who was like her and unlike her. Mouse was a little girl with scoliosis, or a twisted spine, that made one leg longer than the other, so she had to wear built-up Oxfords and walk with a limp,

She made friends with a guy who murdered his wife and was locked up in the nearby facility for the criminally insane. They wrote letters, back and forth.

My twelve-year-old self was nothing like that. I was called Captain Crunch, because I was the captain of my hockey team, every other year, and I had the greatest girlfriend, Sherry, when Sherry, Baby by Franki Valli and The Four Seasons was a big hit. We used to go skating together all the time and we skated to that song as well as other new hit records that Sherry took to the arena to add to their stack of ancient waltzes. The Skaters Waltz. Les Patineurs Valse. Die Schlittschuhläufer-Walzer. A dozen others. And then they played The Boogie-Woogie Bugle Boy and all the waltzers split up and took off, fast, to race around the rink.

Nobody could keep up to me.

Swan has some advice for her younger self in regard to her father. The murderer was a father substitute for her, as she had an absent father. He was busy with work.

My father was long gone by the time I was twelve and my hockey coaches were my father susbstitutes. Two of our hockey coaches were also our phys ed teachers. My first male teacher was our Grade Six homeroom teacher, a guy I liked a lot but the other kids .... Not so much! They called him Lurch because they said he looked like the monster on The Adams Family, on TV, and he walked funny. He was a World War Two veteran and he got shot in the rear end, so he walked with limp for the rest of his life. He moved up to Grade 7 with us and was our first French teacher.

What do I have to say to the 12 year old, Grade 7, bantam version of me?

You go, boy! You skate your heart out, work extra hard at every hockey practice and in every game. And you hold on to that girl as long as you can! -- You know that she is the best girl in town -- the prettiest, most athletic, most charming -- but you think there will be a lot of girls like that in the other towns around and all over the world.

Newsflash, little buddy: She just might be "it"!

PS Forget about skiing. -- Next year you will break your leg.

The doctor will say forget about skiing and hockey, but you make a dramatic recovery. Several years later, you will get hit from behind, playing hockey, at the same time as somebody dives at your feet, and another guy slides into you, feet-first.

So, you might want to think about skiing and hockey. You could save yourself, and a lot of other people, a lot of pain, in the years to come. Those injuries will come back to haunt you, later in life. And you won't meet anybody like Sherry. But, knowing you, at age 12, you're going to do what think you should do, or what you really want to do, which is play hockey, go skiing, and skate with Sherry.

Good luck, guy!

While I was writing to my younger self, I got a call from a colleague at work who was supposed to go to school and supervise some of the senior kids while they decorated the stage for Hallowe'en. They were making a Haunted House.

However, she couldn't be there on time, so she asked if I would be able to let them in at 4:00 p.m. and stay with them until she got there.

Sure, I said.

The kids will be happy to see you! she said.

Okay, I said. I'll be there early and see you when you get there.

Merci, she said.

At school, I talked to our librarian, a former nurse, and she said I should get a foot roller, now that my foot is waking up, so I went to Walmart but they didn't have them. Walking out of Walmart, I almost hit my limit for the day. I almost went down. I had to sit down before I fell down.

So, I came home, driving through a little snowstorm, and was happy to get back home.

Our little storm is nothing compared to what they are expecting in the most heavily populated part of the U.S.A.

Wally Lamb said: Glass half empty: We may be out of power for the next week or more. Glass half full: A reprieve from all those insufferable Linda McMahon for Senate ads.

'Frankenstorm' prompted Red Cross warnings in Ontario.

Forecasts of a hybrid storm blending remnants of a post-tropical storm Sandy with already severe weather over Ontario are prompting warnings from relief workers to brace for an emergency by stocking up with at least three days' worth of supplies.

The Canadian Red Cross issued a release Sunday urging Ontario families to ensure they have enough food and water to sustain themselves for at least 72 hours as the weather system, which is lashing the U.S. east coast and North Carolina, continues to lumber towards Ontario with wind speeds of up to 130 km/hr.

Blackouts and flooding could affect many parts of the province, according to Environment Canada. Evacuations are also possible due to the system, which has earned the moniker "Frankenstorm" — a reference to the mix of meteorological factors comprising it.

CBC weather specialist Craig Larkins said Sandy, which is still a Category 1 hurricane, is 1,300 kilometres wide, making it the second-largest tropical storm in the Atlantic since 1988.

— It could actually still take the No. 1 spot.

That was the weather.

Here's the sports news: The San Francisco Giants beat the Detroit Tigers 4-3 on Sunday night to complete a four-game sweep and win their second World Series title in three years.

My friend in Edmonton sent me an uncharacteristically cheerleaderish message on
Facebook: onward! upward!
health! Light! blessings!
Peace! healing! productivity!
Strength! wellness! purpose!

Day 61: Frankenstorm

Although I hit the sack on Sunday night feeling happy about my own small victories, I was also feeling anxious about the huge storm that swept through the Caribbean and was heading for New York, with a disaster area reaching Canada, including southern Ontario. There would be a disaster area over 800 square miles and there were predictions of 12 inches of rain, 2 feet of snow and sustained 40- to 50 mph winds.

The intensity of hurricane overall kinetic energy forecasts a 5.2 for Sandy's waves and storm surge damage potential. That's on a scale of 0 to 6, putting it up with historic storms, such as Katrina.

The disaster zone was expected to stretch 800 miles (1287.5km), from the Mid-Atlantic to the Great Lakes.

NBC was reporting wave heights of around 30ft (9m) just 100 miles (162km) off the New Jersey coast.

The term Frankenstorm was, of course, an allusion to Mary Shelley's gothic creature of synthesized elements.

Yesterday there was an earthquake on the West Coast, up north, in the area of Haida Gwaii, formerly known as The Queen Charlotte Islands, followed by aftershocks that were expected to last a week. A magnitude-7.7 quake -- the strongest in Canada in more than 60 years -- struck the Haida Gwaii archipelago, and it was followed by a 6.3 aftershock.

"It's good to live in Alberta," our Science teacher said. "We don't get hit by these earthquakes and hurricanes."

It was a night of nightmares, for me, as scenes from disaster movies like 2012, The Day After Tomorrow, and Perfect Storm merged with the ancient Maya text that emerged from the jungles of Guatemala confirming the so-called "end date" of the Maya calendar, Dec. 21, 2012.

Many people were predicting the end of the world or at least the end of the world as we knew it for December of 2012. It was called The 2012 Phenomenon.

That was the date the doctor and the physiotherapist gave for me going back to work. My dreams were about "work" not being there for me to return to.

When I woke up in the morning, I checked for news about Hurricane Sandy, now known as Frankenstorm. Weather forecasters were predicting that Hurricane Sandy would merge with another weather system as it moves, bringing a "Frankenstorm" to Eastern Canada, southern Ontario, and the U.S. just in time for Halloween.

Love Tree posted this note: Man has falsely identified himself with the pseudo-soul or ego. When he transfers his sense of identity to his true being, the immortal Soul, he discovers that all pain is unreal. He no longer can even imagine the state of suffering. ~Paramahansa Yogananda

Hurricane Hazel is the only hurricane or tropical storm to have ever struck the region. Forecasters predicted that Sandy would bring rain to Ontario and Quebec, possibly turning to snow in Ontario.

I believe I was conceived during Hurricane Hazel.

Hurricane Sandy gained strength overnight and is expected to sweep through Southern Ontario and Central Quebec by Monday afternoon, bringing with it raging winds and heavy rainfall that could uproot trees, down utility lines and cause flooding.

Our weather was overcast and cool, a few degrees below zero, with some snow overnight. In the morning, my friend in Edmonton phoned. Let's just call her Edmonton. She sounded positive and upbeat about her health and my recovery, so that helped get the day off to a good start. After that I dropped in at the school, to drop off the doctor's note, plus the physiotherapist's assessment letter, and I planned to zip in and out but some students saw me and pretty soon there was a crowd gathering and a group hug with a lot of kids wanting to know when I'd be back, if I would be coaching basketball, and if I would be back for the second semester or before then, and what was wrong, anyway.

When I told them four weeks of physiotherapy to build up my muscles after a backbone bulged and hit a nerve that controlled knees and ankles and feet, I believed it, but wondered if it would be six weeks, or longer.

December 21, 2012 came to mind.

And then, as Bill Murray says in Stripes, depression set in.

Acutally, I felt great after seeing students at school, plus a few colleagues, and while driving to Bonnyville for physiotherapy. I felt good on the treadmill, walking faster. They had new batteries for the muscle stimulator, so it was more powerful, and felt even better on my back. After that, we tried a new exercise, sitting in a leg press machine, pushing 100 pounds with both legs, then either leg, and then just my toes. It was easy, except for when I used only my toes on my left foot. I could do it, but it was a real challenge, and it left me sweating and tired.

Next up was leg lifts with ten pound weights strapped to my ankles. That was still very difficult.

After that came walking toe to toe, with eyes open and then with eyes closed.

Although I couldn't do it at first, I was able to do it with practice.

But then the physiotherapist said he wanted to talk to me, instead of putting me on the decompression machine. We sat down and talked about GBS.

First he asked me about my feet feeling hot and cold instead of being numb.

He said that was a good sign and so was the improvement to do the various exercises. However, the leg lifts had not improved. And he said he had been doing some research as my symptoms were slightly unusual. They were a lot like the symptoms he sees all the time when a bulged disc hits a nerve or two. But because those symptoms were always accompanied by a lot of pain, he wondered if there wasn't something else going on with me.

He asked me if I had any infection at the start of these problems. I told him I had the flu for a week but spent the rest of the summer hiking and getting in shape and felt great when it was time to go back to school. But then I got a big zap of back pain.

He said GBS started with an infection.

No cuts, bites, or any other problems?

No.

Well, he said, Maybe you should have that lumbar puncture, after all, as well as the EMG.

Isn't GBS a progressive disease? I asked. And aren't I getting better?

Yes, he said. Usually someone with GBS goes downhill fast.

I reminded him that, after the back pain, taken away by a chiropractor, I went down week by week, from staggering around one week to walking with canes the next week to using a walker for about the same amount of time, then crawling around at home for a couple of days, and then back up to the walker, canes, staggering, and now walking well for a mile or two, depending on physiotherapy exercises done the same day.

Yes, he said, well .... Maybe you have GBS and it has already taken you down and you are on your way back up. Sometimes it is much more severe and includes paralysis, including paralysis of your breathing apparatus, and sometimes it is mild.

The EMG and spinal tap would confirm it, or discount it, he said.

I told him again that I feared nothing -- except grizzly bears and the spinal tap.

He said there is a risk and some pain in the spinal, but there is a risk in everything and the pain is not that great.

And then, as they say, depression set in.

The thought of the lumbar puncture operation, or spinal tap, really brings me down. The EMG doesn't sound bad.

He said the two of them together can really determine what you have.

He mentioned MS, so I told him the specialist said he was certain it was not MS, ALS, cancer, Lyme disease, AIDS, STDs, or anything on that scary list of possibilities.

Did he mention GBS? he said.

When I got home, I went online and researched GBS some more, of course.

Here's what I found out: Guillain Barre' Syndrome (Acute Idiopathic Polyneuritis) (Ghee-yaw Bah-ray) Syndrome, also called acute inflammatory demyelinating polyneuropathy and Landry's ascending paralysis, is a disorder of the peripheral nerves, those outside the brain and spinal cord (peripheral nerves and spinal roots are the major sites of demyelination in GBS patients).

It is typically characterized by the rapid onset of muscle weakness and often, paralysis of the legs, arms and breathing muscles.

The cause of Guillain-Barre' syndrome is not known; and why the disorder only occurs in certain patients is still not known. Research to date indicates that the nerves of the GBS patient are attacked by the body's own defense system against disease-antibodies and white blood cells. As a result of this autoimmune attack, the nerve insulation (myelin) and sometimes even the covered conducting part of the nerve (axon) is damaged.

The rapid onset of (ascending) weakness, frequently accompanied by abnormal sensations and pain that affect both sides of the body similarly, is a common presenting picture, and quite often, the patient's symptoms and physical exam are sufficient to indicate the diagnosis.

A lumbar puncture may be performed to find elevated protein levels in the cerebro-spinal fluid to confirm the diagnosis.

The severity of Guillain-Barre' syndrome can vary greatly. In its milder form, it may cause a waddling or ducklike gait, and perhaps some tingling and upper limb weakness that may briefly, for days or weeks, impair a patients lifestyle.

Some primary care physicians have described patients who complained of mild brief tingling and/or limb weakness accompanying or following a viral illness, such as a sore throat or diarrhea. Such a set of symptoms may represent a very mild form of GBS.

In contrast to such mild forms, at the other extreme a GBS patient may become almost totally paralyzed and fraught with complications.

That came from Angelfire, or angelfire.com.

That didn't sound like my story.

There was a link to an article comparing and contrasting GBS with CIDP, by David S. Saperstein, M.D., Phoenix Neurological Associates, Phoenix, AZ.

He said, GBS may also be referred to as acute inflammatory demyelinating polyneuropathy (AIDP). This emphasizes the acute nature of this disorder: symptoms come on abruptly and progress rather quickly. Symptoms stop progressing, often within 2 weeks, and usually not more than 4 weeks. After a period of weeks to months, patients then begin to experience improvement. Although the majority of patients with GBS will do rather well, not all patients will recover fully and may experience chronic weakness, numbness, fatigue or pain. Once symptoms stabilize, there is rarely any further deterioration.

Chronic inflammatory demyelinating polyneuropathy (CIDP) produces manifestations similar to GBS, but there are important differences. Symptoms tend to come on more slowly and progress for a longer period of time.

It sounded like CIDP was out, but I fit Saperstein's description of AIDP or GBS. But I didn't fit the more detailed descriptions of GBS.

So, I Googled 'no pain neuropathy' and got a good answer from the MedHelp website at medhelp.org. There was a question and answer that went like this:

Q. Is it possible to have peripheral neuropathy with no pain? Symptoms include numbness in feet and legs (both sides, numbness covers more of left leg than right, began in shins and has spread from toes to knees); some loss of muscle control; jerking of feet and legs while asleep; occasional falls.

A. Peripheral neuropathy can present without actual pain. In fact, some patients can actually lose the sensation of pain in advanced stages of certain peripheral neuropathies, such as diabetic neuropathy.

The National Institute of Neurological Disorders said the same things about GBS. Guillain-Barré syndrome is a disorder in which the body's immune system attacks part of the peripheral nervous system. Usually Guillain-Barré occurs a few days or weeks after the patient has had symptoms of a respiratory or gastrointestinal viral infection. Occasionally, surgery will trigger the syndrome. In rare instances, vaccinations may increase the risk of GBS. The disorder can develop over the course of hours or days, or it may take up to 3 to 4 weeks.

It said, The first symptoms of this disorder include varying degrees of weakness or tingling sensations in the legs. -- But that's not how it started, for me.

First I had a back pain, then I had a chiropractic adjustment, and that was followed by a week of muscle spasms and contractions, which stopped but left my legs weak, along with my ankles and feet.

After a few weeks, or about 24 days, I started to get stronger. I used the walker, again, then just canes, and then on my own.

My knees got stronger first. After 60 days, my feet were stronger.

My balance was lost, but is returning.

If it's GBS, I'd say it's come and gone, or is on its way out.

Why can't we just say it's the result of an old car crash that put my back out of kilter for years that led to back pain that eventually turned severe and a chiropractic adjustment caused nerve damage to my knees and feet that is healing with time and physiotherapy?

That's the diagnosis I want! -- Not the spinal tap!

Why can't we stick with that?

And now for the weather. Hurricane Sandy was turning into an epic storm and it was going to hit sooner than anticipated -- around midnight in New York City.

A Facebook friend, Andrew Freund, a theatre director, posted this question: Is this a King Lear storm or a Tempest storm?

Lindsay Walker said, i think that depends on the mental stability of the old men we see wandering about in it

Katherine Trowell said, Possibly a, "All Hell's Breaking Loose Storm" . . ?

Martin Avery said, Tempest. "Frankenstorm," the monster mix of Hurricane Sandy, arctic air, and early winter storm, killed the entertainment industry in New York City, but the Letterman Show still went to air!

Linda Ferguson said, It could be a Twelfth Night storm - if you're by the lake, keep an eye out for shipwrecked girls who look just like boys.

Yoga Dawg said he got sent home from work early in the day and planned to spend his time painting.

It struck me that the epic storm called Frankenstorm was like GBS: It hit fast, led to paralysis, could hit your respiratory system, and might prove fatal.

New York City's transportation system was shut down. There was going to be a lot of wind. Fatalities were expected.

The good news was that most people with GBS -- close to 90% -- made a full recovery.

Edmonton called after work to shoot the breeze and got lost while yakking and driving. She must have pulled over to make the call. With some help from Mapquest.ca, I got her turned around so she could go to the West Edmonton Mall.

She asked me how I was feeling, so I told her about the school visit and what the physiotherapist said. And I told her that even though he thought I had GBS and needed a lumbar puncture to confirm it, I didn't think I had GBS and I still didn't want a spinal tap.

My foot felt different every hour or so, all night. As I watched live streaming of the storm of the century on The Weather Channel, online, I was aware of my feet. First they felt swollen, then full of pins and needles, then sore, and then, when I sat on my exercise ball, with my feet on the floor, they felt not bad. -- Still tingling, but in a good way, I thought.

Time and again, I touched them, with different things, in different spots, and I could definitely feel everything. -- I had feeling everywhere in my foot, in every part, even the little toe.

And so, after all this, I concluded that the main population centre of North America was getting hit hard by the biggest storm ever but I was making a comeback.

On Facebook, there was a new posting of an old quote by Simone de Beauvoir: "The body is not a thing, it is a situation: it is our grasp on the world and a sketch of our projects."

Day 62: Sandy

Sandy, one of the biggest storms ever to hit the United States, roared ashore with fierce winds and heavy rain on Monday, forcing evacuations, shutting down transportation and interrupting the presidential campaign. Early reports said there was widespread flooding through New York City, in some cases well inland. Police confirmed at least two people were killed by the storm in the city, and deaths were reported as far away as Toronto as well. The storm's target area includes big population centers such as New York City, Washington, Baltimore and Philadelphia. U.S. stock markets were closed for the first time since the attacks of Sept. 11, 2001, and will remain shut on Tuesday.

That was the news at four in the morning on Tuesday.

While I watched The Weather Network, live streaming, on my computer, last night, as superstorm Sandy rolled in, Edmonton called and she sounded quite different. She had just seen a movie called Flight, starring Denzel Washington, at the West Edmonton Mall, and she said it was profoundly moving.

She also reported that, right after the movie, she went to the washroom, was distracted, and somehow her iPhone fell into the toilet. She has two cell phones -- one for work and one for home. Her home phone, with her contact information for family, was underwater, and was no longer working. But she was still laughing.

Ontario told me she was going to Cuba with a group of women for a holiday in the sun.

It wasn't winter in Ontario, yet. A lot of flights were canceled due to the epic storm.

She told me that somebody I used to know had to postpone a trip to the Caribbean for a writing workshop. Ontario said it was an overpriced adventure based on Shades Of Grey, the bestseller called mommy porn. The person running the workshop lived in the hills northeast of Toronto, and that was one of the areas hit hardest by the wind and water from the enormous storm.

There was no news from Alabama.

Australia said the temperature jumped up to 30 degrees, all of a sudden, so it felt too hot to go for a walk, and she was waiting for it to cool down a bit before she headed it.

I got the impression Australia hadn't been getting my e-mails, so I sent her a message on Facebook about it. I sent a test e-mail.

As for Cold Lake .... It was quiet at four in the morning, there had been a dusting of snow, and that was about it.

My feet had calmed down. The words to describe the way they felt last night do not exist in my vocabulary or experience. Let me try. That pins and needles feeling came and went and moved around, was intense for hours, and was accompanied by a painful feeling, like a sprain or even a break. It felt as though my feet had been injured, fell asleep, and were waking up painfully. It didn't hurt a great deal but they were so sore I did not want to try walking on them. But when I did walk a bit, it didn't hurt, it just felt odd, like walking on bean bags.

This morning, early, they still feel that way, but not so intensely.

I tested my toes for sensitivity and strength, and they passed the tests.

Go, toes!

In other news, possibly related, I just had a glass of cold orange juice, and it tasted fantastic. It usually tastes okay. It's the same orange juice, so .... If the orange juice hasn't changed, I have.

And so, after all this, I feel I can safely say .... The renewed fears of GBS can be forgotten.

But, then, I'm an optimist.

But you know what Mahatma Gandhi said: "Man often becomes what he believes himself to be. If I keep on saying to myself that I cannot do a certain thing, it is possible that I may end by really becoming incapable of doing it. On the contrary, if I have the belief that I can do it, I shall surely acquire the capacity to do it even if I may not have it at the beginning."

Flight was a 2012 American mystery drama film by Robert Zemeckis and starring Denzel Washington as a pilot with a drinking problem. Rotten Tomatoes gave it an 87. The Hollywood Reporter's Todd McCarthy wrote that the film "provides Denzel Washington with one of his meatiest, most complex roles, and he flies with it."

Get it? The movie is called Flight, Denzel Washington is a pilot, and the movie critic said he flies with it.

Speaking of movies .... I've been looking forward to seeing a new movie called Rise Of The Guardians, scheduled for late November, because it looks interesting, and I think my Grade Threes might go crazy over it. However, it features The Sandman, along with The Easter Bunny, Santa Claus, and so on, and The Sandman is called "Sandy", and that name will have different connotations after the superstorm called Sandy. In the movie, the good guys, including The Sandman, fight nightmares, according to the ads and promos, but the name Sandy could cause nightmares for a lot of kids who lived through the storm of the century with the same name.

Up again at seven, I checked the news and discovered the weather picture had changed. President Barack Obama had declared a major disaster in the hard-hit New York and Long Island region.

A massive six-alarm fire destroyed 50 homes in a flooded New York neighbourhood in Queens, on a peninsula jutting into the Atlantic Ocean.

Southern Ontario was expected to bear the brunt of the storm today, with powerful winds being more of a concern for forecasters than the rain. Tens of thousands of customers lost their electricity in both Ontario and Quebec.

Some of the strongest gusts from superstorm Sandy in Canada would hit southern Ontario, especially if you're either over higher ground northwest of Toronto or near the south shore of the Great Lakes, according to The Canadian Press.

The destructive storm had already wheeled through the northeastern U.S. — causing flooding, widespread power outages, and at least 16 deaths, including one in Toronto — and was now walloping parts of Canada with strong winds and heavy rain.

For forecasters, the sheer size of the storm is what sets it apart. Strong winds were reported from southern Ontario and eastern lower Michigan, all the way through Quebec, into parts of the Maritimes and across a large part of the north eastern states.
Andrew, in Toronto, posted this not on Facebook: Woke up to no power, and this time the culprit is a tree that looks like it just blew up in the wind . . . not expecting power, nor internet or cable any time soon.

As for me .... I had a little epiphany. -- I decided I wanted a second opinion from a physiotherapist!

Thinking long and hard about my feet, legs, and back, I went back in time about five years to the moment I was in a big car crash. My little car, an Aveo, was hit from behind by a pig truck. I had stopped for highway construction in the country, below the little town of Blackstock, a bit above Port Perry, in Durham Region, north east of Toronto, as I saw a signman waving a big stop sign. The truck that came up behind me didn't see the sign, or my car, or anything else, and hit me at full speed, pushing my car about 100 yards. My car was totalled but I walked away without a scratch and feeling fine.

I checked in with a doctor at the hospital and all he said was watch for signs of concussion.

However, after that, I had a bit of back pain, and I found a good orthotics person but she was frustrated because she couldn't fix my pain with her orthotics.

The car crash was five years ago. The winter before last, I tried skiing and skating, at the start of winter, but I found both sports very difficult, all of a sudden. My legs no longer had the spring needed for skiing. When I played hockey, my legs ached, and I had to stop a few times. At Christmas, after a few practice sessions, I skated with Sherry, and it went well enough, although my feet felt a little numb.

We got together in our old hometown after re-connecting on the internet, via Facebook, and I met her father, then we went skating, and I took her out for dinner. -- We had a great time doing those things and yakking about the old days, when we were kids, and went skating together all the time, and catching up some of the things that had happened since then.

That spring, I found an acupuncture person in Port Perry, who did Western or medical acupuncture, attaching the needles to electrodes, and cranking up the electricity, and my feet and legs were fixed.

That was when I felt inspired to move out west and found a job in Alberta.

Last winter, I did not ski or skate, but I swam, went to the gym a lot, danced, and hiked in the provincial park a lot.

At the end of the school year, I was exhausted, caught a summer cold, was in bed for a week, but was okay after that, and spent the rest of the summer getting in shape again, going to the gym, walking and hiking, so that by the end of the summer I was hiking 12 kilometres, fast, carrying weights, feeling strong and healthy and ready to head back to work.

In the middle of the summer, I had a little bit of pain in my legs and my feet felt funny, so I tried acupuncture and went to a chiropractor for the first time in my life. Acupuncture took away the pain in my legs. A chiropractic adjustment hurt and I was sore for 24 hours but felt fine after that.

The week before classes started, while I was back at work, getting things ready for the new school year, I felt a huge zap of pain in my back. It was unlike anything I had felt before. Much stronger. So I went back to the chiropractor.

A second adjustment took the big back pain away, but the next day I was walking funny. My legs and feet muscles contracted and spasmed so I looked like a spastic when I walked. At work, I sat down a lot, and experienced more and more difficulty walking.

At the end of the week, I had chiropractic adjustment number three and told the guy I was in a lot of pain so he said, Maybe you should go to the hospital, for the pain.

At the hospital, in the emergency unit, I saw four doctors over three weeks, as I lost the ability to walk. First I staggered, then I needed two canes, and then I needed a walker. For a couple of days, I could not walk at all, and I just rolled and crawled around at home. I was off work by that time.

At the Edmonton Clinic, a neurology specialist ruled out all the big scary things like cancer, MS, ALS, etc., but said it was a little odd that I couldn't walk but didn't feel a lot of pain. So, he scheduled me for more tests and told me to activate short and long term disability at work. He set me up for more bloodwork as well as a lumbar puncture and an EMG.

At the Cold Lake hospital, the doctor said he really didn't think I needed the spinal tap.

By that time, I had found a family doctor, in Lac La Biche, and he agreed that the EMG and spinal tap were not needed.

Also, I had several sessions with a reflexologist, who I felt did fantastic things for my pain and nerves.

After 24 days of going downhill and losing the ability to walk, I've had about 40 days of improvement. I can now walk a couple of miles, my balance is returning, and the numbness in my feet is going away. First my knees came back, then my ankles and feet muscles, and now my toes are returning, too.

So why did the physiotherapist think I had GBS?!

I wanted a second opinion.

When this adventure started, I made an appointment with a physiotherapist in Cold Lake, but I forgot about it because I found one in another town.

When was the Cold Lake appointment for physiotherapy?

Not until the end of October.

The 30th.

When was that.

I checked the calendar.

Today!

What time?

I'd have to call them and find out.

Canadian Health & Sports Rehab Inc. was within walking distance from where I lived. -- Didn't it make more sense to go there than drive to the next town, over half an hour away?

My reflexologist said to stay with the same physiotherapist.

But I wanted a second opinion.

I did not want the spinal tap or GBS.

As they said in Ghostbusters .... Who ya gonna call?

There was no answer, so I left a message, and no call back, so I walked over at 9 to find out I had an appointment at 9:15 but they had to bump me until 11.

While I was there, I ran into the grandmother of a three of my students, getting physio for a sore back, and I told her about the gifted reflexologist, whose speciality was taking away pain. She said, I've heard of her.

They call her The Goddess, I told her.

Yes, she said, I've heard that, too.

She told me I should go to the school the next day to see the parade of kids in Hallowe'en costumes. I told her I was thinking about it, but my last few visits to the school had caused havoc.

Yes, she said, I was there when you showed up at the volleyball game. Everything was calm and then it all shifted, so I looked around to see what was going on and noticed you walking in. And then the screaming began.

They all love you and miss you, she said.

I made a mental note to give her grandchildren higher marks.

-- Just kidding.

My marks are real, not inflated; I don't give bonus marks and brownie points.

So why do they like me?

I have no idea.

They call me The Jolly Rockstar. Or Santa Claus.

They don't think I'm a rock star. -- Last year, somebody found a can of Rockstar, the energy drink, in my desk drawer. -- I liked to keep a can of coffee-flavoured caffeine and so on around in case I felt a need for a pick-me-up.

I've never been a coffee drinker and I like cold drinks more than hot.

Teachers aren't supposed to smile until Christmas. -- That's 'old school'. -- But I laugh a lot.

What can I say? Kids are funny!

After checking my appointment time, I walked down the block to the drug store and the book store, past shopkeepers shovelling snow dust off the sidewalks, around one thousand steps.

My hamstrings were tight, my gluteal muscles were sore, and I walked like a duck.

Even so, I felt great to get outside at the start of the day in a friendly little community.

Wally Lamb must have slept in or something. At 9:30 he posted this little gem:
Superstorm update: One mammoth pine tree down in back yard; didn't hit the house. Phew! Power out at home base but not here in office. Hope everyone fared as well as we did. Spent last evening making fun of my wife Chris about wearing her coal miner's headband light, but I was secretly envious.

Why do I have a sneaky feeling Wally will write a book about Superstorm Sandy? And it will be another big bestseller.

Is it a sneaky suspicion or a sneaking suspiscion?

The correct use is actually, and only, "sinking" suspicion.

This is the most commonly used version on Google & Yahoo because it's the correct one.

"Sneaking" is the second most commonly used version, because it sounds like the correct one.

"Sneaky" is by far the least commonly used version but it is wrong.

Funnily, though not surprisingly, "sneaky" is the version I most commonly hear in conversation.

So, let me just say, I have a sneaky feeling Wally Lamb is writing a novel about Superstorm Sandy and it's destined to be another big hit for him.

Say! I just noticed this book has reached page 300! It's time for a celebration!

Everybody asks me, Are you writing, while you're off work?

And I say, No, all I do is go to appointments with doctors, specialists, physiotherapists, reflexologists, chiropractors, acupuncturists, and nurses, and do physiotherapy homework, and sleep.

The truth is, I started this book as a journal, because I always keep a journal, and to keep things straight, as a lot of information was coming in, about my injury, and then I had a lot of appointments and medication to keep straight, and then I thought I was going crazy, when I lost the ability to walk, and when doctors gave conflicting advice, and nobody agreed about what was going on, and most said, "There's something going on, but I don't know what it is, so I'm going to send you to someone else."

The book grew as my adventure went on much longer than I anticipated and it filled up with info plus related items from my life and inspirational quotes. Around page 100, I thought this book might be useful for somebody else going through the same thing. Around page 250, I thought it might be a big bestseller like No Laughing Matter.

-- Call it No Laughing Matter 2.

At page 300, I have no idea if we are half way through or almost at the end. Of course, I wish it was at the end. But if the physiotherapist is right, and I have GBS, need an EMG and spinal puncture, lol, plus a long recovery time, it isn't even half over.

I want it to be over! I want this dark chapter to be over so I can see the light!

"Is this a comedy, or a tragedy?" one of my writing students used to ask all the time.

Facebook says, This may be a dark chapter but it is not you whole life story."

-- Sounds like something from The Hobbit, or C.S. Lewis, or a Taylor Swift song.

Day 63: Part 2: Breathe

Oh. My. Goodness.

It looks like I was right about one thing: The new physiotherapist, named Angela, might be an angel.

It looks like I was wrong about some important things, like how to BREATHE.

And it looks like GBS is out of the picture.

And I'll be back at work sooner rather than later.

WOO-HOO!

Now .... To get a new doctor ....

How do you fire your physiotherapist?

So, Angela got me to lie down on her table, made me laugh, saying, Lie down and tell me everything," as though she was a psychiatrist, and she noticed, right away, that I was breathing funny.

-- Oddly.

She mentioned it and got me to do a bunch of mechanical tests -- stand up, balance on one foot, balance on the other, walk away, walk toward her, touch my toes, bend backwards, lie down, push against her with my legs, feet, toes, one side of my body after the other ....

First she asked me what was wrong and how it started. Because I had written the story beforehand, I was able to give it to her quickly. Plantar fasciitis for several years starting in 2000; car crash, hit from behind, 2007; orthotics for back pain didn't work; acupuncture worked; healthy year, summer cold, in shape again, big back pain, chiropractor, pain gone, legs gone, staggering, canes, walker, crawling, walker, canes, walking better, numbness going away, getting stronger.

So, you're getting better, she said. That's good.

You're breathing wrong, she said.

She got me to breathe for her, nodded her head, and then got me to breathe while looking at myself in the mirror, from the side, standing up.

What do you see? she said. Describe what happens when you breathe.

Chest goes out, belly goes in, I said.

Exactly, she said. And that's wrong.

What?! I said.

Your belly should go out when you breathe in.

She explained the umbrella breath. When you breathe in, your ribs and stomach should open up like an umbrella. And, believe it or not, there's an umbrella down below that should open up, too. And they should both close, when you breathe out.

She got me to practice that.

Imagine you are picking up a bean with your anus, she said. Imagine you have a vagina. Everybody has two sphincters. They should clench as you breathe out and unclench as you breathe in.

Think of two umbrellas, opening and closing, as you breathe in and out.

Here's the rules for breathing:

Ensure your ribcage is over your pelvis. Line it up!!!

INHALE

Draw breath in to lower side ribs.

Soften the belly.

Allow the pelvic floor to lower.

EXHALE

Blow air out slowly.

Lift up and in with the pelvic floor. Lift your beans.

Do not tighten stomach muscles.

How does that feel? she said.

Good, I said.

Sit down, she said.

She showed me how to sit down properly, with an arch in the back, and asked me if I had a good chair.

How does that feel?

Good. Easy.

I told her I used a ball.

She recommended a chair with lumbar support and a wedge, but said a ball is not bad.

She showed me how to lie down and sleep, too, with a pillow between the knees, feet tucked up close to the bum, and head back a bit.

Good. Easy.

She pushed my feet up and coached me through the breathing routine one more time.

Could you .... I said.

What? she interrupted me. Move in?

I was going to say 'Go over the breathing routine one more time,' I laughed.

After breathing, sitting, sleeping, we discussed walking.

When you stand, where is your weight?

On my heels, I said.

It should be on the balls of your feet and your heels.

Lean forward, she said. For you this, may feel like you're on a ski jump.

It did.

That's how you should stand and walk she said.

Try breathing in that position.

Good. Easy!

When I figured out how to breathe, like that, I could move on to exercise number one, which was a simple push-up with a back bend. Arch the back. Hold it for five seconds. Do that three to five times, a few times every day.

P.S. Do not bend over for the next few days.

We also talked about the gifted reflexologist. She agreed that Mecelle was great.

And she's so funny, she added. And she has such a warped sense of humour.

Oh, I said, I haven't seen that side of her, yet.

Just wait! she said.

And that was that.

I booked three more sessions.

On my way home, I walked to the bookstore and bought the new J.K. Rowling novel for my cousin as a Christmas present. And I ordered No Laughing Matter by Joseph Heller.

I told the cashier about GBS and Joseph Heller.

That sounds horrible, she said. Omigod!

A physiotherapist told me he thought that's what I've got, I added.

Oh no! she said. What are you going to do?

Fire the physiotherapist, I said.

We both laughed.

When I got home, I phoned the office of a group of doctors that one of the emergency docs told me might be getting a new guy around the end of the month.

After that, I wanted to go get a new chair. Shopping online, I found one with lumbar support and a wedge, at Staples.

Everything looked different, up-beat, and optimistic, again.

Woo-hoo!

In three to five days, things are going to be different, Angela said. Better.

You might get some back pain and some pain referred to your legs. That's good, she said. That's temporary. We have to get your back moving and arching and change the way you walk, sit, sleep, and breathe, and you'll be back in the game.

-- Called the doc. They said there was, in fact, a new doctor in town, but he was taking over somebody else's practice, so they weren't taking any new patients.

Dang!

Oh well. -- We'll see!

Suddenly, I felt tired, and I wanted to try sleeping the way I had just been shown. It was so much more relaxing and re-energizing, I could barely believe it. And that was it. -- I was hooked on breathing the right way.

Every breath you take
And every move you make
Every bond you break, every step you take
I'll be watching you
-- Sting

P.S. After trying the sleeping position and practicing umbrella breathing, I felt like going out. Tuesday night was 'wing night' with my colleagues and I try to make it but have missed a few when I could not walk. They have seen me use canes, and stagger, and walk like a duck, but this time they saw me walk the way I used to.

Well, it wasn't exactly the way I used to walk.

Now I lean forward so my weight is on the balls of my feet as well as my heels. And I practice umbrella breathing while I walk, so my posture is better.

The story about the sphincters and the man with the fake vagina made them laugh, a bit, but they could not believe that breathing differently, or breathing, sitting, sleeping, walking, etc., differently could make that much of a difference. They liked the other physiotherapist's idea about GBS and said, "Well, you have to find out what caused this problem!"

Breathing wrong caused it, I explained, along with those other basic things. -- It compacts your spine and that leads to a lot of other problems. The disc bulges, hits nerves, and there goes your legs and feet.

They liked the GBS explanation better.

-- Not me!

After wings, I still felt like walking, so I went to the Walmart superstore to look for a wedge. The physiotherapist recommending getting a good chair with lumbar support and adding a wedge, which will make you sit with your back arched.

First I went to Staples and found a great chair they said was going on sale the next day and then I went to Walmart to look for a wedge.

Although I did not find a wedge, or anything to use as a wedge, I found a couple of other things.

What I found was that I could walk and walk without pain or getting so tired that I had to sit down before I fell down.

While I was out, I practiced another technique the physiotherapist taught me. The trick is to inhale before doing things like standing from a sitting position or getting into or out of a car.

Is that a problem for you? she asked.

Oh, yeah.
-- But not any more.
The catch-phrase for that little exercise is, Blow before you go.
Exhale and then stand up or sit down.

Despite my colleagues disbelief, I decided to stick with the new-found techniques and the new physiotherapist.

I wanted to tell everybody about these discoveries. Shout it from the rooftops. Intervene with every person I saw in the parking lots who looked like they were struggling to get in or out of their car. Post it on the internet. -- Something!

PPS: There is a large amount of irony in this. As a Drama teacher, as a qigung leader, and as a Zen meditation instructor, I teach people how to breathe.

"From the diaphragm," you see frequently to young actors on stage. "Breathe from the diaphragm. -- Then speak."

Meditation is all about breathing and counting breaths and the way you sit is given a certain amount of importance. -- I believe I was sitting the right way and breathing the right way, and teaching those things, when teaching qigung and meditation. -- What about the rest of the time?

I spend a lot of time writing and a preparing lessons and marking student assignments .... And I try to sit with appropriate posture .... But I've been breathing wrong for a long, long time. I tell people to use their diaphragms but most of the time -- almost all the time -- I suck in my gut and fill up my chest.

Angela, the physiotherapist, said it's a North American thing. Most people in the Western world breathe the wrong way. And that's why millions of us have compacted spines and back pain and many have problems walking.

Unlike my colleagues, I find this fascinating, liberating, inspiring, and a great source for optimism.

This book might be just about over!

Day 64: Hallowe'en 2012

Oh what a night!

After wings with colleagues, getting 'the latest' on what was happening at school, I climbed into my car, using the new method (blow before you go), and drove home with a sweat-shirt rolled up for lumbar support, and those two little adjustments made a huge difference, which made me happy.

The things my colleagues said about work and GBS did not exactly fill me with joy. But it was good to see them. And the lumbar support plus the 'blow before you go' exercise made me feel so good that I quickly put the things they said behind me.

"I'd want to know what was going on with me," the Science teacher said. "You should have the spinal tap and the EMG."

"You'd risk a spinal tap, even though you were getting better, just to find out what you used to have?" I said.

"Sure," she said. "I'd want to KNOW!"

-- Not me, I said.

If I had a mild form of GBS and was now recovering, or if I just needed to re-learn how to breathe, sit, walk, and sleep, and so on .... I just wanted to focus on getting back in the game.

Was this new physiotherapy a trick or a treat?

Edmonton called, sounding upset, but she did not want to tell me what was going on, right away, she wanted to hear my news, so I warned her it was happy happy joy joy, and she said, Good, I could use some good news.

So I told her the story about the fake vagina and she laughed long and hard. The part of the story where the physiotherapist climbed on the table with me to test my leg strength and I said, Could you .... And she finished the sentence, Move in with you?

She was funny and upbeat and I sure did appreciate that, especially after the last guy made me worry about GBS.

Yes! Edmonton said.

So I told her it was all about breathing, with a little adjustment to sitting, walking, sleeping, one easy exercise, and I was supposed to avoid bending over for a few days, and I'd be fine.

Then she told me her story. The day before, she dropped her iPhone in the toilet, and just before she called me, she had another mishap like that, she said. She was on the treadmill, at work, after work, walking and watching TV at the same time, and there was something on the news related to one of the stories she was working on, so she lost her concentration, hit the wrong button, the treadmill speeded up, and down she went, off the end of the treadmill, scraping skin off her palm and her arm, and landing on the floor, all alone, with nobody to help her get up or do anything else, if it was more serious.

I asked her if she wanted me to call her a cab or an ambulance.

At first, she thought she broke her arm. -- She had a lot of trouble getting her coat on. So, she just wore it over her shoulders. But when she opened the door to go outside, she discovered the temperature had dropped, and she thought she might freeze to death.

There was a story in the news yesterday about the coyotes of Edmonton, as there was an estimated two thousand of the wild carnivores in the river valley, and somebody's dog had just been taken down by one of them, so she had a vision of falling, again, outside, in the cold, and being devoured by a coyote that came up from the valley.

The way she tells stories is dramatic and her voice is perfect for radio, so she had me laughing, and that made her laugh, but at the same time she said it was no laughing matter, and then her phone died, so we were cut off.

That made me worry about her for an hour or so until she got home and sent me an update on Facebook.

Ontario said 'Bon voyage' as she was on her way to Cuba. There was no word from Alabama about scuba diving in the Turks and Caicos. An old friend was taking a group to the Dominican Republic for an erotica writing workshop. A one book wonder I mentored as a writing instructor. -- That was a mistake! And now she's just somebody that I used to know. That workshop sounded like trouble, to me.

And I was in Northern Alberta, wondering why anybody would fly the day after Superstorm Sandy or travel to the Caribbean during hurricane season. Travel is hard at the best of times. Why go looking for trouble?! I wished them all 'bon voyage'.

After physiotherapy and a little bit of walking downtown, at Staples, in Walmart, and Original Joe's, I thought I'd be ready for bed and sleep, but I was not right about that. Instead of sleeping, I watched A Knight's Tale, even though I've seen it several times, and noticed that I had a headache, which is rare, for me, and then I fell asleep with The Answer Man playing one more time.

In the morning, I felt pretty much the same as before. During the night, I woke up several times, for no reason, and each time I practiced the new breathing routine for a few minutes until I fell asleep again. My legs were stiff, in the morning, but my back wasn't sore, and my feet felt a little less sensitive than the day before. There wasn't a big change. I did not feel like jumping out of bed and getting ready to go to school to see the parade of little kids in Hallowe'en costumes.

Frankly, I didn't feel like going back to Staples to get the chair they said was going on sale today, half price, either. I would have to get the walker out of my back seat and lift the chair in and out and I was supposed to avoid heavy lifting. -- Was that heavy lifting?

What I was looking forward to was reflexology with Mecell.

What could I do to pick myself up and get going? Eat an apple? Have a hot shower? Listen to music? It was a good thing I didn't have to do a lot of exercises for physiotherapist number one. Just doing the simple push-up with a back extension recommended by physiotherapist number two, and getting up again, would be enough of a challenge.

Superstorm Sandy was still in the news but on the way to becoming old news.

Wally Lamb posted a question: Is it possible to be in love with your generator?

Somebody responded with: Sounds like a hot relationship.

My comment was: Lots of electricity!

The storm had weakened significantly, but continued to creep north, from southern Ontario, into Quebec, dropping ice, snow, and freezing rain. Environment Canada said the wintry precipitation will likely continue into the night across broad swath of northeastern Ontario.

Walk Off The Earth had a new EP scheduled for release on Hallowe'en, I discovered, and they were planning to be in Edmonton on November 22nd. I put on a series of 15 music videos by WOTE on YouTube. -- If that doesn't get you going, well, good luck to you!

-- This morning, it didn't work for me.

It's snowing outside.

There's a pain in my backside.

Frankly, I'm fed up and need some fun. I'm sick of being sick and I wish I could go for a run.

-- Maybe what I need is to get going, go outside, think outside the box ....

Of course, I know I should be happy I don't have severe GBS, for instance, that I'm not going downhill, that I'm making a comeback, I'm recovering, getting stronger every day ....

But I've lost that comeback feeling.

What was it they said about carpe diem in The Dead Poets Society?

Gather ye rosebuds while ye may,
Old Time is still a-flying:
And this same flower that smiles today
To-morrow will be dying.
(Robert Herrick, 1591 - 1674)

On Facebook, Positive Thoughts posted a sad note: **Unbelievable damage and losses all around. Please join me to send healing light and love to those who affected by the hurricane Sandy. Many Blessings to All!**

Day 64, Part 2

While walking to the car, looking at the walker taking up the back seat, thinking about taking it out of the car, pushing it inside, getting the new chair from the store, putting the walker back, I had a sudden inspiration that simplified the whole deal and made me feel good.

Why not return the walker?

Was I using it? No!

Would somebody else make use of it?

No doubt!

Also, it could be a symbolic move, representing my recovery, and my determination to walk the way I did before, get back in the game, and get better!

First I went for a reflexology appointment, told Mecell about Angela, and physiotherapy. Reflexology got me feeling better again. Next, I drove over to the Community Health Services building and dropped off my walker. And then I went to Staples to get a great chair with lumbar support for one hundred dollars off.

Those three moves did a lot to change the way my day was going!

I felt so good about the way things were going, I started shopping for a monitor for my computer. -- Maybe it would help me breathe better if I was looking at a big screen in front of me, instead of looking down at a small screen.

The prices I found online but while I was at Staples I noticed they were having some sort of sale. A one thousand dollar screen was being sold for six hundred and something.

Before I headed out, I made lunch, and called the Primary Care Network. Mecell the magic reflexologist informed me that Cold Lake's nurse practitioner, Elaine Wall, who was a local legend, was now her "family doctor" and she suggested I get in there, too. So ....

Finally I have a family doctor in Cold Lake.

Instead of driving for an hour and a half in both directions to see a doctor in Lac La Biche, I could see the famous nurse practitioner in about ten minutes.

Instead of driving to Bonnyville and Lac La Biche for a physiotherapist and for doctor's appointments, I can do it all in Cold Lake. And it only took about 14 months.

Who was it who said, "The better it gets, the better it gets"?

Abraham Hicks said, The better it gets, the better it gets, the better it gets ....

In the evening, I went back to Staples, to look at computer monitors, and then I went to Canadian Tire, to get another air mattress to put on top of the other air mattress, to make it easier to get into and out of.

There were no signs of Hallowe'en anywhere.

One of the cashiers at Canadian Tire was in costume, but that was it.

One of my Facebook friends said she expected 100 kids but only had 30, so she had mountains of candy left over.

In my e-mail, I got an invitation to join NaNoWriMo, or National Novel Writing Month, that was set to start in about five hours. There were 2000 people in Edmonton signed up for NaNoWriMo and close to 1000 more outside of the big cities, including one in Bonnyville and one in Cold Lake.

Could I write a novel this year, during NaNoWriMo?

I didn't think so. I wanted to keep my focus on getting better and getting back to work.

If I was going to write a novel this month, I'd write about a guy with GBS during Superstorm Sandy in NYC or Northern Alberta. Quick, before Wally Lamb beat me to it!

Day 64: Wally And Sandy

Will you still need me, will you still feed me, when I'm 64
-- The Beatles

On Hallowe'en, I went to bed early, around eight o'clock, feeling very tired, after reflexology, getting the new chair and a new bed, a little e-mail, only around 1500 steps, a little driving, with lumbar support, using the "blow and go" trick, the umbrella breathing technique, walking on the balls of my feet as well as the heels, sitting differently, sleeping differently, and trying to keepmy ribcage over my pelvis. Also, I avoided bending over, and I did back extension exercises, arching my back. My glutes were sore, my legs felt heavy, and I was grumpy in the morning, but reflexology changed all that. Returning my walker felt momentous. The new furniture and techniques felt like they were working. But by eight o'clock, I felt ready to crash, hit the sack, try the new sleeping method on the new bed, and get a rejuvenating rest.

Although I slept soundly, I did have a dream, or two, that I remembered, and I kept writing a novel, in my mind, for Nanowrimo. It was called Wally And Sandy. I did not want to write it, but it was writing itself.

At five in the morning, I got up, feeling better than before, or possibly better than on any day in the past 64. My back was less sore, or sore a lot less often, my feet were bothering me less, my legs were still stiff and I staggered around a bit, when I got up, like a zombie on Hallowe'en.

While I was up, I checked the news on the internet to see what was happening to Super Sandy. I read somewhere that Sandy may have caused up to $20 billion in damages, making it one of the costliest superstorms in history.

According to the giant insurance company Munich Re, weather and climate disasters contributed to more than one-third of a trillion dollars in damage worldwide in 2011. And this year's total would no doubt be greater. The U.S.A. suffered through a summer of drought, crop disasters, high temperature records, and wildfires, followed by the superstorm, or Frankenstorm.

There was growing evidence of links between climate change and sea-level rise, heat waves, droughts and rainfall intensity, and, extreme weather events. But tornadoes and hurricanes appeared to be in a different category, like earthquakes.

True, he United Nations Intergovernmental Panel on Climate Change said the intensity of tropical cyclones (that is, hurricanes) will increase as a result of warmer waters. And our atmosphere and oceans are, so tropical storms are getting "wetter."

It is quite certain that sea levels have risen over the last century, and continue to rise, in response to changing climate. And storm surges ride on these elevated sea levels, amplifying flooding losses where they strike.

Sea surface temperatures along the U.S. northeast coast are about 2.8 degrees Celsius above average, which helped to intensify Sandy.

Sea levels along the U.S. northeast coast are rising up to four times faster than the global average, making the region more vulnerable to storm surges and flooding.

In the summer, Arctic Ocean ice and the glaciers of Greenland were melting at record rates. Some weather experts said that was because of an atmospheric weather pattern known as a "block," a persistent area of high pressure, and that was most likely the reason Sandy moved inland rather than out to sea.

Superstorms like Sandy are the start of a "new normal", some people believe, and we can expect more and more extreme weather events with climate change or global warming or planetary overheating -- whatever you want to call it.

Is it possible to fall in love with your generator?" Wally posted the question on Facebook and it got dozens of funny responses. "How old is your generator?" one person asked. "Absolutely," several people said. "There could be a lot of electricity." "Are their sparks?" "It could heat things up." "Sounds hot." And so on and on for the rest of the day.

Wally was a weather watcher for an insurance company in Hartford, Connecticut, where he had lived his whole life, and his wife, Kathy, worked as a reporter for the New York Daily News in NYC, which was about an hour away, on a good day, when the commuter trains and subways were working, and there were fewer cars on the road, and there were no new severe weather events to deal with.

For an American man in his mid to late fifties, Wally was in great shape. He went jogging, a lot, to stay in shape, he didn't smoke, he drank a lot of water but very little alcohol, he ate well without eating too much, and he felt great. After thirty years of marriage to the same woman, he was still pretty happy with the relationship, and he liked their kids, who left quiet Connecticut for wild and crazy New Orleans, but stayed in touch with Skype, Twitter, Facebook, and e-mails.

Wally had just one big worry, really, but it was something that helped him at work. He was sensitive to the weather. He was, in fact, what some people call a "weather sensitive". He did not need to watch the local weather on TV, he knew it in his bones. Everybody who knew him, knew he could predict the weather better than any meteorologist or weather person in the region. "Just ask Wally," they all said.

However, Wally's ability was growing, and as the weather got worse, with climate change, it turned into a disability.

One morning, at breakfast, he told his wife, Kathy, that his coffee tasted funny and his breakfast cereal felt funny in his mouth.

Funny? she said. Off, he said. Odd. Unusual. -- Weird.

That's strange, she said, but she did not pay a lot of attention to it as she had to hurry, as usual, to get into the city, make her connections, catch the train, and take the subway to work.

Kathy was several years younger than Wally and had gone back to work after taking a couple of decades off to raise their kids. So, at age 50, she was working as a reporter. After decades of working for weekly publications, she had done the impossible, and made the jump to a major daily newspaper. -- That almost never happened. It wasn't a jump, it was a leap. A flying leap. And it meant that she was working with, and competing against, a lot of reporters who were half her age. So, she had to work hard, cranking out an average of three stories a day.

Fortunately, her paper preferred articles that were short. Unfortunately, she liked writing long. She was in the habit of writing long articles for weekly newspapers and it was her style, her desire, her habit, that was hard to break. She enjoyed writing long, going deep, making connections between this and that, but her paper wanted her to write short. Her editors liked her work a lot, but they had to say, "Keep it short," all the time, to keep her on track.

She was covering weather-related news and Wally was watching and feeling the weather as always, for his insurance company, so they were both well aware of the tropical hurricane called Sandy that was heading their way. It was when Sandy left the Carribbean and headed up the East Coast of the U.S.A., and appeared to be heading for New York, or at least New Jersey, that Wally said, My coffee is off and whatever I put in my mouth for breakfast feels strange.

Oh? his wife said. Well, I hope you aren't getting sick. And she kissed him on the cheek, said, So long, Sweetie, as she always did, and she ran out the door, to get to work.

By lunchtime, Wally was feeling so unwell that he went to the walk-in clinic at the hospital. He phoned his "family doctor", but he was booked solid, as always, so Wally called a cab and went to the hospital, where there was a nurse practitioner working in a unit called the Primary Care Network, and sat in a chair with two dozen other people waiting to get seen by her or a doctor.

She said, We're going to keep you overnight.

That surprised Wally because he knew that hospitals were in a hurry to get people out of the building, these days, and you only stayed overnight if you were very sick.

When Kathy came to visit him in the hospital that night, she could barely believe her eyes. Her husband, who jogged around four miles the day before, suddenly looked as though he had been sick for months, and he was going downhill unbelieveably fast.

"GBS," their doctor told her. "Guillain-Barre Syndrome."

What the heck is that? she asked. But then she remembered. Joseph Heller Syndrome, they called it in some places in the U.S.A.

Heller, the famous author of Catch 22, was working on his novel called God Knows, and jogging every day, to stay in shape, on the West Coast, when he went down, fast, with GBS. He wrote a book about it, with a friend, called No Laughing Matter. It was a funny book by a famous humourist but GBS was, as he said, no laughing matter.

He was paralysed. Very quickly, his nervous system shut down, apparently, and since your nerves control your muscles, he could not walk, at first, and then he could not move.

And if it progressed to his respiratory system, Wally would not be able to breathe, so he would die.

Oh. My. God. she said.

What should I do? she asked the doctor.

Go home, get some sleep, make sure your house is ready for the storm that's heading our way, go to work, visit when you can, the doctor said. GBS hits fast and hard, over three or four weeks, usually, but then there is a recovery period that can last from a few months to a few years. It's a rare disease, affecting just one in one hundred thousand, but about 90% live and almost all of those recover completely. -- Some patients are left chronically weak and some have nerve damage that effects the way they walk for the rest of their lives. The good news is that your husband is in your hospital and we can hook him up to a ventilator, so he won't suffocate to death, when he stops breathing, as the machine will breathe for him, and he is very likely to live and to recover fully, in three or four months or years.

"He'll be looked after, day and night, by a team of nurses," the doctor added. "And I'll monitor him, as well, of course."

"Our medical insurance is paid up," Kathy said.

"That should cover most of it," the doctor said.

"Nurses?" Kathy said.

Okay, that's it for that story. I really don't want to write that novel. I just want to keep my focus on my recovery and getting back to work.

Besides, it's too predictable. As the storm comes in, the guy gets worse. Kathy covers the storm for her paper and has to deal with the effects at home, not to mention the impossible commute after trains, buses, and subways are shut down. Meanwhile, Wally is nursed through a near-death experience by a team of three women and one of them, in particular, grows close to him.

It's like The English Patient only it's set in the U.S.A. during Superstorm Sandy, the Frankenstorm that hit on Hallowe'en.

Somebody else can write that novel.

Day 64, Part 2: November

It was the first day of November. It was ten below and snowing lightly.

Later that morning I noticed an update on Facebook from my old friend Mel Malton, who is now an Anglican minister. Here's what she had to say about Hallowe'en: Mum and a princess and a superhero. "We don't like the scary costumes," Mum says. "We've got another superhero..." "Who's that?" I ask. Little superhero points upwards. Grin. "Me too," I say.

Mel was a murder mystery writer when I knew her but she morphed into one of the first female Anglican ministers in Canada.

Go, Mel.

We started the Muskoka Novel Marathon together. She liked my idea and knew just who to call to get it going. She hooked us up with the Huntsville Festival of the Arts and the Muskoka Literacy Council, fast. One summer, she got us the building beside the Anglican church to use as our venue. It was a great building, but it was not air conditioned, and a heat wave hit Muskoka, so we were afraid everybody in the marathon was going to die. But everybody survived and had a good time. -- It felt like a minor miracle, at the time.

Love Tree, this morning, posted this dire warning: The trouble is, you think you have time. And they attributed it to Buddha. But I smelled a rat.

Right away, I checked it out with a website called Fake Buddha Quotes. It quoted Buddha saying, I never said that! And, sure enough, it was a fake Buddha quote. On the website, it said, This is another one from Jack Kornfield's Buddha's Little Instruction Book (1994), which isn't a collection of Buddha quotes, but is Jack's rather lovely interpretation of Buddhist teachings.

I let Love Tree know about that.

My friend Mel Malton got more and more interested in her religion after we met and I got interested in Zen Buddhism. And then I got less interested in Buddhism and developed my interest in Zen. Even so, I learned a few things about Buddhism. And I didn't think The Buddha would ever say anything like, The trouble is, you think you have time.

Even so, maybe the trouble is that you think you have time.

On the other hand, there is an old Celtic saying that goes like this: When God made time, He made enough of it.

That reminds me .... The last time I was in the drug store, I noticed two new publications by TIME magazine. One was on Alternative Healing. The other was on Global Warming.

It made me think: Is there a connection?

Maybe the over-heated planet needs alternative healing!

Love Tree posted a response on Facebook saying, Thank you Martin.

They put up a new posting, from Marianne Williamson, this time, that said: As we let our own light shine, we unconsciously give other people permission to do the same. As we are liberated from our fear, our presence automatically liberates others.

Marianne Williamson was a spiritual activist, best-selling author, a lecturer, and a founder of The Peace Alliance, a grass roots campaign supporting legislation currently before Congress in the U.S.A. to establish a United States Department of Peace.

She was also the founder of Project Angel Food, a meals-on-wheels program that serves homebound people with AIDS in the Los Angeles area.

She has published ten books, including four New York Times #1 bestsellers. Her book, The Age of Miracles: Embracing the New Midlife, spent five week on list. She has been a popular guest on television programs such as Oprah and Politically Incorrect.

Years ago, I met her at Trent University, where she was leading a long weekend workshop withe Robert Bly. -- It was Robert Bly I wanted to see. His book called Iron John was on the bestseller list at the time. -- In those days, I was thinking about becoming a drama teacher, as well as an English teacher, and they convinced me to go for it.

Before physiotherapy, I ordered a pizza from the place next door, so it would be ready when my appointment was over.

What a great neighbourhood, I said to myself.

At physiotherapy, I was told my balance was better and I reported success with the "blow and go" technique, said I was sleeping better, and I also told her that I could now pick things up with my toes.

Excellent, she said.

My glutes hurt, I said.

Good, she said.

She checked my balance and breathing and said both were better.

She got me on the table to show me the clam shell exercise. It's a subtle exercise, co-ordinated with breathing, designed to work the muscles that run down the leg beside the sciatic nerve bundle.

You don't want to overdevelop it and cause it to swell, because that can have a bad effect on your sciatica, but you want to engage it and use it, otherwise you are overworking your glutes, and that hurts.

She tried me on another exercise, but it hurt a lot, so she backed off, saying, Stick with the clamshell until next time and we'll do the other one later.

She wrapped a thin sheet around both my ankles, while I was on the table, lying on my back, and pulled it, then relaxed it, and pulled it again, to give me some traction. -- That felt good.

She put a 'stim' machine on my lower back, with me lying down on my stomach, and cranked it up high, and that felt good, too.

And that was it.

She give me a computer print-out of the clamshell exercise and I said, See you on Monday!

After physio, I walked next door, picked up a hot pizza, and headed home.

It was snowing somewhat heavily.

All the school buses had been canceled, I discovered, so it was a "snow day" for Cold Lake. A lot of snow was expected.

A snow day on November 1st, right after Hallowe'en is just wrong. Even up here. But that's what was happening in our little city.

I asked the physiotherapist if I'd be skating by Christmas. And she said, Sure, I can't see why not.

Woo-hoo!

Angela showed me a nerve stretching exercise and coached me through it but it proved to be very painful so she scrapped it.

Have you ever heard of a nerve stretch?

We've all heard of stretching warmed up muscles. How about stretching a nerve?

Nerve stretches do not really 'stretch' the nerves, but they help to improve the movement of the nerves through the joints and muscles by improving their ability to slide and glide.

Like our muscles, the nerves in our bodies can also become tight. Nerve stretches can reduce the tightness in the nerves and also help relieve pain that is associated with tight nerves.

The sciatic nerve runs from the lower back down the leg to the foot and a person with tight sciatic nerve can experience numbness, pain, or pins and needles anywhere along the length of it.

To stretch the nerve, you lie on your back, facing up, then hug one knee to the chest, by holding the thigh under the knee. Straighten the knee to about 90-100 degrees and lift the lower leg towards the ceiling. Bend your ankle and point up and down for 10 times before returning to the start position.

-- Don't try this at home, kids. Try it with a physiotherapist at your side.

After I did the exercises, I watched a couple of old Kevin Costner movies: For Love Of The Game followed by The Upside Of Anger. So I saw him play his last baseball game and then I saw him as a retired ballplayer.

It was a good double-header.

Edmonton phoned and then we chatted on Facebook for a bit, while the movies were playing. She told me she had a biopsy and they asked her if she wanted to see what they took out of her. Being an intrepid reporter, she said yes. And she said it looked like spaghetti.

-- I may never eat spaghetti again.

She thanked me for the Jackie Lawson e-card I sent her the day she flew off the treadmill, saying it meant more to her than I know. So I told her that was how I felt about her phone calls, especially while I was undergoing this adventure in healing.

After that, I noticed a posting about death and regret on Facebook. Somebody did a survey of nurses who saw a lot of people die and they reported on deathbed statements of regret. Here's the top five:

1. I wish I'd had the courage to live a life true to myself, not the life others expected of me.
2. I wish I didn't work so hard.
3. I wish I'd had the courage to express my feelings.
4. I wish I had stayed in touch with my friends.
5. I wish that I had let myself be happier.

None of those would be on my list. That's not what I'd say on my deathbed. -- How about you?

Day 65: It's Global Warming, Stupid!

On the second day of November, after a late night, watching baseball movies, I woke up a little before six to the sound and lights of snow plows clearing the streets, the sight of Facebook exploding with news about global warming, and with an awareness that my condition had altered.

Cold Lake was under a heavy snow warning, a big American financial magazine called Bloomberg had a satellite picture of the superstorm on the cover with the headline in big bold letters saying IT'S GLOBAL WARMING, STUPID, and my back, legs, and feet were a lot less sore.

Environment Canada issued a snowfall warning for Cold Lake and a freezing rain warning for Edmonton, so a big piece of central and northern Alberta was under the weather, and we had a "snow day" that got people talking about an early ski season. Bloomberg used a baseball analogy, saying, It's difficult to blame any one home run hit by Barry Bonds on steroids, but because of steroid use, he hit a lot more home runs more often. And I was happy I had avoided steroid injections for my pain, which appeared to be fading away.

Yesterday, Environment Canada's map of Alberta showed severe weather warnings stretching from the Rocky Mountains to Saskatchewan with freezing rain across the province and the Cold Lake area in red for a heavy snowfall warning. -- Good thing I didn't drive to Bonnyville!

Superstorm Sandy results were reported this way in Bloomberg: At least 40 deaths in the U.S., economic losses expected to climb as high as $50 billion, eight million homes without power, hundreds of thousands of people evacuated, more than 15,000 flights grounded, with factories, stores, and hospitals shut, and Lower Manhattan dark, silent, and underwater. Eric Pooley, senior vice president of the Environmental Defense Fund said, "Now we have weather on steroids."

My feet did not feel numb anywhere and the area with that pins and needles feeling had shrunk away from my heel so it was just in the front half, and that weird feeling of walking on bean bags was gone. Hallelujah! It was a night of solid sleep, without waking up, and only one weird dream, about Tarzan, slugs, and some guy getting crushed by an alligator but eaten by a plant, so I didn't practice my new clamshell exercise with umbrella breathing, much, and I woke up feeling pretty good, for a change.

-- There is nothing like a morning of information overload!

In other news, the Cold Lake Ice, our Junior B team, was promoting their big game on Friday night with a special offer of free admission for the first fifty kids who showed up at the new arena in the Energy Centre. "Great News for the kids!!" their Facebook post said. "The first 50 kids through the door for Friday's home game versus Wainwright get in for FREE, But they must be accompanied by a parent. Puck drop is at 8pm at the Energy Centre. Make Sure to bring the whole family and take advantage... Go ICE Go!" The Cold Lake Ice took over first place, over-all, in their league with a win over Wainwright last week, so their rivals could climb back into a tie for first or the Ice could leave them behind.

Part of me wanted to go to the game and part of me suddenly saw hockey as a game that could leave you with a back injury meaning you might never skate or walk again.

Saint Shirley, up in Lac La Biche, sent me an e saying, WELL, JUST FOR YOUR INFO I TELL MY GRANDKIDS THAT SNOW IS A FOUR LETTER WORD!! I HATE IT. NEW TIRES THIS YEAR... $600 SLIPPING AND SLIDING ALL OVER .., NEARLY WENT INTO THE DITCH ... AM NOT HAPPY ... BUT GLAD U ARE!!!

YES MY DAUGHTER GOT HERE VERY SAFELY, GRANDSON CAME UP FROM VANCOUVER TOO, SO THE HOUSE IS FULL OF LAUGHTER AND JOY AND FUN AND LOVE ... MY HEART IS FULL

I had told her about physiotherapy, so she said, HOW CAN 2 PEOPLE DOING THE SAME JOB COME UP WITH 2 TOTALLY DIFFERENT TREATMENTS AND DIAGNOSIS? WHAT IF YOU DIDNT KNOW YOU HAD A CHOICE .. THANK GOD WE LIVE IN CANADA!!!

I told her the physio in Cold Lake says I am recovering and all I have to do is re-learn how to breathe, walk, sleep, sit, etc., so she said, I LIKE WHAT SHE HAS TO SAY BECAUSE IT IS SO POSITIVE. IT SOUNDS TO ME, THAT WITH HER ENCOURAGEMENT YOU WILL SOON BE CLOSER TO THE OLD MARTIN....

HOW CAN HAVING COLD FEET BE SUCH A MONUMENTAL RELIEF? she added. WHO WOULD KNOW THAT HOT AND COLD BODY TEMPS WOULD MEAN SO MUCH ... I AM DELIGHTED FOR YOU... WHEN IS THE PROGNOSIS ON RETURNING TO WORK? ANYTIME SOON???? I KNOW YOU CAN HARDLY WAIT, AND I AM SURE YOUR STUDENTS WILL BE THRILLED TO SEE YOU AGAIN.

For a while, there, I was feeling in synch with the weather, from Global Warming to Superstorm Sandy, but now I felt we were going in opposite directions. Cold Lake survived its severe weather warning, no problem, the superstorm was fading away but leaving an enormous amount of damage behind, and it looked like the world was waking up to what I'd been worried about for a long, long time: our polluted planet's wild weather.

Locally, Kinosoo Ridge Snow Resort was predicting their ski season would open in less than a month. "Come celebrate more than 27 years of Snow Resort Operations at Kinosoo Ridge Snow Resort, Cold Lake's premier winter destination," it said on their website. "Located 14 km northeast of Cold Lake on French Bay, with its natural and machine made snow, triple chairlift and T-Bar operations, Kinosoo Ridge is gearing up for the winter season. This winter Kinosoo promises great snow, lots of activities and most of all FUN!!"

One of their events was a bag jump for snowboarders. They could slide down a snow-covered hill, get launched on a rail, fly through the air, and land on a huge airbag, to cushion their fall.

Ah, Cold Lake, Alberta .... We create and celebrate global warming!
-- I love it here!

It was dark, ten below, snowing, and early in the morning, when I woke up, so I went back to bed, did the clamshell exercise with umbrella breathing, fell asleep, and got up at eight thirty, when the sky was the same colour as the snow.

It was overcast but not grey. The clouds looked strange, the way it did when there were forest fires to the west of town, and the smoke made the sky an odd shade of white for a few days. It is almost the colour of snow, so the whole world looks quite white, at the moment.

And I'm thinking it might be a good day for getting snow tires on my car.

Umbrella clamshells are easy to do, unless you've screwed up your back and muscles and nerves and so on. You lie on your side with a pillow between your knees. Your legs are bent so your heels are in line with your buttocks and close to your buttocks, Practice the umbrella breathing technique and as you exhale you lift the upper knee slightly.

Repeat that, co-ordinated with your breathing, five times, rest for a few breaths, and do it again, a couple more times. Then roll over and do the same thing on the other side.

I did it 100 times.

My glutes were sore from overuse so I had to do a glute stretch. You like on your back with your knees bent, place one foot over the opposite knee, grab behind the leg that is on the ground and pull it toward you until you feel a gentle stretch, hold that position and relax. If you push on your knee, you will increase the stretch, but don't do it if it hurts. You want to feel it but not hurt it.

I wasn't supposed to do that for a couple of days.

Facebook told me the Rolling Stones did a show in London and were heading for Newark, New Jersey, and that was the start of a tour to celebrate their 50th anniversary, so Rolling Stone magazine did a poll to select their top ten songs and came up with this list: Gimme Shelter, Sympathy For The Devil, Pain It Black, Satisfaction, Wild Horses, Jumpin' Jack Flash, Can't You Hear Me Knocking, You Can't Always Get What You Want, Tumbling Dice, and Angie, with Gimme Shelter in the top spot.

Personally, I'd put Satisfaction at number one.

The Rolling Stones were never my favourite band. When I was a kid and they were hot, I liked Bob Dylan and The Beatles a lot more. Ahead of the Rolling Stones, I liked The Dave Clark Five, Herman's Hermits, and a whole lot of other bands.

Whoever thought the Rolling Stones would outlive so many performers from the Sixties and go on tour to celebrate their 50th anniversary in 2012? Keith Richards was still alive and kicking? Unbelievable!

Would they include Edmonton on their tour?

Would I go to see them?

No.

Walk Off The Earth was going to be in Edmonton on the 22nd.

Would I like to see them?

Yes!

There was a crazy reason I didn't like The Stones, back in the day. There was a guy in town, where I lived when I was a kid, who looked like Mick Jagger. He was incredibly popular. My best friend liked him. The girl I liked -- Sherry -- liked him, too. And he lived right across the street from her.

Everybody liked him.

Well, kids liked him. Adults did not like him. Our teachers, wolf cub and boy scout leaders, even hockey coaches, never mind the priest at St. Paul's Roman Catholic Church, in Gravenhurst, looked at the lanky kid who looked like Jagger, saw the long hair over his eyes, but not over his ears, or long at the back, detected a very bad attitude, and thought he was nothing but trouble.

He lived right across the street from the Catholic church. Behind their house full of kids was the Cosby Funeral Home.

Girls liked that bad boy. And guys liked him, too.

Years later, I heard he committed suicide, in his mid-twenties.

But I never heard the back story.

We all thought he was going to have a great life and be as popular as a rock star.

We were dead wrong.

Another guy his age from our little town took his life around the same time. He was a teacher's kid, adopted while the teacher was working in the Arctic. He and his wife came back from the Far North with an Inuit boy a year older than me. We used the word Eskimo, in those days. -- I liked him more than the Mick Jagger lookalike guy.

After high school, he worked his way up to the position of manager of a bar in Waterloo, Ontario, that was very popular with undergrads at the University of Waterloo, so I saw him a few times when I went to visit my sister, while I was still in high school. He was very cool and looked happy while he had that job.

Some day I'll hear his back story, too. I hope.

Why am I thinking about those two guys my age from my old hometown who committed suicide today?

The Rolling Stones make you think about things like that, I guess.

We used to talk about which band was better -- The Beatles or The Rolling Stones. The guys like Rod, who looked like Jagger, and were a year older than me, liked The Stones. I liked Bob Dylan and The Beatles.

The Beatles music was poppy, bright, cheerful, melodic, and all about harmony.The Stones music was darker, raw, cynical, and flirted with nihilism and misogyny. The Beatles music was generally idealistic. The Beatles sold a lot more records.

Dylan was something else.

I always thought Bob Dylan and The Beatles should get together to make a supergroup.

But that's just me.

If The Beatles were on tour this year .... We would all be there.

RIP John Lennon and George Harrison.

Day 65 PM: Canadian

While my car was at Canadian Tire, so they could replace sumer tires with snow tires, I walked next door to Mark's Work Wearhouse and Stables, back to the car, over to Pizza Hut, to find out it wasn't open, yet, and over to Tim Horton's. It was like a long winter hike for a guy recovering from nerve damage, but I enjoyed walking in the cold and feeling very Canadian.

Edmonton called me while I was walking, so we talked while I walked. She was excited because the Twitter account she created for her grandfather, a World War One veteran, long deceased, was getting a lot of followers, fast. She wanted to tweet his war story of valour from November 1st to the 11th, Remembrance Day. It was like Tweeting from the trenches, she said, and it felt odd, speaking for the war dead on Twitter, but it was also cathartic.

She was planning a trip to Texas to see family for American Thanksgiving.

And she asked me about NaNoWriMo, but I said I was keeping my focus on recovery, physio, walking, and getting back to work a.s.a.p.

In the back of my mind, I was harbouring a secret desire to write a Cold Lake novel and put this place on the literary map of Canada. But what would a Cold Lake novel contain? CFB Cold Lake, the City of Cold Lake, the Cold Lake oil patch, the Cold Lake Air Weapons Ranger, Cold Lake First Nations, the Colk Lake Ice, the lake called Cold Lake, and, possibly, the Cold Lake News.

Novels need conflict and this place hasgreat potential for that. But, frankly, I couldn't see it. Cold Lake looked and felt like the epitome of harmony, to me. -- Nobody wants to read about that!

Cold Lake Murder Case: the novel about cold murder cases set in an isolated northern oil patch. Cold cases of cold blooded murder.

Sounds good. -- Too bad I'm not a murder mystery writer and I don't do commercial fiction.

Well, I have written a dozen murder mysteries, but I haven't published any of them. How about Fifty Shades of Grey in Northern Alberta. Hot Nights In Cold Lake. -- I don't think so!

My legs hurt, on that hike, at the back of my thighs, so I had to step a few times, to rest, but after a short break at Timmy Ho's, it was all good again, as they say.

How about an experimental novel set in Cold Lake. That's my genre. New Wave literature. -- That's an expression I haven't heard for a long time.

Tom Wolf was interviewed by Jonothan Goldstein on CBC Radio, apologizing about his new book, which he said was full of reasons why America was still a great nation. He said it was very unpopular to write positive things about America these days, so he was flying in the face of convention, again. Goldstein asked him about his trademark white suits. How many do you own? Isn't it hard to keep them clean? What about when you eat spaghetti or something messy like that. Wolf, the author of The Tangerine Flaked Streamlined Baby and The Electric Kool-Aid Acid Test, laughed and answered amiably, then apologized to all of Canada for saying so many great things about America in his new book.

Canadian Tire was swamped with people who wanted their tires changed, after a couple of weeks of snow on the ground, so they were backed up and hadn't touched my car in over four hours. After re-booking for the next Tuesday, I got my car key back, drove to the drug store, and then went home, exhausted.

At the drug store, I ran into one of my Grade Twelve students and her mom. They said they were worried about me, being off work for so long, but nobody had told them what was wrong. After I filled them in, they shook their heads in disbelief. "We've lived in Cold Lake forever," the mom said, "so we've always had the same doctor, and we never get sick, so we have no idea what it's like out there for other people."

The cashier at Shopper's Drug Mart said, "It's good to see you walking, instead of using canes or the walker. And it looks like you've lost some weight, too. I'm glad things are looking up for you!"

It never ceases to amaze me how much good a few kind words can do.

Friday night, I heard from an old friend, now teaching Media in a high school in Toronto. Let's call her Toronto. She said: OMG! Are you OK? How badly did you get it? Are you home or in the hospital? Do you have someone there to help you out? What were your symptoms? I want to know the whole story. Sending the very best warm wishes for you and the recovery. I'm sure some of your meditation techniques will quicken the recovery. I swear by Bel Ruth Naperstak tapes. Did me wonders before my major surgery and after two years ago. Do you need assistance with walking or anything? Can't imagine you paralyzed with anything. Hope it wasn't that severe.

After I brought her up to date, she said I should find a doctor who specializes in GBS.

I know you have so many people who have been a part of your life and who you have 'healed' who would want to do ANYTHING they could to help you. I really hope you are not doing all of this by yourself. Even the rescuer needs to be rescued sometimes. Nuf said. Don't want to make this morbid, just want to send concern.

When I met her, she was working in the media, mostly for CITY TV, and freelancing as a story producer, reporter, and producer, and she used to phone me at odd hours to go and get her. She had anxiety attacks after working to exhaustion and would not be able to drive home. I'd get up in the middle of the night, take buses, subways, and taxis to get wherever she was working, and then drive her home in her car, so she got home and her car wasn't stranded somewhere.

After I got her home, to her house, sometimes we would watch a movie or talk about the television industry. I suggested to her that she might like working as a teacher, instead, teaching media to high school students, as it was a different kind of stress. And she liked that idea, took it, went back to school for a year, and became a Media teacher in a big Toronto high school.

Somehow, she reminded me of an old movie by Woody Allen called Annie Hall and the Woody character who says, "I feel that life is divided into the horrible and the miserable. That's the two categories. The horrible are like, I don't know, terminal cases, you know, and blind people, crippled. I don't know how they get through life. It's amazing to me. And the miserable is everyone else. So you should be thankful that you're miserable, because that's very lucky, to be miserable."

Edmonton said she agreed with Woody's philosophy but I told her that I didn't because I believe that "the better it gets, the better it gets, the better it gets ....." She said the choice is between horrible and miserable but I said I choose ecstatic.

She said, I hope you're right.

Day 66: Ice, Ice, Baby

This morning, I woke up at six, feeling tired, and checked Facebook, then e-mail.
Facebook told me our hockey team had a big victory:

"The Ice got their first win in the brand spanking new rink the Events Centre!! A 4-1 win over the Wainwright Bisons! Zach Zarowny was the 3rd star; Dusty Hyde, 2nd star; and Jordan Keeping, with 2 goals, was the first star. Ice will look to do it again Saturday night versus the Lloyd Bandits, 8 PM puck drop at the Energy Centre."

My e-mail finally had a note from my sister in Australia, so I hit "Reply" and wrote back, so it looked like a conversation.

Subject: I'm baaaaaaaaaack!!!!!! :-)
welcome back!

G'day Marty!
Good morning! It's Saturday morning at six.

Sorry I haven't been responding to your e-mails bro', but I haven't been getting them, yours or any others either. Merde!

that's what i wuz sayin'!

Will forwarded two e-mails recently and asked me if I got them and I said no. I sent something to my home computer from work and didn't get that either so had a good look at my in box and saw I hadn't got anything for several days. Will went into our Adam account today and discovered that the box that said "Put everything on line" was no longer ticked, for some strange reason, so that seems to be the problem. He ticked the box, sent me a test e-mail and I got it, so he sent everything to my in box -all 83 e-mails! I've been wading through them reading yours and deleting others as I go.

i'd be ticked off! lol

So, hello!!!!!!!!!! Thanks for your e-mails Marty. I hope your are continuing to make progress with your recovery. I just read up on GBS and hope to hell that's not what you have! You said you don't have many of the symptoms so feel that is not what's wrong. I hope you're right!

joseph heller, author of catch 22, wrote a book about it. bad. would have died if not on a respirator. i may have had a mild form of it, with severe nerve damage. i lost my knees, ankles, and feet, for 24 days, and my recovery is taking longer. heller was paralyzed all over. the physiotherapist has seen thousands and thousands of people with nerve damage and said i wasn't like them. he has had a few people with GBS and said i was like them. but his program of physio was killing me, and he wanted me to go for the spinal tap, etc., that would have set me back several more weeks, at least.

It's good you are happier with the female physio. Is she going to help you with all that re-learning you mentioned?

Angela Plaquin reminds me of you, when you wanted to be a physio! she told me i need to get a fake vagina! lol. she got on the massage table with me and said, Want me to move in? She is so funny! She taught me the "blow and go" method and we laughed about that, too.
She has taught me how to breathe, sit, stand, and sleep, and that has been life-changing. walking comes next. instead of working me until my nerve-damaged muscles are fatigued and i fall on the floor, like the last guy, she is sensitive to the pain i feel, and backs off new exercises as soon as i say, That hurts. -- what a relief!

Oh merde, I have to go and shower and get dressed now as Will just reminded me we are going to a bbq and it's almost time to leave! Where did the time go??
Sorry Marty, must dash now but will write again soon. Thinking of you with much love,
Janisistar
xoxoxoxoxoxoxoxxoxoxo

She has this closing salutation on all her e-mails: Don't wait for perfection in your life to be happy. Be happy in spite of life's imperfections.

It was still dark as night out, at 7:30 in the morning, so I was tempted to go back to bed, but I wanted to do an errand or two.
        At the drug store, I got Dr. Ho's version of the TENS machine or the electrical device used in physiotherapy, but the batteries weren't included, so I wanted to get some and try it.
        Both physiotherapists used it on my back, and it felt good. One of them said I shouldn't get one for home use and the other one said I should, but noted that the model they use professionally is a little different, as it uses two types of current at the same time, and those sold in stores or online only use one.
        She also told me how it works, explaining the theory about tricking the pain centre in the brain. She said the hypothalmus, in the human brain, monitors pain, and it contains an image of the human body or your body's nervous system like a little alien with elongated fingers to represent the most sensitive places. Electricity can mask or confuse or overwhelm the feedback the brain gets from the hypothalmus, so you don't experience pain, for a while.

TENS is widely used around the world for a variety of painful conditions. There are an estimated half a million users annually in Canadian hospitals.
TENS is thought to work in two ways:

On a high frequency, by selectively stimulating certain 'non-pain' nerve fibres to send signals to the brain that block other nerve signals carrying pain messages.

On low frequencies, by stimulating the production of endorphins, natural pain-relieving hormones.

The device is usually used for 15 to 20 minutes, several times a day, and is controlled by the user rather than a health professional. Pain relief may be rapid and last for days.

Unlike many pain-relieving drugs, TENS isn't addictive and seems to have few side-effects.

Of course, as Angela said, TENS takes away pain, which is a symptom, and does not fix or cure or heal the problem that is causing the pain. -- But pain relief can be very helpful.

When I woke up again, around ten thirty on Saturday morning, the sun was up, finally, but hidden behind white clouds the colour of snow again. That didn't bother me because that's good weather for recovery. It encourages you to rest in bed but allows you to go outside to get the things you need, like batteries for your pain relief machine, and it's a good day to get a little exercise walking outside and then inside, at home, again.

Somebody might look at our snow-covered landscape, the overcast skies, and think it was horrible, or miserable, but it made me feel ecstatic.

There's a simple little writing exercise designed to get more ecstasy into your life. You keep an ecstasy journal. If you aren't feeling ecstatic, it can be hard, even painful, to get the ecstasy journal going. The instructions are simple: Write about the last time you felt ecstatic.

Some people have to think long and hard and go back years and years before they can recall a moment they can describe as ecstatic. However, once you've found one moment of ecstasy, you can find another and another. As you keep your ecstasy journal, day after day, you notice more moments of ecstasy in your daily life.

Eventually something magical happens, or it feels as though something magical is happening as it appears as though you are attracting more and more ecstasy into your life.

Instead of thinking of yourself as someone who is miserable or that life is horrible, you think of yourself as someone who has experienced ecstasy and then as someone who experiences a little bit of ecstasy every day and then you realize you are an ecstatic person who lives in ecstasy.

It's a huge transformation and an easy exercise can take you there.

This may sound like strange advice, coming from somebody who can barely walk and is off work and who recently used canes and a walker, after nerve damage took away his legs and feet.

Even on those days that I was crawling and rolling around on the floor, I was laughing at myself and felt ecstatic. When I went to the emergency neurology unit in the city and the specialist told me not to laugh because it was serious, I couldn't walk, I found a lot to laugh about and feel ecstatic about.

Today, day 66 of this strange adventure, I feel frustrated that it has lasted so long, and my recovery is going so slowly, but I'm ecstatic that I am recovering.

My reflexologist said my recovery is amazing and the guy who runs the Wellness Centre agreed with her. Edmonton, my friend in the city, calls it miraculous. Those things make me feel ecstatic.

The last time I felt ecstatic was this morning, just minutes ago, day dreaming about going back to work, seeing my students again, and going skating at Christmastime, and getting my old life back in even better shape than I was before. The last time I felt ecstatic is now, as I write about ecstasy.

Speaking of ecstasy .... On Facebook, Wally Lamb posted a link to a hilarious video on the Huffington Post's comedy page for Canada with Chris Rock giving great advice to white voters in America. He says Barak Obama is just white and even Romney is blacker than Obama. -- It made me laugh!

http://www.huffingtonpost.com/2012/11/03/chris-rocks-message-for-white-voters-kimmel-brooklyn-video

And, of course, one good link leads to another, so there's lots of comedy there, and the better it gets, the better it gets. But the Chris Rock video with advice for white voters was pretty hard to beat.

There was a good hockey news story in the Edmonton Sun on Saturday. The NHL season had just been wiped out until the end of November, at least, and the big game between the Toronto Maple Leafs and the Detroit Red Wings, outside, in a football stadium, was canceled, by Oilers fans were having fun following their team's young guns playing in the AHL with the Oklahoma City Barons.

It was a good re-hab story, too.

It took all of 38 seconds for Taylor Hall to score a goal in his American Hockey League debut.

Hall had not played a game since last March, when his NHL season came to an end due to a shoulder injury. The first overall pick in the 2010 NHL Entry draft needed off-season surgery and has spent the past seven months rehabbing.

He had been skating with the Edmonton Oil Kings, waiting for the green light from doctors to play again and had been "out" for seven and a half months.

Jordan Eberle scored a pair of goals, while Teemu Hartikainen added the other in the victory over the Minnesota Wild's top farm club, the Houston Aeros.

Hall started the game on the Barons' top line with Eberle and Magnus Paajarvi while Ryan Nugent-Hopkins got the night off.

Love Tree posted this quote from Louisa May Alcott: "Painful as it may be, a significant emotional event can be the catalyst for choosing a direction that serves us - and those around us - more effectively. Look for the learning".

Or, as Taylor Hall might say, if you get hurt, go to re-hab, and get back in the game.

It was not the first time Hall got hurt in the NHL.

There was a famous Japanese proverb that went like this: "Fall down nine times, get up ten."

Afternoon, Day 66: For Love Of The Game

On Saturday, I felt like flying to Vietnam for some warm weather and Zen Forest pancakes, so I went to the Wok Box, which features food in the style of the Far East.

Mongoloian Beef and Broccoli with spring rolls, for me, please. If I order the pho soup and a noodle dish, and close my eyes, I can pretend I'm in the Zen Forest, again. When I open my eyes, I see a tree-covered hill with fresh white snow a couple of blocks away. -- It looks like it would be a cool place to ski.

In front of me is Walmart, where I just walked for exercise, looking for triple A batteries, and finding a hundred dollars worth of groceries at about half the price as the drug store.

Did I say groceries? This time the list included Hallowe'en candy, on sale, which I will give to my Grade Three kids as rewards for good work, when I go back.

While I was in the frozen food section, of the Grade Three girls surprised me by popping up and saying, "Hi!" And then she introduced me to her mom, who does not speak English.

There's a quarter of a million francophones in Alberta. -- Who knew?!

On the radio, DNTO had a special on Lost causes, featuring an interview with the girlfriend of a Leafs fan and then a guy with Huntington Chorea and then a guy who day-dreamed his life away imagining winning, finding, or making five million dollars real fast. The dreamy millionaire wannabe said he would spend his days at a gym, working out, getting in shape, and dedicating his time to other healthy things, like nutrition and health care.

And I said to myself, Hey, man, that's my life right now, for the next few weeks.

My fortune cooke gave me this message: You will soon receive help from an unexpected source.

The sweet garlic sauce made the broccoli, beef, and rice taste fantastic.

Energized by that delicious experience, I hatched a fun idea: What if I went to the Cold Lake Ice Game, even for a few minutes?!

Sitting through a whole game would be hard on my system in recovery. But how about seeing some of the game? -- It would be exciting just to step inside the new hockey rink with a good junior game going on, to see other guys skating fast, to hear the roar of the crowd, for the love of the game.

From the Wok Box, I drove over to the Energy Centre, which is a big, beautiful, rec. centre with a college and a high school attached to field houses for indoor sports and a very new hockey rink. It's a great place but I wish it wasn't such a long walk from the parking lot to the front entrance.

From the front entrance, 200 steps from the handicapped parking spot, to the hockey rink entrance, it's another one million or so steps. Just in case my leg muscles gave out on me, I took my canes with me, and used them once I got half way to the arena.

It is a work of art. It looks just like my favourite old arena, in Bracebridge, Ontario, except everything is new, there are red seats instead of blue benches, the seating rows are sloped less steeply, the lights are brighter, and it's not as cold for the fans in the stands. It made me feel ecstatic just to walk through the doors.

But then I saw there was no entrance to the seats except up a flight of stairs.

So I tried the stairs. There was just eight low stairs and I managed them alright. If I could get a seat by those stairs, I could go to a hockey game and have a good time.

Unless I got jostled by a crowd, like Forrester in that movie about the reclusive writer, Finding Forrester, based loosely on the life of J.D. Salinger. -- What a nightmare.

Was I ready to go to a game like that? Facebook said, The ICE take on their rivals from Lloyd tonight as the Bandits make their way to town for a battle. The ICE look to make it 4 straight wins and their 2nd straight on home ice tonight.

By the time I walked out of there and took my groceries in from the car, my glutes hurt so much I was ready to drop.

Or use my new Dr. Ho's Pain Therapy machine!

After inserting the batteries, plugging in the wires, and attaching the electrodes to the sore part of my glutes, I cranked the machine up high and was surprised to feel it as strongly as the machine used by the physiotherapists. -- It felt great.

Of course, I had to try it on my sore feet!

After twenty minutes of the pain relief machine, and an hour-long rest, my glutes felt good. Next I tried the machine on my feet. My expectations were high as I had been anticipating the electric massage would be good for weeks.

However, I could not feel the machine on my feet at all.

They were not numb, as they were before, but I guess they were not yet sensitive enough to feel the electric current buzzing my nerves.

The muscle my physiotherapist wanted me to focus on was one I had never heard of. The piriformis is a flat muscle, pyramidal in shape, situated partly within the pelvis against its posterior wall, and partly at the back of the hip-joint.

The muscle passes out of the pelvis through the greater sciatic foramen. The piriformis is very important in the gluteal region. The piriformis muscle is part of the lateral rotators of the hip.

The piriformis laterally rotates the extended thigh and abducts the flexed thigh. Abduction of the flexed thigh is important in the action of walking because it shifts the body weight to the opposite side of the foot being lifted, which keeps us from falling.

When the piriformis is weakened, the glutes take over, or try to compensate, so they get sore and you walk like a duck.

That's what was going on with me, anyway.

To strengthen the piriformis, you do the clamshell exercise, co-ordinated with your umbrella breathing technique.

It's easy to do this exercise incorrectly. It is not a power move, like something you would do in weight-lifting. It is a small move that should be smooth. -- If your thigh or leg moves in a jerky manner, you are doing it wrong.

It was hard for me to feel this muscle, or to feel I was in touch with it, or communicating with that part of my body, as it was a small muscle I had never heard of and could not picture, and it hadn't been working for a while.

I Googled 'piriformis' and got the picture.

After all that fun, I did not feel like going out again, to see the Cold Lake Ice at the new hockey arena in the Energy Centre, so I looked for a movie to watch on Netflix, but then .... The internet went out. Or off. Or wherever it goes. So, I got my computer to read this book to me, from the beginning, and tried to sleep.

Day 1000

Okay, it only feels like Day 1000. What is it? Day 66? Or Day 666?

It's Day 67.

Ever have one of those days when there is nothing on Facebook that you like? Every have one of those nights when the internet goes out or off or whatever happens to the internet when you can't get online? And when you get hooked up in the morning, there's nothing there that you like?

Do you ever get up in the middle of the night and test your toes, only to discover they are numb, in places, and then you freak out because you're afraid you're having a relapse and will lose your feet and legs, again?

Ever have that dream about eating marshmallows and then, when you wake up, your pillow is missing?

Ever lose the ability to walk and then doctors can't tell you what's going on but a physiotherapist tells you he thinks it's GBS?

-- Man! I hate it when that happens!

This morning we have freezing fog, outside, but the sun is supposed to come out this afternoon and the temperature should go up to two above.

If you know the difference between ice fog and freezing fog, you could be a Canadian.

-- My back is sore and my feet are bugging me and I'd like to just go back to bed and start over, fresh, but I've already logged to much horizontal time and things would be the same, anyway, when I got up again.

Who needs chocolate?!

Where are those Hallowe'en candies?

-- Good thing I left them locked in the trunk of my car.

Dr. Ho's pain relief box has three modes. The first one, they say, gives the sensation of thumb and palm kneading, finger tapping, soothing pincement and gentle chopping.

I don't know what pincement means.

It says it is a good mixture of deep and soothing massage and is highly preferred by most users.

The second mode gives the sensations of pounding, chopping, oscilating, and shaking techniques and may be described as a very deep penetrating massage.

Actually, it says "May be describe as a very deep penetrating massage."

The third mode has a sensation of hands on muscle squeezing, lifting, and letting to, it says.

Modes 1, 2, and 3 are described as medium, strong, and gentle.

It's on my lower back, right now, and I am wishing it was stronger. -- A lot stronger. -- One hundred times stronger.

Now for the good news: Last night the Cold Lake Ice defeated the Lloyd Bandits on home ice by a score of 5-4. The Ice are on a 4 game wining streak and sit first overall in league standings with a record of 8 wins 1 loss. Next Home game is Saturday November 10th versus Vegreville at The Energy Centre.

While I experimented with the electric massage and pain relief, I watched Captain America get zapped with the Vita-Ray and get transformed from a little guy to a super-hero. And then I wished Dr. Ho's little machine was a million times stronger!

But there was some good news: Today I could feel the electricity massaging my feet.

Alabama returned, after a week away, and started several games on Scrabble.

So. That was the weekend. Maybe I overdid it, walking around the Canadian Tire area on Friday and the Energy Centre on Saturday, especially going up and down those stairs. On Sunday, my leg muscles felt sore. You would think I had just finished running the New York City Marathon. That winter fog, outside, never cleared up, all day, and the Cold Lake sun did not come out. The high point of the day was experimenting with the little electric massage machine.

Day 68: Jagr Day

The famous #68, Jaromir Jagr, former NHL superstar, was quoted in the Edmonton Journal, saying, "Everybody knows that the first month of the season is nothing like a rose garden for the NHL. Only a few make money. They face huge competition in American football and baseball. The baseball competition is getting into its playoffs, so one competitor is gone, and the interest will switch to hockey. It doesn't bother the that they're going to lose the first month and a half."

Jaromír Jágr was an owner, now, but still playing, in the Czech Extraliga. He should be an Edmonton Oiler. He's the owner of Hockey Club Kladno, also known as Rytíři Kladno, or the Kladno Knights, in the city of Kladno, in the Central Bohemiana Region, in the Czech Republic. Kladno is part of the Prague metropolitan area.

He wears #68, famously, because 1968 was the year of the Prague Spring, the year that Dubcek (the Party Secretary of the Communist government in Czechoslovakia) decided to open the borders economically and culturally to the West. The Soviets saw this as defiance to the Warsaw Pact and invaded the nation overnight, posting tanks in Prague. They arrested Dubcek and took him out of power and put in their own puppet regime.

That year is especially important to the Czech Republic as it was the time that serious resistance to the Soviet machine was started.

Jagr chose this number to show his pride in his nation and a reminder of what the Prague Spring meant to the world.

His grandfather died in prison that year, after participating in the Prague Spring.

Jagr is one of my all-time favourite hockey players. It was rumoured he was going to join the Edmonton Oilers a few years ago and I'd still like to see it happen. He would fit in well with the young guns, be a great mentor for them, and he can still play the game better than 90% of the guys in NHL.

Prague is one of my favourite cities, not only for its place in history, and for Jagr, but because it was Kafka's city. Franz Kafka was my favourite writer for decades and remains one of my all-time favourite writers, right up there with Hermann Hesse, Margaret Atwood, Alice Munro, Vonnegut, Paolo Coehlo ....

As for this, ahem, hockey player .... Everything is changing. After walking in the snow on Friday, climbing stairs at the new hockey rink in the Energy Centre on Saturday, resting a lot on Sunday, and using Dr. Ho's little electric machine on my back, legs, and feet, this Monday morning, my feet feel quite different. Numbness is giving way to pain.

And that's a good thing, I believe.

My back feels okay, my legs are less stiff and sore, my feet are less numb and tingly and feel tight and sore. Maybe that means the numbness, pins and needles, the feeling of walking on beanbags, is on the way out. -- I believe it is!

Excite? You bet! In fact, I'd do my happy dance, but my feet are sore and my balance is still quite bad. -- Maybe I'll do a version of the happy dance anyway, holding on to the kitchen counter. -- Woohoo!

How I would love to go skating!

A Facebook friend posted a Pooh cartoon this morning. It's a famous one as it's very Eckhart Tolle. What day is it? asked Pooh. It's today, squeaked Piglet. My favourite day! Pooh said.

Somebody else posted a poster I'd never seen before, which said, You can't start a new chapter in your life unless you stop re-reading the last one.

Although I get the metaphor, about living in the past, I have to disagree with that saying, on the literal level. Personally, I love living in the past and the future, despite spending several years in the Zen Forest. When writing a book, especially this one, I re-read the last chapter and the previous chapters again and again and again. Well, I get my computer to read them to me.

And I have a friend who has a problem with her memory, after her temperature went sky high, and she almost died, so she journals her way through the day, and re-reads her journal constantly, as she cannot rely on her memory.

And she appears to be fearless in regard to moving on with her life.

In fact, she just got back from scuba diving in the British Virgin Islands.

It's 9:45 and it looks as though the overcast skies are turning into yesterday's news. There's some blue sky out there and the sun is shining through some leftover clouds. -- And you know what they say: The better it gets, the better it gets!

After physiotherapy, what will I do? -- Something celebratory ....

Maybe I will go back to the Energy Centre and climb some more stairs! Cue the Rocky music! I'm going the distance! Play The Eye Of The Tiger! Gonna Fly Now!

Do you know the words to "Gonna Fly Now"?

Trying hard now
it's so hard now
trying hard now

Getting strong now
won't be long now
getting strong now

Gonna fly now
flying high now

gonna fly, fly, fly...

In the USA today is the last full day of campaigning, and predictions on the presidential race are rolling in — most of them favouring President Obama.

The Globe And Mail says Obama would be best for Canada. Even John Ibbitson, who wrote that book called The Promised Land, praising Harris, the Ontario premier who tried to privatize education, was saying Obama was the one Canadians understood.

-- Wait a minute .... Was he praising Obama, now, or was he just calling Canadians 'stupid'?

The Edmonton Sun says Romney is the one for Alberta. They say Obama had his chance and he failed. All the Sun papers across Canada backed Romney.

The Edmonton Journal wasn't as biased. They reported that, at President Barack Obama's ancestral village in Kenya, witch doctor John Dimo tossed some shells, bones and other items to determine who will win Tuesday's election.

After throwing the objects like so many dice outside his hut in Kogelo village, Dimo, who says he is 105 years old, points to a white shell and declares: "Obama is very far ahead and is definitely going to win."

Yesterday I watched a documentary on Netflix called Media Malpractice which looked at the coverage of the last U.S. presidential election and claimed the American media made Obama president. The media loved the Obama story, the media embraced Obama early, and the media does not like to be proved wrong. It gave a lot of credit to Oprah and gave a lot of time to Sarah Pallin, complaining about media bias.

In the final days of the election, Obama was getting support from Bill Clinton and Bruce Springsteen, not to mention the Frankenstorm, Hurricane Sandy, the superstorm.

He also enlisted an army of A-list performers and public figures -- from Stevie Wonder and Lady Gaga to Billie Jean King, from Jay-Z to Crosby, Stills and Nash -- to promote his re-election. The Obama campaign provided a who's-who of 181 actors, musicians, authors, athletes, mayors, Congress members, and more that fit any and all demographic groups.

How could he lose?

Romney, the first Mormon on the presidential ticket, the Republican supported by The Tea Party, was the underdog, and we love underdogs, so some pundits said the race was very close and he could overcome the odds.

I've been following American elections since 1960, with Kennedy Versus Nixon, which was called a "cliffhanger".

In the next Canadian election, it looks like we will have Harper Versus Trudeau.

And it makes you wonder: What will North America look like with Obama and Trudeau?

What would North America look like with Harper and Romney?
As 2012 draws to a close ....
Wait a minute! -- It's only November 5th!

Jagr Afternoon

Love Tree's Facebook post today: Talking about our problems is our greatest addiction. Break the habit. Talk about your joys. ~Rita Schiano
The Attitude of Gratitude asks, What are you grateful for today?

Where do I begin?
My Scrabble score went up to 1638.
Physiotherapy was brilliant.
The sun is shining.
My fortune cookie at the Wok Box said, You will soon receive pleasant news of a personal nature.
That came after pysiotherapy, a trip to the gas station, and new shoes.
Appara did not have winter Crocs, so I got something similair and better: lined, slip-on, winter shoes.
At physiotherapy, Angela was in top form. She told me she had big plans for me as soon as she saw me.
Uh-oh, I said. I mean .... Good!
She asked for news, so I showed her my balance, slightly improved, and told her my foot felt normal from heel to arch, and the balls and toes of my feet kept changing, but felt less numb and tingly, overall.
Good, good, and good, she said.
But then I confessed I overdid it on Friday, walking in the snow, and Saturday, climbing stairs while checking out the new arena.
Isn't it beautiful?! she said.
My glutes hurt, I said.
I'm okay with that, she said. Let's check your clamshells and I'll show you a new nerve stretch after I get you to balance on the parallel bars.
We improved my clamshell so it was an even smaller and much smoother movement, coordinated with my breathing.

Between the parallel bars, I stood on a square board on rockers and slowly, very slowly, went down on one side and then the other, without touching the bars on both sides of me. And then I did it backwards and forwards, one foot in front of the other. Just thirty seconds for each exercise.
The nerve stretch was actually for my sausage casing, the sheath around m nerves. Angela said my sheath was constricted.
On your back, grab the back of one leg, lift it, straighten the knee, lift the toes and lower them. Move the foot like that, toes up and down, thirty seconds each leg, three times.
Your legs will be sore after this! she predicted.

Are you coming back tomorrow? she asked.
I could, I said.
Book it, she answered.

Edmonton called while I was in the waiting area. My students' grandmother came in as I was going out. "Baby steps," she said. "That was a good movie," I said.

So, on Tuesday I had physio at 2:15, it was wing night around 3:00, and Canadian Tire for winter tires at 4:30. Wednesday was reflexology at 10:30. Thursday was physio at 11:00. -- That was my life, for the next few days. -- With exercises from physiotherapy and some walking. -- And I loved it.

Angela got me to look in a full-length mirror as I practiced surfing exercises.
    My proprioceptivity is off. I don't know where my body is in space. -- That sounds odd, but when she says, lie on your side and bend your knees so your hips are stacked, I can't do it without looking down at my legs and hips.
    "I feel as though I don't have very good communication with that part of my body," I said.
    "Of course," she said. "You don't. -- But it's improving."
    That guy in the mirror in front of me looked very familiar but like no picture I had ever seen.
    He was taller, not as stocky, with thinner legs, a flatter belly, and a longer neck.
    Maybe all that decompression made me taller! I said to myself.

At the Wok Box, I had Ginger Beef: sweet ginger sauce, battered beef strips, fresh peppers, and onions.
    It's not the Zen Forest diet, that's for sure, with beef and onions, but the rice and peppers were on the Zen Forest menu, and ginger was recommended very highly.

The temperature shot up to 10 above while Muskoka dropped to 5 below.
    Go, Cold Lake!

Martyrs' Shrine posted: On Behalf of the Shrine Staff we would like to thank all pilgrims who visited the Shrine for the privilege of serving you this season. May the Lord, through the intercession of the Holy Martyrs' and St. Kateri, bless and guide you, give you health in body, mind and spirit, answer all your prayers and fill you with God's peace. We look forward to seeing you, your families and friends again next season.

On Netflix, while I did my exercises in front of open window doors, with the sun streaming in, I watched a movie called The Age Of Stupid, made up mostly of news footage from our era, with a future archivist trying to figure out why we used all our fossil fuels and destroyed our planet when we knew better. He blamed it on consumerism fuelled by capitalism and the American lifestyle fueled by oil.

It was a 2009 movie, made before the tar sands of Alberta changed the game. The movie was made when the concept of peak oil was in vogue, or the idea that we had used up half the oil on the planet, and oil was running out fast. Of course, the oil sands of Cold Lake and Fort MacMurray changed all that. Technology was developed to make it possible to take oil from the tar sands.

The movie called The Age Of Stupid argued that we should use the planet's fossil fuels to fund a transition to a lifestyle that used alternative sources of energy, while pointing out that if the whole world consumed at the speed of Americans we would need four to six planets to supply the demand. And, of course, we have just the one planet.

There's no place like Earth.

After going to bed early, after doing my exercises, I got up, after a couple of hours, and had a strange experience: I walked to the kitchen without any pain in my back, without walking like a duck, and with little stiffness in my legs or pain in my feet.

And not only that .... I stepped on something and it hurt a lot. I really felt it!

You know that strip that holds down linoleum under a doorway? I've stepped on that thousands of times without feeling a thing. This time it felt like stepping on a tack.

I experimented by stepping on a few other things.

My foot has gone from numb to hypersensitive!

How will I be able to sleep, after this?

It reminds me of that old story about the princess and the pea.

Woo-hoo!

While I was up, I checked Facebook, of course, and this time I found a link to a YouTube video that has gone viral, shot at a game in the Canadian Lingerie Football League, when a brawl broke out among the fans in the stands during a game between the Regina Rage and the Saskatoon Sirens. And there, in the middle of the little video, was a big guy with the number 15 on his jersey and the name Pilon, and right behind him, also wearing #15, with the name Momma Bear printed on it, was my reflexologist, Mecell Pilon!

It had half a million hits!

She told me there was a lot of tension in the stands but said they stopped a fight. But the video shows there was a brawl with punches tossed and clothing torn.

And those lingerie league football players looked like they did not have enough equipment -- or clothing -- on! Somebody could get hurt out there, as well as in the stands.

Don't people realize how fragile they are? How vulnerable? Do they think they are immortal? Made of kevlar? You don't want to hurt your back and lose your legs!

Those two late night experiences gave me a good idea, combined with other things in this book. I've been thinking about a title for this book. The Lulu Titlescorer just said The title

Fall Down Nine Times, Get Up Ten has a 44.2% chance of being a bestselling title! -- That's a very high score on this title checker!

Edmonton liked a different title: Zen And The Art Of The Lingerie Football League.

My working title was Cold Lake Zen Healing. -- I liked that title, even though it sounded rather flat. It sounded like a sumo wrestler walking down a short flight of stairs after losing a fight.

Day 69: Yinyangs

It was 2 below, going up to 2 above, but the sun was shining, so it looked warmer. On Huffington Post, CP had an article featuring an interview with an Environment Canada Researcher saying warming permafrost, reduction in summer sea ice extent, increased mass loss from glaciers, and thinning and break-up of the remaining Canadian ice shelves, is increasing evidence of an accelerating response to global warming in places on Earth so cold that water is solid — either ice or snow.

Apparently the climate models projected that we'd see these changes, but they were projected further into the future.

In other words, the Arctic was warming and melting.

Cold Lake had a winter like southern Ontario, last year, but winter came early this year and shows no sign of letting go.

Huffington Post published pictures of famous Canadians with short reports on who they wanted to win the American election, and why, with everybody aligned with David Suzuki, in support of Obama, except Conrad Black, who said he liked Romney.

There was also a big picture and a message from the Queen on Facebook called, appropriately enough, A MESSAGE FROM THE QUEEN To the citizens of the United States of America from Her Sovereign Majesty Queen Elizabeth II.

It said, In light of your failure in recent years to nominate competent candidates for President of the USA and thus to govern yourselves, we hereby give notice of the revocation of your independence, effective immediately. (You should look up 'revocation' in the Oxford English Dictionary.)

Her Sovereign Majesty Queen Elizabeth II will resume monarchical duties over all states, commonwealths, and territories (except North Dakota, which she does not fancy).

Your new Prime Minister, David Cameron, will appoint a Governor for America without the need for further elections.

Congress and the Senate will be disbanded. A questionnaire may be circulated next year to determine whether any of you noticed.

To aid in the transition to a British Crown dependency, the following rules are introduced with immediate effect.

1. The letter 'U' will be reinstated in words such as 'colour,' 'favour,' 'labour' and 'neighbour.' Likewise, you will learn to spell 'doughnut' without skipping half the letters, and the suffix '-ize' will be replaced by the suffix '-ise.' Generally, you will be expected to raise your vocabulary to acceptable levels. (look up 'vocabulary').

2. Using the same twenty-seven words interspersed with filler noises such as "like' and 'you know' is an unacceptable and inefficient form of communication. There is no such thing as U.S. English. We will let Microsoft know on your behalf. The Microsoft spell-

checker will be adjusted to take into account the reinstated letter 'u" and the elimination of '-ize.'

There were a dozen more rules, including:

4. You will learn to resolve personal issues without using guns, lawyers, or therapists. The fact that you need so many lawyers and therapists shows that you're not quite ready to be independent. Guns should only be used for shooting grouse. If you can't sort things out without suing someone or speaking to a therapist, then you're not ready to shoot grouse.

5. Therefore, you will no longer be allowed to own or carry anything more dangerous than a vegetable peeler. A permit will be required if you wish to carry a vegetable peeler in public.

7. The former USA will adopt UK prices on petrol (which you have been calling gasoline) of roughly $10/US gallon. Get used to it.

8. You will learn to make real chips. Those things you call French fries are not real chips, and those things you insist on calling potato chips are properly called crisps. Real chips are thick cut, fried in animal fat, and dressed not with catsup but with vinegar.

9. The cold, tasteless stuff you insist on calling beer is not actually beer at all. Henceforth, only proper British Bitter will be referred to as beer, and European brews of known and accepted provenance will be referred to as Lager.

10. Hollywood will be required occasionally to cast English actors as good guys. Hollywood will also be required to cast English actors to play English characters. Watching Andie Macdowell attempt English dialect in Four Weddings and a Funeral was an experience akin to having one's ears removed with a cheese grater.

11. You will cease playing American football. There is only one kind of proper football. You call it soccer. Those of you brave enough will, in time, be allowed to play rugby (which has some similarities to American football, but does not involve stopping for a rest every twenty seconds or wearing full kevlar body armour like a bunch of nancies).

13.. You must tell us who killed JFK. It's been driving us mad.

15. Daily Tea Time begins promptly at 4 p.m. with proper cups, with saucers, and never mugs, with high quality biscuits (cookies) and cakes; plus strawberries (with cream) when in season.

God Save the Queen!

It came with a PS: Only share this with friends who have a good sense of humour (NOT humor)!

It turned out to be a post by an American columnist named Michael Yon.

Wally Lamb's Facebook posting said, May the better man win today--by which I mean the candidate who stands for inclusiveness, advocacy for the underclass, women's health and reproductive rights, regulation of corporate greed, economic sanctions before bombs, and the responsibility of the wealthiest Americans to pay their fair share.

That picture of the Queen reminded me of the new hockey rink in Cold Lake that looks like my favourite old arena in Bracebridge, Ontario, and what it was missing: a huge picture of the the Queen.

    The new hockey palace in Cold Lake also reminded me of another great arena in Ontario, the home of the Jr. A Peterborough Petes, which has a few more seats but otherwise looks very similair.

    Both Bracebridge and Peterborough have arenas in which the dressing rooms for players and skaters are a few steps below the ice surface, so you have to walk up a few stairs with your skates on, and the ice surface in the Peterborough arena is unlike any other I've seen anywhere as it is above street level so everybody has to walk up a few stairs to get to ice level. It's very theatrical and gives both the ice and the game an elevated status like a stage or a church. It is a hockey cathedral.

    So is the new rink in Cold Lake.

    The Cold Lake Energy Centre arena was a community project. They got a good deal of funding from Cenovus.

The old arena in Cold Lake, in Cold Lake North, looks like the old tin barn of a building in my hometown, right behind our house, where I spent so much of my time as a kid. They built a new arena in the Seventies and renovated it last year, so now it looks fancy and new.

    Gravenhurst is the home of the South Muskoka Shield and Bracebridge is the home of the Phantoms, in the Greater Metro Junior A Hockey League, not affiliated with the Canadian Junior Hockey League or a member by Hockey Canada. -- Some people call it "the bastard league".

    The Cold Lake Ice, Jr. B, could probably take the Phantoms and the Shield. I'd bet they would win easily.

    Gravenhurst used to be the home of the Sr. A Indians and Bracebridge was the home of the Jr. C Bears. -- Those were the days!

The Cold Lake Energy Centre is the home of the Jr. B Cold Lake Ice, back to back champions of their league, now in first place, and for them -- and us -- THESE are the days!

Stalling?
What do you mean 'stalling'?
Do you mean to suggest I am not reporting on my ability to walk, balance, or feel with my feet or toes, this morning? Are you suggesting that I am afraid to try walking, et cetera, because I am afraid last night's success was a dream?
Well .... You might be right.

After doing my morning exercises -- umbrella breath, clamshells, nerve stretches, back bends -- I felt good.

This just in: If Canadians were electing the next U.S. president, Barack Obama was the clear favourite, by about 78% to 12, according to a Forum Research Inc. Poll, reported in the National Post. But even though Canadians love Obama, there are many who say Romney would actually be better for Canada.

In the morning, I felt pretty good, but after doing my exercises in the a.m. and the p.m., my glutes hurt.
Angela said she liked the way I walked, when she saw me in physiotherapy. She tested my balance and the tightness of the muscles in my left leg, then gave me an adjustment by pulling my leg back as I tried to push it forward. She got me on the surfboard for a while, in the parallel bars, and showed me a leg stretch that helped the pain in the butt.
After that, it was time for acupuncture, with needles in The Four Gates, for pain relief (hands and feet) plus the "third eye" and then needles added at my heels to help with my hamstrings.
Four Gates usually tackles pain as well as feelings of frustration—basically anything that suggests things aren't flowing as smoothly as they should be.

Angela said I'd likely feel good for a while, get revisited by the leg pain, maybe tomorrow, and then it would go away.

Right after physio, I went home, typed up my notes, and immediately felt my eyes grow heavy as though I was being hypnotized or falling asleep. But I couldn't hit the sack -- I had to take my car to Canadian Tire for snow tires.

Angela's prediction came true. Waiting around for about four hours for snow tires was painful.

It hurt to walk, when I got to Canadian Tire, so I had to take a shopping cart to lean on, to make my way over to Pizza Hut, and back, for dinner.

It was painful sitting in the hard chair in the waiting area at Canadian Tire, watching U.S. election results on my iPhone.

When I finally got home, I posted a prediction: Obama, President, and Trudeau, Prime Minister.

Not long after that, CBC Radio announced Obama was President. BBC followed, and then the American networks all said the same thing.

The Democrats would control the Senate and the Republicans would have a majority in the House.

So, the most expensive election in history and ... nothing changed!

Amy Tan, the American novelist, posted this note on Facebook: Clink! Clink! That is the sound of our champagne glasses celebrating the re-election of President Barack Obama!!! We are thrilled, relieved, and grateful.

Ianthe Brautigan, Richard Brautigan's daughter "liked" it.

Toronto said, Phew!! Now I can go to sleep and not have nightmares. Oh no, I can't. I still have 2 tests to create for tomorrow. And marking to do....

Edmonton said, That's great. -- She's a former Texan.

Alabama said, That's horrible. -- She's a Republican.

The Edmonton Sun said, Obama's Challenges Just Beginning.

The Edmonton Journal said, simply, U.S. President Barack Obama won a second term Tuesday.

The Globe And Mail said, Obama wins second term: Obama vows 'the best is yet to come'. It featured stories called America Keeps Faith With Obama and Romney's Campaign One Of Blunders And Missteps.

The Globe also had a story called @BarackObama gets 'Four More Years' and a Twitter record. It said that just moments after the U.S. television networks projected an Obama win, the President's Twitter account posted a photo of Michelle Obama embracing her husband, with a simple message: "Four more years".

Within an hour that tweet was being discussed as the most popular tweet of all time, garnering more than 300,000 retweets.

Social media literally heaved under the weight of millions of people taking to Twitter, Facebook, YouTube and countless other platforms to share their view.

TIME magazine did not appear to be impressed. Their big election story went like this: A hurricane couldn't stop it. Two billion dollars couldn't buy it. A weak economy couldn't swing it. Americans re-elected Barack Obama on Tuesday, affirming the goals of the

President's tumultuous first term and giving him a second. This wasn't 2008. Not as many states went his way. Fewer of his supporters wept.

Bobbie Ann Mason, the American novelist, posted an E-mail from her friend in Paris. She was in the French Resistance in WWII. "Félicitations!" she said. "yes, HE can. et il a gagné. Affection. Michèle".

After watching election coverage, all sorts of responses, after winter tires and pizza after acupuncture, I could not sleep at all. So, I watched movies until about five in the morning. Mr. Baseball, Coach, and A Thousand Words, again.

Mr. Baseball had Tom Sellick as a baseball star, sent to Japan, falling in love, and making a comeback. It was thin but fun and it got me going. The soccer movie called Coach was even thinner but it got me even more. After all that, I finally fell asleep during A Thousand Words.

Does acupuncture energize you and make you sentimental? It never had that effect on me before!

-- Maybe it was the election drama.

Day 70: Reflexology For Acupuncture

They say acupuncture does a lot of things for you, and it also releases toxins. And reflexology moves those toxins out of your body. -- That was my plan for the day.

At nine in the morning, I woke up feeling odd. The dream I was in felt real, so I was a little disoriented for a few minutes. My feet felt very funny. And I was hungry as a big league baseball player in Japan.

It took me half an hour to give my groggy head a shake and then I realized that pain in the butt was gone, that ache in my glutes had finally vanished.

That alone was a cause for celebration.

Last night, after a painful time at Canadian Tire, I came home feeling "down" despite the exciting election news. And I felt bodily tired but mentally awake, while watching those movies. I was quite emotional. And this morning I am in a much better frame of mind.

But can I walk?

My feet feel sore. You would think I was in those soccer games in the movie last night. But my back felt fine, for a change. My butt muscles were no longer aching. My thighs felt normal. My shins felt as though they had stopped some soccer balls. My feet felt .... sprained.

After my morning shower, the pain was back again.

In the shower, I was thinking, I feel as though I've been drugged. -- I don't know what drug would make your body feel good for an hour, then make you feel sleepy, and then bring all your pain back again for a few hours, and make it go away again, but keep you awake all night, and after a few hours of sleep in the morning, feel good, but more aware of your feet, until after a shower, when you started to feel the way you did before you took the drug the day before.

Reflexology was two hours of heaven. Mecell the magic reflexologist was running a little late but I didn't care because I was lost in time, sitting in a comfy chair with good lumbar support, focusing on my breathing exercise, meditating while staring at big, soft, snowflakes drifting down like feathers.

When the alternative practitioner came out, I asked her if she knew she was a star on YouTube and was surprised when she said, What are you talking about?"

So I showed her the short video on my iPhone and told her it was posted everywhere and went viral so it was approaching one million viewers, and that was incredible.

She laughed and shrugged her shoulders. She looked aghast when she saw the brawl on video, as it looked worse than in real life, she said, from her perspective.

In her comfy recliner, I almost fell asleep. First she told me all about the state of my health, from looking at my feet, and asked me what happened about four days ago. That was the day I climbed the stairs in the hockey rink. She said, You've been out of alignment and in pain ever since then, haven't you?

She gave me an exercise like the Crocodile Twist, which I found on the internet, for putting hips back into alignment, but warned it could hurt a lot.

As I zoned out, in the big chair, she told me stories about her life as the wife of a guy who is part of the Snow Birds team.

It was the first time she mentioned the Snow Birds.

Until that time, she just said her husband was in the military.

Apparently that is what she always says and that is not the way some or most of the wives attached to the men in the Snow Birds team present themselves.

But she likes to meet people and have fun, she says, and she finds she gets along with more people in a much better manner, that way.

She always makes me laugh.

She also told me she has a new theory about what happened to me as she is getting more and more men as clients with the same ailment. She thinks it might be viral, an airborne virus, in this area.

"Hope it's not something in the air from Fort MacMurray," I said.

She told me she intended to do some research and find out if there was a virus or something in the air that hit men but not women and hit them hard.

She had one guy with the symptoms who spent six hours in emergency at the hospital in Cold Lake as they thought he was having a stroke. She had other guys of all ages and occupations and son on with the same symptoms: sore back, legs and feet gone, numbness, and a lot of pain.

I was the only one who did not get the pain.

She took away the pain I was feeling in my feet, shins, thighs, glutes, as always, and I floated out of her office.

Appointments are supposed to be one hour, but she kept me in there for over two hours, and then sent me home to sleep, after a snack, or lunch.

On the way home, I stopped at the No Frills grocery store for milk and a few other things, then went straight home.

There were snowfall advisories for Edmonton and Northern Alberta and that meant it might be a good time to hibernate in Cold Lake.

Winter weather hit the Edmonton area in full force on Wednesday, prompting a snowfall warning and causing traffic delays, road closures and hundreds of collisions. -- That storm did not reach Cold Lake.

Us. President Barack Obama's acceptance speech in Chicago, in the evening, was incredible.

Did he say global warming?!

Oh, my! -- He did.

Well, he said "the destructive power of a warming planet".

And he also said this: "America, I believe we can build on the progress we've made and continue to fight for new jobs and new opportunity and new security for the middle class. I believe we can keep the promise of our founders, the idea that if you're willing to work hard, it doesn't matter who you are or where you come from or what you look like or where you live. It doesn't matter whether you're black or white or Hispanic or Asian or Native American or young or old or rich or poor, able, disabled, gay or straight, you can make it here in America ...."

Day 71: The Fun Has Just Begun

Fall down seventy times, get up seventy one.
    Okay, it's not that bad. It just feels that way.

In my e-mail, late at night, there was a nice note from my old friend Karen Hood-Caddy. She was a Life Coach and novelist who did a lot for the MNM in its early days. She was also Findhorn's rep in Canada.
    Findhorn was the alternative community in Scotland, famous for many things, including talking to plants and getting things to grow up north in sandy soil beside the North Sea.

Oh my goodness, Martin!!" she said. "What a LOT to be going through. I just wish I'd known earlier and I could have prayed for you or at least held your health in my morning meditations. I will do that now.

    Sounds brutal!!

    I hope all your spiritual work has helped and that somehow this will be a doorway for you to emerge in a different place for yourself. But yikes, how scary!!   Oh, Mars! I feel for you!!!! I care for you!! Please get better. And please stay in touch. I am a resource for you in my heart and in my soul!"

The morning after, my feet felt different. Again. It felt as though I stepped in glue and it has hardened on the bottom of both feet, from arch to toes, and up the side, a bit. That annoying tingling feeling, or pins and needles, is gone! -- What a relief. That ache or soreness has faded, too.
    Yesterday I asked the alternative practitioner who practices reflexology to please tell me it was not unusual to have the feelings in the foot change frequently, that I'm not completely crazy, that it's normal to feel a lot of changes in your feet when the nerves are regenerating, or coming back to life.
    How often? she asked.
    It's constant, I said. Every hour is quite different.
    That happens, she said. There's a lot going on down there. As your nerves come back to life, and you develop different muscles in your legs and feet, you're going to be aware of a lot of changes.
    What kind of exercises does your physiotherapist have you doing? she wanted to know.
    So I told her about nerve stretches, surfing on the balance board between the parallel bars, and back extensions.

Of course, I did not have to tell her that my legs feel different to the touch because she was aware of that. They felt quite different to me. -- Before the injury to my back and nerves, my legs had muscles you could see and feel. My legs were hard and squared and carved. And after the injury, the angles were all gone. My legs were thinner, soft, and rounded.

Now they are firm and squared but not carved.

Walking, this morning, went well, with less pain in the glutes and next to nothing in the back. -- And I love that.

Frankly, I'm fed up with focusing so much on my feet, my legs, my back, myself. -- How I long to be other-centred again.

My brother sometimes gives me a hard time for being other-centred. He says, You should be selfish and self-centred like me and everybody else.

And he is not even kidding.

Yesterday I phoned him, to get our sister's mailing address in Australia, as it wasn't coming up on the internet. He told me he just a tee-shirt made with a star over the heart, our last name under the star, and an Aussie flag on the shoulder. Her nickname is "Star", apparently. I call my sister JanisiStar.

He didn't ask me what I was sending her but said it would be good if she got things from both of us at the same time.

Online I found a gift delivery service in Australia that delivered hampers to the door in her city. Apparently 'hampers' are popular in the land down under. A hamper is a basket of goodies, usually with wine and gourmet snacks. -- I sent wine and cheese with crackers and a cutting board and cheese knife with a note that said, Congratulations! Felicitations! Bon voyage et bon chance!

My sister lived in Montreal for a decade or so, teaching deaf children, and now she lives in a place where the weather is the opposite of Montreal's and she had a second career as a relationship counsellor. She is turning sixty, soon, and is leaving the place where she has worked, at Relationships Australia, to do something else. First she wants to focus on her health and then she might launch a private practice, she says.

After I sent her a hamper, I went online to find out if I could do something like that in Canada, for my cousins in Lac La Biche and my friend in Edmonton.

It was a lot easier to get something delivered in Australia.

My brother told me his back was sore, again, and he had a bad couple of days, so I told him about the "Blow And Go" exercises, that sounded too simple, but was oddly effective. He tried it while we were still on the phone and was amazed to discover it worked.

Last night, after chatting online with Edmonton, I felt energized, so I decided to go out and do something. It was eight o'clock and Walmart was still open so I took a poster I had made awhile ago and went to the superstore to buy a big, black frame for it. I had been thinking about doing that for months, had shopped around a bit, and had wanted to get it, but it was so big it was awkward to carry and would have required a separate trip, I thought, so I waited until I could handle it.

It was a short drive, a long walk, and I got it, brought it home, put the poster in the frame, and thought it looked good.

Then I played Scrabble for a while, online, and fell asleep with a movie on. -- A sports movie with Owen Wilson, Jack Nicholson, and Reese Witherspoon.

-- I love Jack Nicholson and Owen Wilson.

It was called How Do You Know. Witherspoon was a female jock who just got cut from the American baseball team and Wilson was unbelievable as a major league baseball player with an incredible fastball and salary. But he was funny.

There was a love triangle and family business fiasco and I fell asleep before it was over.

Edmonton said, I thought you saw that before.

Oh yeah, I said. It was funny but forgettable.

Yesterday, the principal of my school phoned to ask me to drop off my keys, so the supply teacher could use them, as they were going to have a lock-down. -- As well as fire drills, schools now practice lock-downs, too, in preparation for emergency situations, such as a shooter in the school.

When I lived in Pickering, with a view of the nuclear power plant, on the shore of Lake Ontario, we practiced fire drills and lock-downs and also had an emergency plan in place in case of a nuclear meltdown. -- The plan was to get all the students on buses and take them to schools up in the hills, on the northern boundary of our board of education.

It reminded me of the old "duck and cover" plan for nuclear bombs practiced in America in the Fifties. -- They thought that hiding under a school desk would protect kids from the radiation of an atomic bomb blast. -- We thought a half hour drive into the forest hills of Durham Region would save our students from radiation if something happened to the nuclear power plant.

The morning after Obama was re-elected, the Dalai Lama was in the news for his message of congratulations. He said, "As you know, it is over a year since I handed over all my political authority to the elected Tibetan leadership, but as just one among the six million Tibetans I want to thank you for your steady encouragement of our efforts to find a peaceful resolution to the problems in Tibet. I am very appreciative of your support for our Middle Way Approach, which I continue to believe is the best way for us to ensure a solution that is beneficial for both Tibetans and Chinese. Given the recently deteriorating situation in Tibet, of which the tragic series of self-immolations is a stark symptom, I hope your Administration will be able to take further steps to encourage a mutually acceptable solution."

For the first time, the Americans had elected a Buddhist to their senate. Mazie Keiko Hirono was the first female Senator from Hawaii, the first Asian-American woman elected to the Senate, the first Senator born in Japan, and the nation's first Buddhist Senator. She called herself a non-practicing Buddhist.

She was a lawyer and a politician.

What is a non-practicing Buddhist?

In my e-mail, there was a nice note from the publisher who discovered Robert Munsch. First, Rachel Sa, the Toronto Sun columnist, posting on my Novel Marathoners page on Facebook, asked if anybody had Anne Millyard's e-mail address. So I sent her an e, with the address, and Rachel sent Anne a thank you note, as her novel was getting launched, and Anne cc'd me on the note she sent in response.

Here's how it all went down:

Hello Anne:

Rachel Sa here. It's been a long stretch since we were in touch, but I wanted to let you know that my first novel for young readers, The Lewton Experiment, is now available from Tradewind Books. I'm happy to tell you that this book is the final incarnation of The Big Box, the manuscript on which you gave me so much helpful advice after I wrote a first draft at The Muskoka Novel Marathon back in 2004.

The Forester has done an article on the launch, here http://www.cottagecountrynow.

Thank you again for your earlier enthusiasm for the story. It's what helped to spur me on to keep at it!

Cheers, Rachel

Anne said, Isn't it absolutely fabulously wonderful that another of the early MNM novels has been published?

I loved that manscript. She had put her whole heart into the true story, and took such risks, and - as we see - put soooo much additional work into the project (as did Cheryl for hers!), she truly deserves this success.

Christmas is coming up. There are a couple of middle readers in my vicinity who will love the book! And needless to say, the Huntsville LIBRARY will need to stock a few copies.

I hope the MNM keeps going forever.

Cheers, Anne

Notes like that made me feel happy I started the Muskoka Novel Marathon. Also, when Rachel was feeling "stuck", I suggested she go to grad school, recommending the International Writers Program at Iowa. She got an MFA at UBC and thanked me for urging her to go.

I forwarded Anne's note, with Rachel's little letter, to Mel Malton, who got the MNM going with me.

Novice Buddhist monks are required to help other people anonamously. It's called 'going in the back door'. So why am I mentioning it? First of all, I'm not a novice, or a monk, or a Buddhist, or even a non-practicing Buddhist. -- I just live like one. And my brother advises: Be selfish! Be more self-centred! Be like me and everyone else!

In the morning, I went to work, to drop off keys and the big, framed, poster with pictures and info about our drama groups winning at the provincial level, left a gift certificate for a shave for our phys ed teacher, who is a great guy, and would likely be shocked to get a present from me, out of the blue, like that.

Marc Hamel, our phys ed. guy, gets a great number of teams to represent our small school, and they have unbelievable success against the other schools in the area, which are all bigger. He has four kids in elementary school, so he is going 24/7, working with young people. So, I thought he could use a break today, as we say, although I knew he would have a hard time fitting a relaxing shave into his schedule.

Physiotherapy went twice as long as usual, close to a full hour, with acupuncture followed by exercises and walking around. -- It was great.

Angela got me to lie down on the massage table, backside up, so she could stick needles in my lower back, butt, and ankles. She was hesitant to use so many needles but I encouraged her to, Go for it.

This is pretty aggressive, she said.

Good, I told her.

She did not hesitate to pull my sweat pants down, so she could prepare the spot to be needled, but then she caught herself, laughed hard, and apologized. She kind of collapsed on top of me, she was laughing so hard. I told not to worry, I just had a shower and fresh underwear, and I had total faith in her as a professional.

She asked me if I could feel the buzz of chi, as she inserted the needles. First, you feel a little pin-prick, and then you feel an odd sensation that gives you the impression you have electrical circuits running through your body and the wires had just been touched or tapped. It is not a vague feeling, it is quite definite.

After lying in the dark like that for ten minutes, she showed me one of the needles. There are short needles and long needles used in acupuncture and she showed me a long needle. It was about three inches long and very thin.

"Okay, get up slowly," she said. "Try standing and let me know how you feel."

"Good!" I said.

She got me to try a back bend and I found I went about twice as far as before.

She sent me to the parallel bars to work with the balance board and later showed me a very gentle hamstring stretch, lifting one leg at a time onto a platform, only a few inches high, and sticking out the butt until you feel the stretch in the muscles at the back of your thighs.

After some time given to walking it off and stretching, she asked me to walk for her, and I took a few hesitant steps. She coached me to try a longer stride, place each foot heel first and then putting the toe down. -- Basic walking instructions.

When I discovered that it didn't hurt, I walked faster than my usual frightened gait, with long strides, so I felt like I was flying around the room.

She gave me a cane but I carried it, instead of using it.

Okay, she said. Look at you go! -- Get outta here!

She said, Book three appointments for next week.

She warned me that I might feel some pain in my leg muscles later on.

Right after physio, I drove over to Sobey's, the grocery store, and I practiced walking like a normal person as I picked up some food for the weekend.

By the time I carried my bags to my car, my legs were starting to get sore, so I headed home.

CBC Edmonton on the car radio said the capital city of Alberta had a huge dump of snow with fifty car crashes the day before and another twenty today, and there were snow warnings for Calgary and most of the southern end of the province. Cold Lake was getting a few soft, dry, floating snowflakes.

The big news story was the announcement of the approval of an oil sands upgrader for the area just north of Fort Saskatchewan, which is just north of Edmonton. It would turn tar sand into diesel oil and other projects.

The up-grader, or refinery, would cost 5 billion dollars and provide 3,000 construction jobs and then a lot of "good jobs" for the next fifty to one hundred years. And they hoped that this up-grader was just Part One, with two more planned.

They interviewed an engineer who said it was an all-Canadian project with engineers taking it from a blank piece of paper and an unsharpened pencil through construction to operation.

He sounded happy and excited and pointed out that the building of the Canadian Pacific Railway -- an enormous engineering project at the time of the Confederation of our country -- cost about one billion dollars, in today's figures. And the oil up-grader would cost more than five times that amount.

As I drove home, listening to the news and weather, I wondered if it would be wise for me to contact my old high school, back in Ontario, and see if we could set something up, so some of the guys and girls graduating or going to secondary school in Ontario could have a contact in the oil patch out west where the new jobs were being generated.

When I got home, I got a message on Facebook from a former student in my old hometown: hey Mr. Avery! I just wanted to let you know that I am in my first semester of college. I'm working towards my diploma in social service work, then the degree, then masters of social work. It will be a long process, but I am happy to report that after mid-terms I have an average of about 90-95%! I have worked very hard for this, and I wanted to share that with a teacher who knew me as a very angry and confused teenager. I am a little late in starting my career, but I am confident that I can make a difference to people. anyways, just wanted to share my success with you! -Kari

Love Tree went crazy and posted half a dozen things on Facebook, instead of the usual one per day:
1. Don't want to spoil the ending for you, but everything is going to be okay.
2. When you raise your vibrations, you begin to see sacredness everywhere.
3. We're all just walking each other home.
4. True strength is standing in the discomfort of the moment and moving forward with love.
5. God has no religion. -- Gandhi
6. Lighthouses don't go running all over an island looking for boats to save; they just stand there shining. — Anne Lamott

After several hours of running around like that, after lunch, my eyes grew heavy, again, as though I was with a hypnotist saying, You are growing sleepy ...."
So I hit the sack and wondered if I would ever wake up.

But I didn't sleep. Instead of snoozing away the afternoon and resting my tired leg muscles, I watched an amazing movie called Soul Surfer.
It's getting to the point that I've seen so many movies that it seems difficult to find anything I want to watch. Netflix has a lot of movies, but I may have exhausted their list, for my kind of movies.
While I was thinking about sleeping or watching a movie, I reflected on the day so far. -- Did I get a cool reception at school this morning? I wondered. One of the Grade 12 girls saw me and left an exam she was writing beside the principal's office to come out and give me a hug.
That was the second time she did that. -- The last time, I told her, I think this is against the Education Act, or something. But she said, To heck with that!
Actually, she has a funny habit of swearing a lot, in English and in French, so she said something like, Screw that! And she gave me a big hug with the principal looking on.
You've got to love a rebel!
Remembering that moment made me think about the last school year and how busy it was and I realized I did not see a single movie from 2011 or 2012, so I Googled 'best movies 2011' and got a list of a dozen movies that all looked good.

The one I picked was called Soul Surfer -- the movie based on the true story of a 13 year old girl in Hawaii who got attacked by a shark while surfing. The story made the news in Canada as well as a lot of other places and I remembered reading about it at the time. Bethany Hamilton lost her arm in a shark attack.

A 'soul surfer' in surfer talk is someone who surfs for pleasure, or fun, not for competition or sponsors or championships. In this case, Bethany was also a 'soul surfer' because she was religious, a Christian, and followed her faith, doing mission work in Thailand after it was hit be a huge tsunami.

Also, she was in tune with the energy of the ocean.

The movie did not do well with critics and Rotten Tomatoes only gave it a 45%, but I loved it. Critics complained it was a good story lost under so much Hollywood cheese, but it really moved me.

Helen Hunt plays the part of the injured girl's mom, beautifully, and Bethany Hamilton did the stunt double work for the actress who played her in the movie.

The script was based on her autobiography, called Soul Surfer: A True Story of Faith, Family, and Fighting to Get Back on the Board.

Now, maybe I'm overly emotional these days, due to my injury and recovery, plus acupuncture this morning, and you have to know that I only cry a few times a year, at big moments, but I sobbed while watching this movie. I talked to the computer monitor, saying, No! No! No! during several key scenes.

What more can I say?! This movie is replacing The Karate Kid at the top of my list of inspirational sports comeback movies.

The Facebook page for the movie has a quote provided by Bethany Hamilton from the Bible: "Whatever you do, work at it with all your heart, as working for the Lord" - Colossians 3:23

Too Christian for you? Too cheesy for you? -- Too bad for you!

Bethany went on to become a professional surfer.

The movie shows some of those moments when she was down and wondering why this happened to her and how it could possibly be part of God's plan. Later she realized that her story, reported in the media, inspired huge numbers of people, especially kids, to keep trying. Her book and then the movie added to the numbers, so the little surfer girl who survived the shark attack went on to inspire millions of people.

-- Including me!

What other movies did I miss in 2011, besides all of them?

At the top of the list were The Tree Of Life and Hugo.

Day 72: The Names Of God

There are 72 names for God, according to the Kabbalah, the number one book of Jewish mysticism, right up there with the Zohar, and they say that just looking at those names will give you strength. Posters are sold and displayed and you can find pictures online.

For each of the names of God, I have had one day on this adventure in healing. If my injury happened to somebody in the time of Lazarus, he might have met a great healer who brought him back from the dead. If it happened to an American named Joseph Heller, they would have kept him in hospital for half a year and then given him a bill for $120,000.00. In any era, a mild form might go undetected. In my case, it means busy doctors send you for more testing and for physiotherapy, and if you are lucky you will find a physiotherapist who gives you the right exercises plus acupuncture. If you are lucky, you will also find an alternative practitioner who does reflexology. It will be up to you to decide which doctor to listen to, which physiotherapist is good for you, and what kind of alternative healing works for you. And it will be up to you to get to your appointments with all those people, to do the exercises and take the drugs, to make sure you get the rest required, and get good nutrition, and it is up to you to stay motivated.

Americans are more interested in self-help books, motivational speakers, and the powers of positive intentions than most Canadians tend to be. Canadians are more stoic, we follow strong and silent role models, would like to be left alone to recover or get better in peace.

Not me!

Hollywood sports comeback stories do it for me, along with slogans and sayings designed to inspire a positive attitude, plus alternative healing.

Even so, after a long time with little progress, you might be tempted to take all 72 of the Lord's name in vain.

-- But not me!

Last night I saw an American football movie, made in Hollywood, that I had hoped would be inspirational, called Any Given Sunday, but it turned out to be dark and disturbing. Soul Surfer lifted me up and Any Given Sunday brought me down.

Edmonton called last night and this morning, which cheered me up.

Sleeping in, this morning, until nine, also helped.

Vivid dreams, last night, also helped, I believe.

In one dream, I was running. I was me, in the dream, and I was recovering from an injury that took away my ability to walk, but I was walking again, and one day I started running. First, I tried jogging, a little, but before too long I was running the way I used to, and it was exhilarating.

In another dream, I was having trouble climbing stairs.

In another dream, I decided it was time to call my mother on the telephone, and that idea made me happy, so I wondered why I hadn't called her sooner, and stayed in touch by calling her frequently. But then I remembered that she passed away decades ago, half a lifetime ago. So I phoned her in my dream so I could talk to her in heaven. But she was out, dancing, having a good time. And that made me feel happy, too.

All in a dream, all in a dream ....

Also, I had a dream about a new novel I could write, based on a recent news story, from the Edmonton Sun and other newspapers, about a bully whose peaceful brother stomped on his head until he was dead. After a lifetime of being a bully, getting away with murder, even, the bully's little brother "lost it", his mother said in court, and stomped him to death, after a blow-out. The little brother was wearing nothing but socks on his feet.

That would make the bloody climax to the ultimate book on bullying. Show the bully's life story, with his brother and his mother, from school days to adult life, and then the meditative peacenik turns into a cold blooded murderer.

Friday morning, it was 13 below and snowing lightly in Cold Lake but there was a snowfall warning for a lot of areas in Alberta. Edmonton already had over 30 cm of snow, which was more than the city usually got in all of November, and it wasn't even Remembrance Day, yet.

The RCMP said there were over 200 collisions as cars headed downtown on roads that had ice and snow. Their advice: travel only if you have to.

I didn't have to.

I had three days ahead of me with no appointments with doctors or specialists or for physiotherapy or reflexology or anything else.

If I was working as a teacher and I got three days off, it would be time to celebrate. I'd be traveling to Edmonton or Jasper, maybe even Banff, for a few days of fun, with a trip to a big bookstore, steak at a good restaurant, some time in the great outdoors, getting inspiration from the magnificent mountain range called The Rockies.

If I was in Ontario, I'd be excited about getting three days to write, and would knock off a novel.

And what do I have planned for these three days, this year? Rest, physiotherapy exercises, vitamins, nutrition, and a movie or two, plus online Scrabble, Facebook, and e-mail.

Boring, you say?

Well, maybe it sounds boring to an outsider. But if you believe you are fighting for your life, or to get your life back, it is the opposite of boring. It is a dramatic life and death struggle to overcome the odds and recover for a comeback story worthy of a Hollywood movie.

Or, since this is Canada, it is a peaceful, orderly, weekend of stoic recovery punctuated by sleep as it snows outside and the temperature is below ten below zero.

Love Tree posted a quote from Albert Einstein: "Each of us is here for a brief sojourn; for what purpose he knows not, though he sometimes thinks he senses it. But without deeper reflection one knows from daily life that one exists for other people."

One of my former students, now a teacher, posted a picture of a smiling President Barak Obama on a cellphone, with this cutline: So you want to go to Canada because I got re-elected and you hate socialized healthcare, marriage equality, and the separation of church and state? Well, I've got Canada on the line for you, and they can't stop laughing, either."

Somebody posted a quote from Mitt Romney, saying, "I'm not in this race to slow the rise of the oceans or to heal the planet." It was pasted over a picture of New York City after the superstorm, with cars half-submerged on flooded streets.

This morning, I don't feel too bad, thank God.

Even so, I think I'll give it a rest, at least until the afternoon.

The movie of the day? Skyfall, the newest James Bond. And the theme is resurrection.

Day 73: Heal Your Life

After a full day of rest, with very little walking, I felt better than the day before.

The new James Bond movie didn't thrill me, so I watched Hugo, instead. -- What a charming movie!

After going to bed early, I got up around three in the morning, with a memory of an adventure in healing about ten years ago. In those days, I had a vague pain in my legs that left me feeling like I didn't want to hike or dance or do anything like that, and that wasn't good because I loved hiking and I was coaching soccer. So, I let someone talk me into going to a Quantum Touch workshop.

QT, as it's called, is incredibly simple, but it worked. The instructor showed us that many people have hips that are out of alignment. It is easy to see and to demonstrate. Just have a look at the way people stand and look at their hips. -- They should be parallel to the ground or the horizon.

The QT instructor found someone whose hips were obviously out of alignment, put his hands on that person's hips for twenty minutes, and showed us the results. That person's hips had shifted back into alignment and they reported feeling better. Some back pain went away and they felt happy, the person reported.

At the same time I explored QT and got certified as a QT healer, which led me to Reiki, and a lot of other things, I discovered Louise L. Hay, the author of You Can Heal Your Life. She is an American publishing and self-help phenomenon whose lectures attract huge numbers and whose website offers help to many more. Hay has more than 50 million books sold worldwide.

Louise Hay established Hay House Publishing Firm and it is now the primary publisher of books and audio books by Deepak Chopra and Doreen Virtue, as well as many books by Wayne Dyer.

Hay believes that in order to heal the thoughts and beliefs that are the true source of the pain, you must face your pain head-on.

Pain comes in so many different forms, she says, and people try to hide from it, hoping that it will go away or that they can cover it up with medication. But ignoring your body just makes it try harder to get your attention—your body is asking for your help.

One way to deal with it is to change your perception of the situation; simply do not give in to it! For example, instead of focusing on the fact that your wrist hurts, try referring to your wrist as having a lot of sensation. This can help you get through the unpleasant experience and allow you to focus on healing your mind and soul. Healing the pain will then follow.

Right now, I am having a lot of sensation in my feet, a little in my lower back, and in my glutes, or butt.

Hay says, What if you looked at every symptom as a metaphor, or as a message that your subconscious was sending you using your body as the messenger? Rather than shooting the messenger full of drugs to suppress the symptoms, another option is to investigate the symptom from an emotional perspective. The body and subconscious tend to be rather literal in terms of communication, so sometimes the simplest approaches and questions yield remarkable results.

Asking simple questions that stretch beyond the usual medical explanations can be quite revealing, such as "If there was an emotional component to this physical illness, what would it be, or what was happening in my life when the symptoms first started?" For cervical spine problems, ask "Who is giving me a pain in the neck? For lumbar spine problems, ask "Who is stabbing me in the back? For calf pain, ask "Why am I afraid of moving forward in my life?"

Sometimes the answers that come back seem too simple to be true, but it is best to reserve judgment until you explore whether the metaphor might actually fit your situation. Once you have a metaphor you can work with, you can experiment with creating a new metaphor that might address the message that your body is sending you.

Then, you can make appropriate adjustments in your attitudes and beliefs, so your subconscious an translate those changes into action at the body level.

The emotions are meant to flow continuously through the body like the qi flowing through themeridians, and an "e-motion" can be considered to be "energy in motion." Symptoms occur when the flow of energy is blocked.

A helpful guide in this process is Louise Hay's famous little blue book, Heal Your Body. It is an encyclopedic list of health problems, probable emotional causes and new thought patterns presented as affirmations. Despite what appear to be oversimplistic generalizations, it often provides useful insights into the underlying causes of symptoms leading to healing.

What metaphor worked for me, in regard to my back?

Years ago, I had a friend who often said, "I've got your back." And it turned out that the opposite was true. She did things behind my back. -- It took a lot of work to get over that, but after a few years I forgave her.

Hay says, for lumbar spine problems, ask "Who is stabbing me in the back?" And for calf pain, ask "Why am I afraid of moving forward in my life?"

Maybe I should ask, "Who is getting on my nerves?!"

While I contemplated that, just before I went back to bed, a little after three thirty, I checked Facebook and got the latest news about the Cold Lake Ice. Their latest post said, The Ice make it 5 straight wins and remain undefeated on the road this season with a 2-0 win over the Vermilion Tigers! Brandyn Garth got the shut out with his first start of the season and Dallas Ansell Scored the game winner.

Their Facebook page also had some information on last year's season, when they met the Killam Wheat Kings in the finals, and won.

One of my students told me the Wheat Kings used to have a different name. Like the teams I played for when I was a kid in Gravenhurst, the Killam team was called the Indians. -- The Killam Indians.

For a decade or so, whenever we played a game on the road, there was somebody in the stands yelling, Kill the Indians!

Our jerseys had the famous logo of the Chicago Blackhawks on the chest.

We lived close to a Mohawk reservation called the Gibson Indian Reserve, now called Wahta First Nations: Mohawk Territory. A few families moved off the reserve and into town and back again, and back and forth, and high school kids from the reserve were bused into town, so we got to know a lot of Mohawks. And we hated it when people in other towns yelled, Kill the Indians!

While I was looking online for the current name of teams from Gravenhurst, I discovered a piece of news. The principal of the high school was retiring at the end of January and the former VP was returning to take over as principal at the start of February. -- She was a very popular person in the role of VP, when I was there as Arts Department Head, and they would be happy to have her back, I was sure, as the P. In fact, I thought I could hear the cheering from three thousand miles away!

Just before I shut down my computer, I checked e-mail and noticed a Jackie Lawson card from my sister in Australia, suggesting a Skype date on Sunday.

So, I sent her a reply that said, Hi Janisistar! Thanks. Sunday Skype sounds super!

And I added this little note: Want to hear a funny? I was contemplating illness as metaphor, thinking about who was getting on my nerves, who stabbed me in the back, who was a pain in the butt. What was the last straw, that broke the camel's back?

That made me think about hip alignment, back pain, walking problems, at 3 in the morning, and remembered it all started when I was working at GHS when we got a new principal. I really liked the old principal! That was when I discovered Quantum Touch, which led to Reiki, the Diploma In Spiritual Healing, the Zen Forest .... And I just got the news that that principal is retiring at the end of January, and the former VP, who we loved, will return as the P at GHS. -- Quel coincidence!

After that start to the day, I went back to bed and slept in until 9 or so, reliving my experience at that school in my dreams and daydreams. It was like that old movie called The Good, The Bad, And The Ugly. And when I got up again, I found these words of wisdom posted by Love Tree: "The process of fighting something only feeds and strengthens what we are fighting."

When I left that school, after six years as Arts Department Head, teaching English, Drama, Media, and Writing, coaching swimming, junior and senior boys soccer, and drama, working with the Outers and OSAID clubs, not to mention the yearbook, serving as the teachers' rep on school council, handling the publicity portfolio for the school and writing a weekly column for the district newspaper, etc., I left the school with a new trophy as an award for Drama that is still in place.

Our little school had declining enrolment in an era when small schools were being closed, so I led a group to get the school an Arts designation so it would be the Arts destination school of the region, and we added Dance to the curriculum, as well as more Drama and Writing.

When the Board of Education wanted our small school to join with the big school in the closest town and a community college in a big, new, building proposed for a location at the north end of town, backing on Lake Muskoka, not far from the airport, I felt highly conflicted. My friend from childhood, the publisher of the newspaper, led the SOS movement to save our school, with support from another friend from childhood who was on the school council for over a decade and was like the mayor of the town as he had made millions in the oil business and was dedicated to public service for the town. On the other hand, the school building was quite old, the facilities needed to be updated, and the size of the student body made timetabling -- or offering all the courses we wanted -- a nightmare. And my school would be joining the school I went to as a kid, and loved.

It was a dynamic, emotional, conflicted time for the school and for the town.   I joined a multi-faith group that brought together several of the local Protestant churches and the Catholic church for weekly services with a lot of music, singing, and prayers, and the leaders told me they believed the town needed that because it was depressed economically, psychologically, and spiritually.

It's funny: That town is in a physically beautiful location just one hundred miles north of Toronto and it is famous for being cottage country for the biggest city in Canada and one of the top tourist destinations in the nation, so it is a busy place in the two summer months, but the reality for people who live there all year round is quite different.

It was hard, leaving that place, so I left behind a Drama award with a distinctive trophy, the Arts students gave me a dramatic farewell assembly, and I was given the podium to make a final statement, but used it only to do a reading of a poem by Hermann Hesse, from Magister Ludi, called "Stages":

.., The Cosmic Spirit seeks not to restrain us
But lifts us stage by stage to wider spaces.
If we accept a home of our own making,
Familiar habit makes for indolence.
We must prepare for parting and leave-taking
Or else remain the slave of permanence.
Even the hour of our death may send
Us speeding on to fresh and newer spaces,
And life may summon us to newer races.

So be it, heart: bid farewell without end.

On Facebook, my old friend Richard Thomas, a former reporter now working in public relations for the school board in Owen Sound, posted the front page of the Owen Sound Sun Times with a picture of his son, Finn, who was starring in a high school drama production.

There was a big picture on page one.

Beside it, there was an article headed Teachers Sanctions Up In The Air.

Inside there was a story on drugs in high schools and the OSCVI principal was quoted, saying there were no more drugs in his school than the others.

Richard Thomas was instrumental in getting the novel marathon I started in Owen Sound going. His marathon novels were published by The Ginger Press. When they published an omnibus edition with three of his murder mysteries in one book, he dedicated it to his writing instructor and marathon buddy.

When I moved up to O.S. from Toronto, I moved my writing workshop from York University to Georgian College and then to the Ginger Press Bookstore, where I met Richard and many other people who became great friends.

In sports, a goalie from Gravenhurst, playing for the Jr. A Owen Sound Attack, was the OHL's player of the month and had his picture on the front of a cereal box. Jordan Binnington backstopped the Owen Sound Attack with a record of 6-1-0-1 in October posting a league-low goals-against-average of 1.96 and league-high save percentage of .940. He made 30 or more saves in six of his eight starts and was twice recognized as first star of the game when he made a season-high 42 saves in a 4-2 win over the Plymouth Whalers, and when he stopped 34 shots in a 3-1 win over the Sudbury Wolves.

Binnington was playing in his fourth OHL season with the Attack. Selected in the third round of the 2011 NHL Entry Draft by the St. Louis Blues, he leads the OHL with a 1.96 goals-against-average and a .940 save percentage in 12 games where he holds a 9-1-0-2 record. He competed for Team OHL in the 2012 Super Series.

The picture of him on the Alpha-Bits cereal box was a great action shot. There was a story in the Sun Times about it, saying the cereal had sold right out, locally, because their goalie was on the cover.

The O.S. Attack also had a guy from Bracebridge. A big defenceman named Keevin Cutting was in his fifth year with the Owen Sound Jr. A team and the Toronto Maple Leafs had expressed some interest in him. He was moving up in the draft and could go in the third round, the scouts said.

It was a rare moment when my old hometown and my adopted hometown had something in common. It was less than two hundred kilometres and under a three hour drive from Gravenhurst to Owen Sound, and a beautiful route from Muskoka past the Blue Mountains and Georgian Bay to the city of 20,000 on the Bruce Peninsula, but not many made that trip.

This is Tom Thomson country, painted by the Group of Seven. He was called the 8th member of the Group of Seven.

Georgian Bay is called the 6th member of the five Great Lakes.

I love this region and can't believe my current students have never heard of these places.

The news from my old hometown made me feel a little homesick, so I planned to go to the new hockey arena and attack those stairs again.

If my legs, feet, hips, whatever, could handle it, I'd be going to the Cold Lake Ice game at night.

In local sports, The Lloydminister Composite Barons had a disappointing finish to a promising season, losing in the Wheatland semifinals to the Cold Lake Royals, 40-14.

Despite finishing first in their division and securing home field for their first game of the playoffs, the Barons were far inferior to the Cold Lake team, who got it done in all facets of the game. Cold Lake broke open a tie game by scoring 33 consecutive points against Lloyd Comp in Wheatland Football League semi final play at Armstrong Field. Day 73, Part 2: Go, Cold Lake Ice, Go!

Although I set out with some doubt and trepidation, things went well on my second attempt at the arena in the Energy Centre that is the home of the Cold Lake Ice.

The eight stairs leading up to the seats? No problem. -- I took them one step at a time.

How about the 24 stairs leading up to the top row of stairs?

No problem, sargeant major.

How about walking around the arena? There is a wide walkway going all around the top row of seat in the rink.

No problem!

What about walking down 32 stairs, back to ground level?

No problem!

While I was playing Rocky, climbing stairs, trying to get in shape, I was watching a little hockey. On the perfect ice below, there was a practice going on for a team of twelve kids who looked like they were in Grade Four or about nine years old. The twelve kids had six coaches on the ice with them, or one coach and five trainers, and they were all doing an awesome job.

The girls and boys, in full equipment, did one skating drill after another, with lots of encouragement from the six men on the ice with them. They did power skating drills and old-fashioned drills and drills I'd never seen before.

One part of the practice really hit home. The coach got all the kids gathered in one corner and told them to watch closely as he demonstrated a small but important skating technique. He showed them how to balance on their skates: not on your heels, as though would put you off balance, and you'd go down, fast and hard, if you got hit, and not too far forward, on the rocker blade of your skates, because you would do a face-plant.

That was precisely the lesson I had to learn in physiotherapy. I was too far back on my heels. I had to learn to lean forward so my weight was balanced between heels and toes. Standing straight felt as though I was leaning so far forward I would do a face-plant. How had I forgotten how to walk?

Watching those kids and coaches, I had a strong desire to get back on the ice, on skates, and skate again.

But, first, I had to learn to walk again.

After my little victory, walking up and down 32 stairs and all around the arena, I walked through the Energy Centre, to the back entrance, to the front entrance, back to my car, and calculated the number of steps.

Only one thousand?

It occurred to me that my stride was twice as long as it used to be.

-- That meant my earlier claims about walking a mile or two were probably way off. Divide them by two.

Was I excited about my latest little accomplishments?

Not only was I walking farther, and taking stairs, but my stride was longer and faster. And I wasn't walking like a duck.

Although I took one cane with me, I only used it a few times.

To celebrate, I drove by the lake and then over to the provincial park, where I went for a short walk.

After all that fun, I felt like heading home and getting some tickets, online, for the hockey game that night.

It was the Cold Lake Ice vs. Vegreville Rangers.

Season's tickets for the Ice cost #160.00. For the Bracebridge Phantoms, it was just #145.00. For the Peterborough Petes, it was $180.00, but you got a Petes t-shirt, 10% off the Petes store, autograph sessions, a chance to watch a warm-up from the penalty box, and more. They had special deals for snow birds, seniors, students, kids, and so on.

Edmonton said hello on Facebook chat, so I told her about climbing stairs and walking around the arena. Here's what I really said: i'm feeling so relieved! was worried about walking and making it back to work. but now i'm thinking ... in two weeks ... i will be ready, freddy!
ps 47 days till xmas !

My sister got a similar message: yo janisistar! 47 days 2 xmas and i just got my pressie: walked to arena, up 32 stairs, around the arena at the top, and down 32 stairs, using longer strides, faster, not walking like a duck, with one cane but barely used it, and then i went for a little walk at the provincial park. rested all day yesterday. now i'm thinking of going to the Junior game at the rink tonight! wooo-hooooo!

And then I told Edmonton, I'm so excited?
    But she said, Why?
    So I told her: Lazarus! i can walk! i can work! i avoided surgery and the spinal tap! made the right decisions about physio, doctors, and alternative practitioner -- reflexology. yesterday i had a lot of doubts.
    it feels like Christmas!

After a long winter nap, I got up to the rockin' music of Vinyl Tap with Randy Bachman on CBC Radio One.
    The show was a particularly good one as it celebrated Leiber and Stoller and other musical talent from the old Brill Building in New York City.
    They played a lot of my favourite golden oldies so what could I do but get up and DANCE?!
    Ruby Baby by the Drifters, Hot Dog by Elvis Presley, Kansas City by Little Willy, Wilber Harrison, and The Beatles. All Lieber and Stoller songs. Peggy Lee singing Is That All There Is? Benny King singing Where's The Girl and Stand By Me.
    My dance steps were simple, but I had a lot of fun moving to the music, for the first time in a long time.
    After that I won half a dozen games of Scrabble and then headed out to see the hockey game.
    It was the most exciting day I'd had in a long time!
    -- How pathetic is that?

The hockey game was great.
    The arena was not half full, which is great for so early in the season. The fans in the stands made a lot of noise every time there was a big hit, a bit of a brawl after a whistle, a big penalty, or a goal. The referees were right on top of the rough stuff. And the game was tied at zero in the first period and tied at zero in the second. And the third.
    There were a few big hits, a couple of fights, lots of rough stuff after the whistle, around the nets. They guys looked big in their uniforms and on skates, mostly, but there were a few small guys, too. They could all skate and pass.
    The Cold Lake Ice looked like a Junior A team, to me.
    I had no problem getting to a seat.
    A couple of guys from my Grade Ten English class came over to say hello.
    I hired one of them to be the trainer for my basketball team.

The other guy was in the senior drama group.

What the Ice needed was a guy from our school on the team, and the rest of our school in the stands, cheering for him and the rest of the team. -- That would make it a lot more fun for everyone!

The ICE looked for redemption against the only team that has defeated them this season as the Vegreville Rangers made their return to the Energy Centre. The first place ICE were in search of their 6th consecutive win.

The game was tied at zero in the first period, tied at 1 in the second period, and took the lead in the final period.

Redemption was complete by the end of the game as the ICE won, 2 to 1.

There were a few huge hits, a couple of fights, lots of penalties, and a lot of rough stuff after the whistle, around the nets. Both teams could skate well and pass the puck well and looked as though they knew what to do in every situation.

The teams were tied in shots on net in the first period but the Ice took over in the category in the second period. The Ice held the puck and dominated the play so much, it looked as though they were playing cat and mouse, and the cat was just daring the mouse to try to do something so it would have an excuse to pounce and devour.

The Cold Lake Ice won their 6th straight hockey game and 3rd in a row on home ice with a 2-1 victory over the Vegreville Rangers.

Tanner Corbeil scored the game winner.

The Ice were 10-1 on the season. Next game was this coming Friday in Lloyd then next Saturday they will return home to play Killam!

(Note: The Killam Indians are now called the Killam Wheat Kings.)

Cold Lake was all alone in first place with 18 points after nine victories in ten games. Vegreville was in second place, in the North division, with 13 points after six wins and four losses.

As much as I enjoyed the game, and celebrated the fact that I could make it to a game, walk that far, climb a few steps, without falling or fatiguing my muscles, or getting tired, I remembered that there is something I like a lot more than watching a hockey game: I wanted to skate and play hockey again!

I've never been big on spectator sports. I'd rather be doing something than watching somebody else playing any game or participating in any sport.

Was I reduced to being a spectator, following spectator sports, or would I be able to get back in the game?

When I got home, I chatted online with Edmonton. I said, Put me in coach, I'm ready to play!

She congratulated me on my latest victory but warned that there might be a relapse or two to deal with on my road to recovery.

Day 74: Remembrance Day

After getting up at 3:00 a.m. and watching WOTE music videos until 4:00 a.m., I slept until 9:00 a.m., then checked out the system. Back: okay. Legs: okay. Feet: sensitive. Walking: stiff. Mood: Up-beat.

There was no hangover from the big day yesterday, with stair climbing, walking in the Energy Centre and Provincial Park, and going to the hockey game.

In the morning, I woke up with a wild desire to shave off my beard, but keep the moustache, for Movember, and go to Humpty's for their egg scrambler.

After that, I wanted to check out CFB Cold Lake, to see if they had a Remembrance Day service. -- No doubt, they did. -- But was it open to civilians?

Who are you gonna call?

Their website said, 4 Wing Cold Lake is the busiest fighter base in Canada. It provides general purpose, multi-role, combat capable forces in support of domestic and international roles of Canada's Air Force. Home of fighter pilot training for the Canadian Forces, 4 Wing attracts Top Gun crews from all over the world to our annual air combat exercise, Maple Flag.

4 Wing Cold Lake is, without question, the home of the fighter pilot. It not only hosts Canada's world-class tactical fighter force training, but also deploys and supports fighter aircraft at a moments notice to fulfill the domestic and international roles of Canada's Air Force.

As an international Centre of Excellence for tactical fighter operations, deployable combat support and leading edge training, 4 Wing focuses on people, leadership, innovation and technology.

Two operational CF-18 Squadrons and two training squadrons, including Phase IV of NATO Flying Training in Canada (NFTC), and numerous premier support units, make 4 Wing Canada's largest and busiest fighter wing. As such, 4 Wing has some of the best and most sought after amenities in the world, including an almost unrestricted 1.17 million hectare air weapons range equipped with state-of-the-art threats and targets.

There was a "Contact" page, but it did not say who to call, locally, to find out about a Remembrance Day ceremony. So, I just drove over.

Mecell, the alternative practitioner and reflexologist, the wife of a military man, part of the Snowbirds team, told me what it was like. She said it was a big ceremony with a lot of military people packed into the J.J. Parr gym on the base -- CFB Cold Lake -- and it was almost overwhelming on all levels: physical, emotional, intellectual, and spiritual.

And she was right.

Being so close to the military airfield, with fighter jets taking off and flying by, is moving at any time, but especially so on Remembrance Day.

My school used to be on the base and it still has a strong connection to CFB Cold Lake, with many students who are the sons and daughters of men and women who are part of CFB Cold Lake.

My thoughts, during the minute of silence, and the rest of that morning, and the afternoon, were about my students, their parents, my family, and my own experiences with war.

My grandfather was a veteran of World War One, my father was a veteran of World War Two, my brothers and I grew during the Cold War, believing we would be veterans of World War Three. My mother did "war work" in an aircraft factory in Toronto, making mosquitoes at DeHavilland -- the wooden airplanes some people say won the war. My father worked on the top secret radar project which some people say won the war. Towards the end of the war, my father joined a Canadian corps that went across the channel from England to The Netherlands and liberated Holland and the Nazi concentration camp in Holland where Anne Frank was held. -- They saw the Canadians coming and packed Anne Frank and the other prisoners of that camp off to Auschwitz just a few days before the Canucks got there.

Anne Millyard liked the book I wrote about my father's war story and said it was a story that needed to be told, deserved a great publisher, and should be promoted heavily.

My commitment to that story was not great, I have to admit, because that wasn't the whole story.

Both my father and my grandfather returned from their wars as victors. But as veterans, their levels were hell. Both men returned from their wars shell-shocked, with TB, and a cross-addiction to alcohol and nicatene. Or "booze and fags", as they used to say. Alcoholics, they used to say, rather than persons afflicted by alcoholism.

My father's story was complicated with irony.

Because he had TB after WWI, my father was sent to a sanitarium to recover. Instead of the TB san in Toronto, where he lived, and where his wife and son lived, with her parents, he was sent to a TB san up north, in Muskoka. The small town with the strange name, Gravenhurst, which sounded like grave and hearse, had four TB sans.

Four separate sanatoriums operated in Gravenhurst over the years.

In 1897 the first sanatorium in the country was built in Gravenhurst at the end of Muskoka Road North and named the Muskoka Cottage Hospital. It operated as a private facility.

A few years later the Minnewaska Hotel was converted into a private sanatorium. It was torn down when the Calydor Sanatorium was built on Gravenhurst Bay around the First World War. The Calydor was later converted into a prisoner of war camp in the Second World War and eventually became the site of the Gateway Hotel.

Around the turn of the century, the Muskoka Free Hospital for Consumptives was built beside Lake Muskoka on what is now the Ontario Fire College property. It provided free housing and treatment to TB patients.

When it burned down in 1920, the National Sanitarium Association spearheaded an initiative to rebuild the hospital at the Muskoka Centre property. The Gage building was used as the Muskoka San for the next 40 or so years.

By the 1960s, when tuberculosis was no longer a major threat, the site became the Muskoka Centre and housed hundreds of mentally disabled residents.

My father was sent to the Calydor and my mother moved to Gravenhurst, within walking distance of the san, to nurse him back to health. The cure for TB, in those days, was bed rest, mostly, with a lot of fresh air. Mostly, he wanted my mother to smuggle in booze and fags, and to get their family going, again.

The Calydor's history was more complicated than the description above.

The Minnewaska Hotel was confused with the Swastika Hotel, not far away.

During World War Two, it was a Canadian P.o.W. camp for German prisoners of war, especially Nazi officers. After the war, the Nazi P.o.W. camp was turned into a TB san. And then it became a resort with a hotel and cabins called The Gateway. At a time when Jews in Canada were not accepted by hotels, it became a Jewish resort. -- A kosher resort.

So, the Nazi P.o.W. camp became a kosher hotel and campgrounds.

My father survived TB. First he got a day pass, to leave the san, then a weekly pass, and then he just had to report regularly for check-ups. He had to get a job in the area, so he joined Brewer's Warehousing. He was an alcoholic, after the war, and he got a job in the beer store. Later, he started a career as a civil servant with the Department of Manpower and Emigration, working in the federal building in the closest town to Gravenhurst, up the highway in Bracebridge.

Gravenhurst, in those days, was also the home of The Northern Book House, which was the largest distributing centre for Communist propaganda in North America. Material from Moscow and the CCCP was sent there for redistribution across Canada and the U.S.A.. They also produced their own publications for distribution. And we saw some of their magazines designed to make life in the Soviet Union sound like a workers' paradise.

I hate to admit it, but I fell for their propaganda, when I was a kid, and really wondered about which side I should be on in the Cold War between the superpowers on either side of us: The U.S.A. or The U.S.S.R.

Complicating matters a little further was the fact that Gravenhurst was also Norman Bethune's hometown. He was the son of the Presbyterian minister, born in the manse, or the minister's house, and he went on to fight against Franco and fascism in Spain, with thousands of other Canadians, where he distinguished himself by inventing a mobile blood transfusion unit, the start of the mobile army support hospital, or MASH, was decorated in Moscow, advocated for socialized medicine in Canada and the U.S.A., and then went to China to join in the communist revolution. He was a socialist and he did not foresee China becoming a totalitarian empire.

Today we wonder how people at that time were not aware of the Holocaust, the huge system of Nazi concentration camps, or the fact that the Soviet Union had the Gulag

Archipelego, which murdered even more people than the Holocaust, or that the Chinese communist revolution would lead to totalitarianism.

My oldest brother was in army cadets, our other brother was in air cadets, and they said I was in space cadets, because I was fascinated with Sputnick, the space race, astronauts and cosmonauts, landing on the moon, and Star Wars. My oldest brother became a beatnik. Our other brother wanted to join the air force but found out he was colour blind and could not get in. I was more like my beatnik brother. After getting all the badges, stripes, and stars in wolf cubs, I left boy scouts after a year to focus on playing hockey. And I left hockey to be a writer and a spiritual seeker.

I grew up with this wild idea, which I kept secret, even from my best friend: We heard about people defecting from the CCCP to live in the West, especially athletes, and I dreamed about defecting from the West to play hockey in the CCCP. Their hockey teams were getting better, year by year, but I could see they needed me! -- Keep in mind that I was just a kid, not even half-educated, growing up in an isolated little town in the middle of nowhere in Canada under the influence of Communist propaganda during the Cold War after my father and his father came back from World War One and Two as shell-shocked veterans with TB and terrible addictions.

My closest contact and first major near-death experience happened while I was a university student, taking a year off to work and travel, before finishing my first degree and going to grad school. First I worked as a labourer/teacher for Frontier College, following in the footsteps of Norman Bethune, working in a prison in Manitoba. From there, I went to Israel and lived for half a year in Jerusalem, just for the experience. As well as exploring religious history in the Holy Land, I wanted to travel around the region, but only one border was open. With my travel partner, I planned a trip up north, to Lebanon, to see Beirut, which was called The Paris of the Middle East. It was a complicated plan because the railway did not run all the way, to prevent trainloads of terrorists from traveling to Jerusalem, so we had to take a series of trains, communal cabs called sheruts, and buses, from Jerusalem, to Tel Aviv, to Haifa, and up to Beirut, in Lebanon.

However, we slept in, for some mysterious reason. And when we woke up, we could not find alarm clocks or watches, so we switched on the radio to find out what time it was. The IBA interrupted their usual program to give coverage of a terrorist attack. The PLO, hiding out in the Bekka Valley, alongside the Red Army, the Baader-Meinhoff gang, and other terrorists, traveled down the Litani River to the Mediterranean, by zodiak, or inflatable raft, landed on the coast of Israel near Haifa, killed an American on the beach, hiked over to the highway, hijacked a bus, let all the Arabs off, kept the Jews as hostages, and then got caught in a dog-fight with the IDF. It was the Palestinian Liberation Organization versus the Israelli Defence Force. They had a shoot-out that ended with the bus getting blown up.

We checked our schedule. -- That was bus.

Years later, I found out all about that terrorist attack. It went down in history as The Coastal Road Massacre. The IDF unit that caught the bus full of terrorists was led by Ehud Barak, who later became the Prime Minister of Israel. The PLO group was led by a young woman named Dalal Mugrabhi, who became famous, or infamous, as "The Mother Of All Female Suicide Bombers". Apparently, in the Palestinian Territory, there are schools, summer camps, terrorist training courses, and so on, named after her, her grave site is a martyr's shrine, and she is held up as an inspiration for new generations of Palestinians.

The PLO in Lebanon, in those days, was headed up by Yassir Arafat. And Arafat got a lot of funding from the KGB, the deadly Soviet spy agency.

Why would the KGB want to kill me? I wondered as I hurried to get out of Israel.

That's what I think about on Remembrance Day.

These days, I look up at the Canadian fighter jets in the skies, flying out of CFB Cold Lake, and I say, happily, but in all seriousness, Thank you, guys!

-- I am happy to teach their sons and daughters.

(And I can hardly wait to get back in there!)

PS: Ricky Ray and the Toronto Argos blew the game open in a record-setting 31-point second quarter against the Edmonton Eskimos. Ray, traded from Edmonton to Toronto at the end of the last season, threw for 239 yards and two touchdowns and rushed for a touchdown.

The game took an ugly turn late in the third quarter when Eskimos' quarterback Matt Nichols replaced Kerry Joseph and scrambled but was tackled and his leg was caught awkwardly underneath him, resulting in a horrifying leg injury, with his foot left pointing in the wrong direction to such a degree it almost made you sick just to see it.

The Saskatchewan Roughriders almost pulled off a come-from-behind victory at the end of their game. The took the lead with less than a minute to play. But Calgary came back to win by a score of 36 to 30. -- What a wild finish!

Drew Tate threw a 60-yard touchdown pass for the win with 20 seconds left.

I could not believe the Calgary Stampeders beat the Saskatchewan Roughriders like that!

Toronto would travel to Montreal to play the East final against the Alouettes while Calgary plays the B.C. Lions. The winners of those game would advance to the 100th Grey Cup in Toronto on November 25th.

Day 75: Happiness

When Calgary beat Saskatchewan in that wild West CFL game, everybody watching saw quarterback Drew Tate throw a sixty yard touchdown pass and they saw his pass receiver, Romby Bryant, make the catch and then point up to the sky.

The morning after, we learned why: Bryant pointed skyward in honour of his good friend Tse Ogisi, who died recently and unexpectedly in his sleep.

The Calgary Stampeders and the Calgary Dinos were on their way to the Grey Cup and the Vanier Cup, according to my Alberta buddy, Andrew Freund, who was living in Toronto but not cheering for the Argonauts, Blue Jays, Raptors or Leafs. He was still cheering for his old hometown teams. He said, The world is lining up nicely for the Dinosaurs to play for the Vanier Cup and the Stampeders to play for the Grey Cup.

Toronto's big league sports teams were all experiencing droughts, but the Argos were winning, so their fans were getting ready for a huge celebration.

In the morning, Love Tree posted two things on Facebook. 1. Happiness is an inside job. 2. Sometimes I go about pitying myself but all the time I am being carried on the great winds in the sky. ~Ojibway saying.

Personally, I felt pretty happy on Monday morning.

Why?

First of all, I had a good weekend, health-wise, with big improvements in my ability to walk, some improvement in my balance, and a reduction in the pain in my butt, legs, and feat. -- I mean the sensations in my glutes, thighs, and the soles of my feet. -- Thank God.

Also, I Skyped with my sister for about three hours on Sunday night.

While I was fighting my way back to work, she was working on her plan to retire early and get away from her workplace, because she didn't love it so much, anymore, and she had health concerns she suspected were related to her job. Although she loved working as a relationships counsellor, still, the agency she worked for had made some major changes and she felt they were going in the wrong direction.

Instead of private rooms for counsellors, they moved to an open area concept.

Counsellors had to report on their activities every fifteen minutes, using a Canadian computer program called Penelope.

And that was just the short list. Her long list left her feeling quite frustrated.

It was a pain, she said, and it was giving her a pain, or pains, in her body.

Everybody expected her to go into private practice, but she said all she wanted to do was look after her health, for a while, and she would think about private practice after that.

She planned to exercise more and explore alternative healing in her area as the mainstream medical system wasn't helping.

Of course, I recommended acupuncture and reflexology, if she could find people who were particularly talented or gifted.

We both lamented the fact that doctors and alternative practitioners could not work together with Chinese and Indian healers for an integrative approach to healthcare.

It was the first time we had Skyped or talked for months.

It felt good to yak with my sister, again.

As well as health and work, we talked about movies and music and motivation, sharing our favourite resources. We had both discovered the TV show called X Factor and she liked a show called Australia's Got Talent and we talked about performers who took the stage looking like unlikely candidates but revealed amazing abilities to sing. And we talked about how that made so many people cry.

When she called, I was watching a movie called Hatchi, with Richard Gere and a dog, that was guaranteed to make you cry.

I told her about Soul Surfer and she added it to her list of inspirational movies to recommend to clients.

An evening with a movie double-header, watching Hatchi and Soul Surfer, was guaranteed to get the tears flowing, and they say it is very good to cry like that because your tears release chemicals and toxins that otherwise stay in your body and have a negative impact on your health.

Take these two movies and you won't have to call your doctor in the morning!

We talked about the big election in the U.S.A., too. She said Australians would have voted for President Obama in huge numbers, like Canadians, and the same was true for many places all over the planet.

She said she only heard part of the president's acceptance speech, so I promised to send her a link that would let her hear the whole thing.

"It's great," I said. "He just made one mention of global warming and did not talk about poor people at all, but it was pretty great anyway."

She explained Australia's new carbon tax and noted that poor people got their money back. She also said hydro or electricity bills were going up and up in Australia. -- I asked her when they were going to get solar panels. -- They live in one of the sunniest places on the planet and the price of solar panels had dropped significantly.

"It's on our list of things to do!" she said.

When I told her a bit about the NHL lock-out, she laughed, remembering how important hockey is to Canadians. She said that if "footy" players in Australia went on strike, and there was no footy on the telly for blokes to watch in Oz, she could open up a practice as a social worker specializing in helping people deal with the loss of their number one passion and pass-time, and probably make a lot of money, until footy was back.

Footy, or football, in Australia, occupied the same position as hockey in Canada, or baseball in America or cricket in India or .... You know.

My sister also wanted to talk about turducken. So I told her about the popularity of the bacon weave.

A turducken is a dish consisting of a de-boned chicken stuffed into a de-boned duck, which itself is stuffed into a de-boned turkey. The word turducken is a portmanteau of turkey, duck, and chicken. The dish is a form of engastration, a recipe method in which one animal is stuffed inside the gastric passage of another.

The thoracic cavity of the chicken/game hen and the rest of the gaps are stuffed, sometimes with a highly seasoned breadcrumb mixture or sausage meat, although some versions have a different stuffing for each bird. The result is a fairly solid layered poultry dish, suitable for cooking by braising, roasting, grilling, or barbecuing.

The bacon weave is the new way to cook massive amounts of bacon at once. The bacon weave is simple to create. It consists of bacon woven together to make a mat of bacon. You can add it to all kinds of things.

My sister said she missed Canadian bacon and peameal bacon and was looking forward to her planned trip to North America next year so she could go to Los Vegas and so she could get some Canadian bacon.

That made me laugh. I told her she might be able to get a turducken with a bacon weave and have the whole thing deep fried.

Note: I do not recommend any of those food ideas!

On a more serious note .... She said she planned to be in the U.S.A. in the summer but the trip might be postponed until the fall. So I reminded her what last summer was like in the U.S., with record-setting heat and drought, so she said she'd look into that.

We talked about Christmas, too, as it was only 46 days away. She told me a friend from Montreal was planning to visit her in Australia at Christmastime as her daughter was spending the year traveling around Oz and she had always wanted to see the country, too, so she was going to rendez-vous with her kid and they were all going to get together for Christmas.

She had split up with her husband, who was devastated, and apparently he offered to pay for the trip to Australia, if they could get back together and travel together, but she said, No, I'd rather do it on my own.

She was looking forward to seeing her old friend for Christmas.

I told her I was thinking about spending some time with our brother, back in Ontario, at Christmastime, maybe taking the train, but it all depended on my recovery, or comeback, and if I would be like Tiny Tim at Christmas.

All I want for Christmas is my legs and feet back again, I told her.

She said she was making a list of 60 thing to do to celebrate turning 60.

Her birthday was right after New Year's.

She was making a bucket list of things to do that year. It would be 2013.

I quoted the American poet, Mary Oliver: "Tell me, what is it you plan to do with your one wild and precious life?"

And I sent her a link to a website with one thousand suggestions for a bucket list for age 60: http://www.squidoo.com/100things.

A lot of the things on my bucket list had already been checked off.

But I still had a few.

One of them was to teach the triangle offence to my Grade 5 and 6 basketball team that would also have some Grade 4 guys on it.

First, I had to get my legs and feet back, and my balance, and get back to work.

Something strange happened on Day 75: Even though my feet, ankles, legs, glutes, back, etc., felt a bit better, again, I felt as though I needed a bit shot of energy if I was going to do anything athletic. After breakfast, I thought I'd just give it a rest and lie in the sun for a bit, reading The Winner Stands Alone, by Paolo Coehlo, while listening to WOTE on YouTube.

Walk Off The Earth is an unconventional, multi-talented five-piece musical phenomena that is currently taking the world by storm. Based in Burlington, Ontario (just outside of Toronto), their brilliant 5-people-playing-one-guitar interpretation of Gotye's "Somebody That I Used To Know" exploded on Youtube garnering well over 35 million views in under 2 weeks. The massive fan response quickly drew attention to their collection of innovative songs and videos spanning their last 5 years and generated an unprecedented flood of media interest. New fans from around the world immediately fell in love with the band's organic, independent, sincere and honest original songwriting, cover interpretations and beautifully filmed videos. The band has built a massive following of dedicated fans from around the world through their unique and heartwarming approach to songwriting, filming and constant, open interaction with everyone their music touches.

However, I closed my eyes after a little while and .... Well .... I woke up at four in the afternoon!

Just before I woke up, I was dreaming about eating lunch at work, and I thought I'd get up to make my lunch, but it was almost time for supper!

What's up with THAT?!

In the evening, things did not improve much. While playing Scrabble online, I listened to the Massey Lectures on CBC Radio, a series on magic and physics, and I looked at the news and weather.

My online Scrabble score went up to 1649, with 222 bingos.

The past week had been one of weather extremes.

In the Prairies, there was between 15 and 30 cm of snow in Calgary and Edmonton last Wednesday and Thursday. The same or more fell on Regina and Saskatoon from Friday and Saturday. Winnipeg was under nearly 30 cm of snow while a small west-central Manitoba town was buried under 65 cm!

Strong winds and frigid temperatures followed.

The Great Lakes area and along the St. Lawrence had mild temperatures with rain and freezing rain for eastern Ontario and southwestern Quebec, and there was a record high temperature in Toronto of 18.4 degrees C, beating a 74-year old record, and record breaking high temperatures were expected in the Montreal area.

Venice was sinking, according to some reports. Apparently, nearly three quarters of Venice was flooded on Monday and tourists swam in St Mark's Square as a wave of bad weather swept through northern and central Italy, forcing the evacuation of 200 people from their homes in Tuscany. Shops, homes and historic palaces filled with water in Venice and authorities said 70 percent of the lagoon city was flooded. High water in Venice reached 149 cm (5ft), the sixth highest level since records began in 1872, forcing residents to wade through waist-deep water.

Tourists in swimming costumes sat at cafe tables under the water.

The big news from our neighbour to the south was a prediction about being energy sufficient in the near future, thanks to fracking, plus alternative sources of energy, and the Americans were the only market for Canadian oil, to date, so .... Alberta needed to find other markets -- and the way to get to them. The United States could overtake Saudi Arabia and Russia as the world's top oil producer by as soon as 2017.

As for me .... frankly .... I felt like Edmonton, Winnipeg, and Venice. -- Not hot. -- Under the weather.

Woke up at 3 in the morning after an oddly positive and useful dream about creating a website to consolidate information about novel marathons. Also, I got the idea that I should focus on my comeback or recovery, in this book, rather than the length of time since I got injured. And I woke up with a desire to find a note on my Facebook page.

Where did the Notes section of Facebook go? I had to go to the Help section, as Facebook has no Customer Support, and in a Users Forum I found a link to Notes. -- What a pain!

The 'Note' I was looking for was an old poem I wrote. Ever since I watched the movie "Hatchi", I've been thinking about this dog I used to like. It was a friend's dog. She called it a borderline collie because it was a wild and crazy border collie that only liked around five people on the planet. Fortunately, I was one of them. Well, I worked hard to tame and train that dog, and eventually had a great relationship with her. The dog's owner lived in the country and liked to keep the collie as a guard dog, so she was happy it would attack people and animals, for her protection.

In the movie, they said the dog called Hatchi was not picked by its own, it was the dog that did the selecting. And that was the case with Phantom, too. "Don't look in her eyes!" her owner said, the first time I met the dog. After numerous meetings, with snacks and toys, and attention, she warmed up to me. And eventually I became her best friend. -- That did not go over all that well with her owner or with her son. -- It was their dog. Why was she friendlier with me than them?

Well, of course, the answer was that I played with her more and taught her things. Border collies are smart and they respond well to learning new things. I taught her some stupid pet tricks and games and she responded by playing with me in a way nobody else dared. She would let me put my hand or my arm in her mouth, playing around, confident she would not savage it. -- Not even her owner would play that game!

She ran circles around us when we took her for a walk in the woods: enormous looping circles that kept bears and whatever other wild things were in the forest far away from us. She always came back at exactly the same time, five minute intervals, as though she had a watch, looked for a pat on the head, from me, and took off again for another five.

That dog loved to play "fetch" so I got a Chuck-it, like a plastic lacrosse stick, so you could throw her ball about a mile, and she would race after it and bring it back by the hour. She was confused by a pet toy called a "flying chipmunk" but learned to go after it and take it out of the air just like the frisbee I got her.

At dinner, she sat at my feet, and when I wrote, she curled up under the desk and kept my feet warm.

When my relationship with the dog's owner ended, after a few years, I was very sad. And I still miss that dog!

Here's the poem I wrote about that relationship and that dog:
I'm The Man The Dog Loves

I'm the one who gets all the snacks for the dog
I'm the one who gets the dog's breakfast and dinner, when I visit
I'm the one who plays catch with the dog, with tennis balls, frisbees, and Flying Squirrels
I'm the one who lets the dog out of the car so she can run
I'm the one who throws sticks for the dog to fetch
I'm the one who exercises the dog's brain
I'm the one who wrote a song for the dog
I'm the one who sings the dog's song
I'm the one who taught you to sing the dog's song
I'm the one the dog rolls over for
I'm the one the dog doesn't bite
I'm the one the dog seems to like
I'm one of the few people the dog loves
I'm one of the few people who loves the dog
I'm the one who taught the dog to catch
I'm the one who taught the dog that other stupid pet trick
I'm the one who makes the dog sit with a treat on her nose
I'm the one the dog likes despite the stupid pet trick
I'm the one who loves to hose down the dog
I'm the one who laughs when the dog tries to bite the water
I'm the one who throws snowballs at the dog and laughs when she bites the snow
I'm the one who throws sticks in the pond for the dog to swim to
I'm the one who swims with the dog
I'm the one who races the dog to the ball thrown in the water
I'm the one who wins the race to the ball thrown in the water
I'm not the one who holds onto the dog and gets pulled around in the pond
I'm the one the dog soaks when she shakes herself dry
I'm the one the dog buffs with her nose when she wants a treat
I'm the one the dog buffs with her nose when she wants a pat
I'm the one the dog buffs with her nose when she wants to go out
I'm the one the dog cries for when I go away
I'm the one the dog cries for when I come back
I'm the one the dog wakes up first in the morning, looking for sugar
I'm the one who makes the dog feel like hiding behind the furnace when I leave
I'm the one whose back seat is covered with dog hair
I'm the one whose car looks like the car of a dog owner
I'm not the owner of this dog
I'm the boyfriend of the owner of the dog
I'm not the alpha female of this pack
I'm the alpha male

I'm not the one the dog obeys the best
I'm not the one the dog has grown up with and loves the most
I'm not the one who takes the dog to the kennel
I'm not the one who takes the dog to the vet
I'm not the one who raised this dog from a pup
I'm not the one who trained this dog
I'm not the one who tamed this dog
I'm not the one who got the electric dog collar and the zapper for the dog
I'm not the one who shocked the dog into behaving well enough to be with people
I'm just the one who made the dog a little sweeter
I'm the one who holds the dog when the dog's owner pulls porcupine quills out of the dog's snout
with needle-nose pliers
I'm the one who laughs on the other end of the phone when he gets the news the dog smells like a skunk, again
I'm the one who believes this dog is the dog of his childhood, reborn
I'm the one whose feet the dog sits at during breakfast, lunch, and dinner
I'm the messiest eater, I guess
I'm the one the dog stares at, if I have a hidden bag of Beggin' Strips
I'm the one who has staring contests with the dog, and loses
I'm the one the dog has never tried to bite
I'm the one the dog has never even snarled at
I'm the one who puts his socks over the dogs eyes while getting dressed
I'm the one who tosses shirts and towels on the dog's head just for fun
I'm the one who plays Bison wrestling with the dog
I'm the only one who puts his hand or arm in the dog's mouth
I'm the one the dog is staring at right now, while I write this
I'm the one who grabs the dog when somebody strange comes to the door
I'm the one who never forgets dog is god spelled backwards
I'm the one who writes poems about the dog

That poem was one I read at a lot of "readings" and it always got the strongest response. People applauded and wanted to talk to me about it.

     The game we called Bison wrestling had me down on all fours with the dog and the two of us pushing against each other, joined at the shoulder, like a couple of buffalos trying to push each other around.

     She would sit with a dog treat balanced on her nose, vibrating with excitement, until I said, "Okay!", and then she would snap her jaws so ferociously it scared everybody watching. "How did you ever teach that wild thing anything?" people asked.

     The answer, of course, as anybody who saw "Hatchi" would know, is, simply, love.

That dog was the double of the dog I had as a kid. Phantom had the opposite temperament as Barney, but the two dogs looked a lot alike.

Hatchi has a way of making you remember and miss the dogs you've been close to. The movie is all about loyalty and the bonds that connect us to pets. And to people, too.

I'm thinking my back started hurting, maybe, when that relationship ended, when my loyalty was not rewarded. That dog's owner was very effusive and said a lot of great things, including the well-known expression "I've got your back", all the time.

Did you ever see the cartoon showing two stick people and one is holding the other's backbone stick, so it has no backbone, anymore, and the cruel one says, "Don't worry, I've got your back!" -- It was like that.

I found the cartoon online and posted it to my Facebook page. By then, it was four o'clock in the morning. -- Time to go back to bed!

Recovery Day 50: Walk-Outs And Lock-Outs

Alberta bragged about having the best education system and the government often said education was a top priority. -- Trying to remember anybody saying anything like that in Ontario.

Sometimes I think about going back to Ontario, to teach, leaving my small class of students with laptops and our SmartBoard to go back in time and write on a blackboard with chalk for a class of 33 or so.

It's a daydream more like a nightmare and I hear myself screaming, No! No! No!

Here's today's headlines from Ontario:

Locking out Ontario teachers 'only tool' boards have to protect student safety: minister. -- Globe and Mail - 1 hour ago

Education Minister Laurel Broten says she opposes lockouts, but 'the only tool that [school boards] have if they think student safety is at risk, ...

Ontario teachers' job action threatens student safety: Principals -- CANOE - 2 hours ago.

Education Minister threatens action against Ontario teacher protests ...
www.theglobeandmail.com

Here's the story: Teachers are opposed to a controversial law, which forces a wage freeze, cuts teachers' sick days and limits their right to strike. Teachers are not expected to walk out, but many have limited their duties to exclude administrative tasks and supervision outside the classroom. -- That means no clubs or teams after school.

In Alberta, teachers were negotiating for a better deal, too, but they were talking about hours, not money, and not a wage freeze, sick days, and the right to strike.

Go, Alberta!

In Ontario high schools with four periods per day, teachers were guaranteed one period as prep time, for marking and lesson planning, and that was "carved in stone". Teachers had fought hard for it, gone on strike for it, and would never give it up without another battle including a province-wide strike.

In Alberta, prep time for teachers was regarded differently. Principals told teachers they were lucky if they got a prep period. Last semester, I had not prep period. Our local teachers federation was discussing the prep time issue with our school board and it was being looked at in other places around the province, too.

Do you want your teachers prepared for class? Or burnt out?

This morning I had a dream about going back to work. First I went in for after-school drama and basketball, for a week, and then I went back to my full day timetable. Basketball was like Hoosiers, the movie, but without the alcoholic father taking over as coach, and with shorter guys, since my team is Grade 5 and 6, with some help from Grade 4. Drama was like The Dead Poets' Society.

My dreams have changed. Instead of nightmares about never walking again, they have shifted into positive scenarios with useful ideas. -- That's more like it! -- That's more like me!

Edmonton called. She said something was going on with the Wild Rose party. There was a coup in the works. And the Environment Minister was hosting a press conference at the Shaw, the conference centre on the hill, overlooking the Saskatchewan River, with too many stairs and escalators, she said. And expensive parking.

I told her she could park right across the street, at the north end, instead of underground at the library, which was often full, or in Chinatown, which was almost a cab-ride away.

On the radio, I read a report on snow removal in the capital city, which was unlike cities to the south, including Calgary and Winnipeg. It's the same story in Cold Lake. Snow and ice is not removed. They don't scrape down to pavement. They maintain the snow pack, or snow and ice pack, with sand on top.

That's what I suspected. -- I didn't know it was their philosophy, I thought it just worked out that way.

We have ice roads.

After four days with almost no pain, I was looking forward to physiotherapy and acupuncture. My balance was a little bit better and my walking ability had improved by strides.

I wanted my physiotherapist to tell me I was good to go to the gym and do everything to build up my muscles again. So far, all she wanted me to do was stretch my hamstrings, nerve sheaths, and arch my back.

I wanted to lift weights, do sit-ups with weights, get on the elliptical machine, push weights with my legs, hang out at the gym all day everyday for the next twelve days or thirteen days.

Edmonton said, Take it easy. You don't want to have a relapse.

Australia said, You want to be stronger than ever when you go back to work.

I said, Just put me in, coach, I'm ready to play, and I can play with pain. I'll get better while I work.

Everybody disagreed with me about going back to work. My reflexologist and physiotherapist agreed with the doc, saying it would be better to go back half-time for a while and then go full-time.

Why? They say it's a huge transition. The physiotherapist said, Your body is at about 50%, so going back to work 100% could be too much and it might set you back.

That was not what I wanted to hear.

Working a half day is not much different than working a full day, the way I see it. -- Maybe I could go in next week, in the afternoons, to do drama and basketball, and get everybody used to the idea that I'm making a comeback.

As the Zen master said in the famous story .... We'll see.

Mecell Pillon gave me a great idea for a book and a movie. While Angela Paquin left me after inserting acupuncture needles, I had more book ideas.

They were both excited about the way I was walking. Mecell said it was amazing. Angela said it was very good progress.

Mecell said the reason my big toes hurt was that they were involved in balance, were being used in the new way I was walking, and they were connected to neck and shoulder muscles, which I was using differently for balance when I walked my new walk.

She said I was warm while walking on a cold day because my circulation was ramped up.

She loved my Rocky at the hockey rink story, about climbing all the steps and then walking more and more.

Angela liked the stair climbing story, said I was hot because my auto-immune system was on high alert, and the pain in my toes was deferred pain from other places.

They both made me feel better, physically and mentally, but I was not happy with their advice about going back to work part-time.

I'm chomping at the bit, I told Angela.

Tough, she said.

After physio and acupuncture, I went next door to order pizza. It was 2 for 1 day, so I got a large veggie for the people who work in the physio office, and dropped it off.

My new exercises, from physiotherapy, are a squat and a leg raise.

For the squat, lean against a big exercise ball, against the wall, legs apart, a little, feet pointing straight ahead, and bend the knees to go down a little bit, co-ordinated with umbrella breath.

For the leg raise, lie on your front, toes down, and straighten one leg, then the other, co-ordinated with your breathing.

They are small exercises but, added to back extensions, clamshells, blow and go, and umbrella breathing, plus nerve stretches, the physiotherapist said that was all I should do, so I wouldn't overdo things, and it would build up the core muscles I needed for walking, stability, and relief of back pain.

However, when I told her I missed going to the gym, she suggested going to the pool on the base and doing some deep water walking while wearing a bouyancy belt.

But she cautioned me not to overdo it, because it feels like it isn't much work, at first, because of the bouyancy provided by the water.

-- I liked the sound of that.

Here's the book and movie idea .....

It's like Patch Adams, in some ways.

When I told Mecell that I wished we had integrative health care, with mainstream Western medicine, alternative practitioners, plus Indian and Chinese techniques, she told me she believes my wish is coming true. And it's not just something promoted by Deepak Chopra, Dr. Andrew Weill, and a few others. It's happening here and now.

Well, maybe not in Alberta, right now. But it was happening in B.C. and Saskatchewan, and was starting to take hold in Ontario, so the rest of Canada would not be too far behind.

She said there was a hospital in Vancouver that had a wing dedicated to alternative healing, with reflexology and massage and several other things.

And she said she made a lot of progress in Saskatchewan, while she was there, getting reflexology recognized by doctors, hospitals, nurses, and health insurance providers.

Picture this on the big screen: A tiny woman surrounded by a dozen skeptical M.D.s, all of them men with their arms crossed, scrutinizing her work as a reflexologist. They gave her their most difficult cases, including screamers and paralyzed people.

But she did not give in to the intimidation. She just did her thing and got great results. The screamer stopped screaming and participated in a conversation for the first time in months. The guy paralyzed in the fetal position straightened out for the first time in half a year or so.

All the doctors were won over except for one and they told him he was out-numbered.

The nurses were worried about losing their jobs to reflexologists, but she assured them their work was still needed, along with what a reflexologist could do.

Health insurance providers were told by the doctors that reflexology lowered patients' pain levels and their need for expensive painkillers and other drugs.

Just as reflexology was getting accepted into the mainstream in Saskatchewan, our reflexologist had to move to another province, and start over again, winning over doctors, nurses, physiotherapists, et cetera, again.

While lying facedown on a massage table with long needles stuck in my lower back, rear end, and ankles, I had a big idea for my classes: I could give each of my students a book I've written for Christmas, to inspire them to read over the winter break.

Recovery Day 51: MAN Syndrome

It was a night of wild dreams. Some good, some bad.

In one dream, I was working as a teacher in my old high school, where I was a student in the Seventies, and I had been off work for a while, but was going back, only I couldn't find my classes, and rooms kept changing, and eventually I was naked, my clothes turned into a big, reddish-coloured, horse, like Secretariat, the race-horse, so I hopped on it and rode it, bareback, out of there, fast.

In another dream I was at some casual writerly gathering with two dozen men and women and Wally Lamb was there. By coincidence, we were wearing the same sort of shirt. It was royal blue, almost purple, with lines of yellow and red, like a subtle plaid. He wore his open, with a tee-shirt underneath, but my shirt was done up, except for the top button. We had our pictures taken together, holding each other's books. I was holding She's Come Undone and he was holding this one, called Fall Down Nine Times, Get Up Ten.

In the last dream I had before waking up, Guillain-Barré Syndrome was ruled out, for the growing number of men in the Cold Lake area who were coming down with a mysterious virus or disease that caused nerve damage and destroyed leg muscles so you couldn't walk. My reflexologist was right: It was a local phenomenon. Because she called it, she got to name it, and she named it after me: Martin Avery Neuropathy Syndrome, or the MAN Syndrome.

Last night I went to sleep at my usual time and this morning I got up at my usual time. The physiotherapist told me to start normalizing my nocturnal rhythms so they would not be all over the map and I would be in synch with the routine for going back to work.

In another dream, I was walking in a heated swimming pool at CFB Cold Lake, for physiotherapy, and then I climbed into the hot tub.

-- That was a dream I could make come true and planned to as soon as possible.

Love Tree posted three interesting items on Facebook:
1. Face the thing you fear the most and the death of fear is certain. ~Cricket House
2. You are the universe expressing itself as a human being for a little while. -- Eckhart Tolle
3. A pair of world-renowned quantum scientists say they can prove the existence of the soul. American Dr Stuart Hameroff and British physicist Sir Roger Penrose developed a quantum theory of consciousness asserting that our souls are contained inside structures called microtubules which live within our brain cells.

Since I had never heard of Hameroff and Penrose, I Googled for a while and found out they are quantum physicists who believe near-death experiences occur when the soul leaves the nervous system and enters the universe, claim two quantum physics experts.

They say consciousness is a program for a quantum computer in the brain which can persist in the universe even after death, explaining the perceptions of those who have near-death experiences.

My Scrabble score jumped up to 1657, with 223 bingos.

And then everything went to hell in a hand basket, as we used to say.

Recovery Day 51, Part 2: Insert Swearword Here

Just as I was leaving for the pool at the base, the phone rang. Just when I was feeling so good that I wanted to get into a swimming pool for deep water walking, as part of my physiotherapy, I had a setback.

Edmonton called to say she was in a fender-bender, but she was alright, and so was her car, but she dinged some guy's bumper, and she was on the way to the doctor. She was multi-tasking, she said, distracted by thoughts about work and life. There were cutbacks where she worked, but she survived. She was going to the doctor for results of tests, not because of the little collision.

We talked for an hour or so, and then it was too late for me to go to the pool. In order to make it to my physio appointment, I had to shower and get changed, quick, and while I was doing that, I bent over, the way I'm supposed to avoid, and that old pain in the butt came back.

After four or five good days, I was doing the zombie walk, again, and feeling a good deal of pain.

Damn!

Angela, in physiotherapy, pointed out that I was walking the old way, again, and coached me through the correct method, one more time, then checked the way I was doing the clamshell exercise, the wall squats, and the balance board.

Just when I was yelling at myself, silently, she said, Don't beat yourself up. This happens. It's hard to develop and maintain new habits. The body takes over and does things that run counter to what we are trying to get the body to do.

She got me to stand in front of a mirror and pointed out that I was back on my heels, a bit, again, and had to lean forward, even though it felt odd, as though I was leaning into the wind. Relax your rear end, unclench it, put your ribcage over your pelvis, walk heel to toe, and don't forget to breathe using the umbrella breath technique.

Those light exercises made me sweat. So I asked her what was going on.

You are working extremely hard to walk properly and to correct what some strong muscles in your body want to do, she explained.

Half a dozen people had warned me about a relapse, and I was determined to prove them all wrong, but I just bent over, quick, the wrong way, to pick up something at home, and -- ouch -- I was back to where I was a week or two ago.

That made me mad. -- I mean angry.

When I left physio, I was dive-bombing the lower emotions, as the poet Richard Brautigan used to say.

In the car, I heard Dan Hill, the singer/songwriter, doing a CBC interview on the subject of prostate cancer. So, I phoned my brother to tell him about it, as his doctor had him worried about that. Dan Hill said men don't talk about their injuries, illnesses, or problems, much, but that was changing, and it was a good thing, because it's important to share information.

He said that so much has changed in the world of prostate cancer detection, surgery, and recovery, it was worth risking being vulnerable to talk about these things, because you could pick up some invaluable information.

An old friend of mine went to high school and to church with Dan Hill, decades ago. My mother liked his music a lot, and I took her to a Dan Hill concert, when he came up to Barrie, to perform at the Gryphon Theatre. I introduced his brother, Laurence Hill, when he came to do a reading at the Pickering Public Library, when I was Writer In Residence, and he was launching his first novel, called The Book Of Negroes, in Canada, and something else in the U.S.A.

Hearing him on the radio, on The Story From Here, interviewed by Matt Galloway, a former student of mine, now a CBC Radio star, made me feel happy.

Although I didn't feel like it, I went to the grocery store and got a lot of things I needed, and on my way home I drove by the lake, just because it is beautiful, and by the time I carried my groceries out of the car and into my kitchen, I was thinking worried thoughts about going back to work.

If bending over to pick something up was all it took to set me back a week or two in my recovery, how was I going to go back to work in a week and a half?

However, after I made some comfort foot, 'mac 'n' cheese', and checked Facebook, I started to perk up a bit.

A new Facebook friend from Muskoka posted a poster with a picture of a ferris wheel on a snow covered street lined with trees decorated with blue lights, and it said, Just because you don't see results after a day or even a week, DON'T GIVE UP. You may not see changes, but every smart choice you make is affecting you in ways you'd never imagine.

-- That was a comforting thought.

Because I just caught the end of the Dan Hill interview with Matt Galloway on the radio, in the car, I found the podcast version of the show called The Story From Here on the CBC website. He said, Men don't talk about their health problems. He said, The surgery for prostate cancer is painless. He said that, after the operation, there were problems with continence and sexual performance, for a while, but you can work on that and get things functioning again.

Dan and Matt both said, Get tested, it could save your life!

On Facebook, I chatted with my cousin's daughter, down in Colombia, where she is teaching, after stints in England, Egypt, and Thailand, and she told me she was sitting at her desk, pulled out a drawer, and pinched a nerve, so she's had a lot of back pain for the past couple of months. Also, a doctor told her she has the knees of 100 year old, with almost no cartilege. So, she's getting a second opinion.

The news form Europe was not good, either. Hundreds of thousands of Europe's beleaguered citizens went on strike or snarled the streets of several capitals Wednesday, at times clashing with riot police, as they demanded that governments stop cutting benefits and create more jobs.

Workers with jobs and without spoke of a "social emergency" crippling the world's largest economic bloc, a union of 27 nations and half a billion people. In Madrid and Barcelona, where the crisis is hitting particularly hard, protesters and police fought street battles resulting in dozens injured and numerous arrests.

There was some good news from the sports world -- in a way. A junior hockey team from Russia was touring Canada to play all-star teams in Quebec, Ontario, and the West. Russia's star was the NHL's #1 draft pick, Nail Yakupov, who will join the Edmonton Oilers when the lock-out ends and the season finally starts.

In the game against OHL stars, he was shellacked by a guy named Tom Wilson, who will play for the Washington Capitals if the NHL ever gets going again.

Tom Wilson is the nephew of my friend Marty Avery, a woman who works in Banff, and must be related to me, somewhere back in time, maybe in Avebury, England, at the time of Stonehenge. So, Tom Wilson, NHL rookie, must be related, too!

Wilson won the Goal Medal with Team Ontario at the 2011 World U-17 Hockey Challenge and was also selected to play in the 2012 CHL Top Prospects Game. He was selected 16th overall in the 2012 NHL Entry Draft by the Washington Capitals.

Marty Avery and I have met a few times, by the way, and got on like long-lost sister and brother. -- Or something!

She found me by Googling herself, as they say you are supposed to do, to check out your online presence, and she kept getting pictures of me, so she sent me an e, came to a writing workshop I was leading, and joined us for a novel marathon, one summer.

At the start of the marathon, I ran an ice-breaker and got all the writers to say a few words about themselves. Marty said, Martin was wondering, one day, what his life would be like if he was a woman, and then "poof", I appeared, born out of his forehead!

She is an amazing woman. I liked her so much I said she could be Marty Avery and from now on I would be Martin Avery, to avoid confusion on the internet.

It took a while for me to get a grip after that set-back. After I watched a bit of The Answer Man, again, then slept, woke up, had a snack, chatted online with my cousin's daughter, exchanged e-mails with my cousin in Lac La Biche, played Scrabble with Alabama, I tried walking, again.

It is strange when you have to think about walking in order to do it correctly, when it doesn't come naturally, and you have to relax your glutes but work your piriformis, balance between your heels and toes, blow before you go, lean forward as though going against the wind, and walk heel, toe, heel, toe ....

But around 11:00, I did it, again, and then headed for bed, hoping that I would be in better shape in the morning.

Recovery Day 53: Water Walking

In Canada, we can all walk on water: we just wait until it is frozen. Most of our lakes freeze over with ice so thick you can drive a car or truck on them.

Cold Lake is always cold so you would think it would be the first lake to freeze over, but it is the last, because it is so big.

Last summer, I did some water walking in Cold Lake. -- It was too cold for me to swim in, but wading, for a short time, worked out okay on certain hot days.

But we are not walking on water or ice today, we are deep water walking, for physiotherapy.

Here's how:

There are two types of water walking: deep water and shallow water.

Deep water walking is done in water where your feet don't touch the bottom. You mimic land-based walking movements such as walking up hills, walking sprints, and backwards walking.   Consider wearing a flotation vest or belt to keep you afloat and holds you in a relatively stable condition in the water so that you can concentrate on your form instead of on keeping yourself afloat. You can use a Styrofoam weight under each arm to keep yourself afloat. Stick them firmly under your arms so that they don't interfere with your arm swing and so they keep you relatively stationary in the water.

Deep Water Walking is essentially an aerobic exercise. So, adding warm up and cooling exercises before and after the the schedule will help you a great deal.

Drink plenty of water. It consumes your body fluids.

Deep Water Walking does not strain the musculoskeletal system. The entire body is cushioned by water. This offers great protection to the joints which bear most of your body weight.

It involves all major muscle groups. Muscular imbalance can be prevented and even corrected because water equalizes tension to the body. Because water is a liquid medium, the body is relatively less vulnerable to stress injuries. It not only reduces but removes muscle soreness for all practical purposes.

Shallow water walking is done in thigh- to chest-high water where your feet still can touch the bottom. Walk the pool back and forth widthwise, driving your arms and legs as hard as you can. Don't worry about speed. You won't move very fast. But you will work up a sweat and tire out quickly. You might want to start with a few short intervals with a minute or two of rest in between and gradually increase from there.

Stand waist deep in water.

Land on your heels first and then proceed to the ball of your toes just as you do in regular terrestrial walking exercise.

Walk back and forth. First take some 8 to 10 strides forward, then take almost the equal number of strides backwards. This enhances your muscle tone.

Push your arms alternatively in forward and backward directions. While doing so, ensure that your hands are straight as far as possible and more importantly, turn your hands in such a manner that they move against water.

Move your arms and legs in opposite directions. When you stretch forward your right leg, stretch your left arm along with it and vice-versa.
Consciously, lift your knees to a higher level. This adds more intensity to your exercise.

Walk back and forth in short steps, long steps, average steps, or step kicks.

Move in a fixed pattern. For example, you may move in a circular or a rectangular fashion. But ensure that you trace the shape in both the directions. This meets the demands of your body in a better manner.

When your intensity increases considerably and you gain in confidence, try taking longer strides. If possible even bound at times by pushing off with your back foot. By doing so, you will bounce up the pool. Try this between the strides if you can. It will add to your overall out put.

Water walking is an excellent aerobic and calorie burning workout; you can burn up to 550 calories an hour walking in the water. Water walking can also help correct muscle imbalances. Walking on land emphasizes the muscles in the front and back of your thighs. Walking in water works your upper body equally as hard as your lower body because water gives all your submerged muscles 12 to 14 times the resistance of air.

Swimming went so well, it was like swimming from hell into heaven, from injury to health, from sickness to wellness.

Ecstatic Man was back!

First, you have to register and get your picture take for a membership card that says 4 Wing on it, so you feel as though you have been made an honourary member of the air force.

At noon, there was a lifeguard, two swimmers, and me, in the big pool, and some moms with tots in the shallow pool. The two swimmers did lanes, so I had half the huge pool to myself.

Go easy, everybody said. Don't overdo it. And I had every intention of doing just that, for about fifteen minutes.

But the more I did, the better it felt, so I kept on going and going and going. Shallow water walking, deep water walking, forwards and backwards, for half an hour, a little break in the hot tub, and then right back in to try the flutter-board.

By then, I was feeling good, my confidence was growing, so I tried swimming a length.

One thing led to another: breast stroke, back stroke, side stroke, treading water, and then more walking, backwards and forwards, lifting knees way up and lifting legs upwards behind me, with more mobility that I've had in months.

Heaven!

But then I discovered, once again, there is a heaven above heaven.

While I was swimming on my back, looking around at the pool, admiring the architecture and decor, the coloured triangles like sails displayed like a mural, I noticed, outside the huge windows on the north side, a couple of school buses pulling up and a lot of kids pouring out.

They were escorted by a big guy in a winter military outfit, so I assumed they were from some other school.

However, after a quick shower, wrapped in my big black towel, I walked into the change room to discover that it was full of kids from my school.

Dozens of boys from my Grade 3 and 4 classes, plus last year's Grade 5s, ran over in their bathing suits to give me a high five and ask me how I was doing and when I would be back. They invited me to stay and swim with them and told me what they were doing in classes without me.

The Grade 4s had finished Charlotte's Web and were doing a play.

After I had a chat with their teacher and a parent who was helping out, they headed for the pool, and I headed out, thinking that was better therapy than swimming for an hour, even, and that hour in the pool was great, great, great.

What do you do for an encore, after a physical and emotional workout like that?

How about steak at the diner by the lake?

Why not?

Hungry? -- I could eat a the legendary big fish of Cold Lake.

The big lake called Cold Lake looked serenely beautiful, in a northern way, with pale blue water to the distant horizon, just a blur, and frozen water covered with snow inside the big breakwall where docks held boats in the summertime.

Across the street was a diner I'd never tried before. Clarke's Eatery looked great inside, decorated with a lot of Coca-Cola collectibles plus lots of Christmas stuff. The restaurant had an Owen Sound tin ceiling. -- I don't know if they call them that, out west. And two walls were packed with memorobilia. One wall had windows with a great view of the wintry looking lake.

Sit wherever you want, the waitress said, so I took a round stool at the counter, and tried to talk them into making me steak and eggs, even though that was on the breakfast menu, and it was after 1:30 p.m. A steak sandwhich was just as good, with fries, white vinegar, and a 7Up.

It was like lunch in the past, like going to Avery's Marina and diner in my old hometown back in the 1950s, with the same chrome and red vinyl stools, the same counter, and an old fashioned milkshake machine.

That song about heaven that goes, "Heaven, I'm in heaven ...." was in my head but the diner's music took over, with blasts from the past, including a little Elvis Presley that sounded as though it was coming from a Wurlitzer juke box in 1962.

As I walked across the street, to the lake, to the car, I had to think about how I was walking, as my body wanted to use my glutes, and that hurt, a bit, but when I focussed, I could relax those muscles and use the piriformis, which I had exercised in the pool.

I felt I was back on track and I'd be back at work on schedule.

The pool cost just $17 per month, and the price went down if you signed on for three months, half a year, or a full year. And I could go to the early bird swim, from 6:15 to 7:30, the Noon Lane Swim, from 11:30 to 1:00, and the Lane Swim from 8 to 9.

There was public swimming from 6:30 to 8:00, but I was told the pool filled up with teenagers at that time.

When I got home, I saw my Scrabble score had jumped up to 1649, with 224 bingos.

What can I say? The better it gets, the better it gets.

P.S. In sports news .... Calgary Stampeder's QB, Drew Tate, who threw that long bomb to win the game against the Saskatchewan Roughriders, did it with a broken wrist, apparently. He would not be in their next game, the semi-finals against the B.C. Lions. -- Tough break!

Recovery Day 54: Sense Of Touch

In order to get in synch with the working world, trying to normalize my sleep patterns, I set the alarm for six thirty, and got up, had breakfast, checked e-mail, and continued something I started the night before.

Netflix tells you which movies and TV shows you should watch, based on your ratings of other things you have watched, and a show called Touch kept moving up on my list, so I thought I would try it, even though I don't watch television.

After teaching media for several years, I gave away my TV set and stopped watching the boob tube. It was quite a move for a guy whose first career was in the media. I used to read three newspapers a day, listen to the radio all day, and catch three or more newscasts at night, while writing press releases and calling reporters, editors, and columnists on the telephone all day.

As part of the first generation to grow up with TV, I loved it, as a kid, but I had a one channel childhood. Our TV only got CKVR TV in Barrie, Ontario, which was a CBC affiliate, so we didn't get many of the American shows, other than Bonanza and Ed Sullivan.

A few years ago, I watched 24, the series starring Keifer Sutherland as Jack Bauer, at the CTU or Counter-Terrorist Unit -- a very fast-paced, tight, dramatic, series full of conflict, cliff-hangers, death, and end-of-the-world style disasters.

The new series Netflix told me I would like starred Keifer Sutherland as a guy named Martin Bohm, a former newspaper reporter whose wife died in 9/11, now busy looking after their son, who is precocious but appears to be autistic, or afflicted with autism, and just might be one of the lamed vav, or the 36 people on the planet who, some believe, are the reason God keeps the Earth alive.

Lamed and vav are two letters in the Hebrew alphabet. In the traditional system of Kabbalist numerology, they represent the value 36: lamed = 30, vav = 6. According to Jewish legends, there always live thirty-six men, the Lamed Vav, who are also called Tzadikim Nistarim or the "Hidden Just Men". They are usually poor, unknown, obscure and survive by the sweat of their brows. No one can guess that they are the ones who bear all the sorrows and sins of the world. They don't even know it themselves. It is for their sake that God does not destroy the world,

The blurbs say Touch is about Martin Bohm, a widower and single father, haunted by an inability to connect to his emotionally challenged 11-year-old son Jake.

The show also stars a British actress with a Zulu name, Gugu Mbatha-Raw, as a social worker who handle's the kid's case.

The show is called Touch and it started in 2012 and was renewed for a second season, but then got bumped by some show called Kitchen Nightmares, or something.

This is why we don't like television: You get hooked on a series that is unusually bright, tightly edited, set in exotic locations around the world, with people speaking many languages, about communication and how people are connected, and it gets replaced by a reality TV show about cooking, or something.

Martin Bohm morphs into Jack Bauer, sometimes, but that's alright with me.

I really liked Jack Bauer.

My brother thinks he's Bruce Willis in Diehard.

I'd like to see Jack Bauer take on John McClane.

-- No I wouldn't.

I'd rather see Martin Bohm get it on with Gugu Mbatha-Raw!

So, I saw several episodes in the evening, back to back, with no commercials, the way I like to watch a TV series, and I saw a few more when I got up the next morning.

Edmonton called, but I had to cut it short, as I had an appointment for physiotherapy.

The acupuncturist was a little under the weather so I put myself on the treadmill and the balance board, she got me going with the big ball and leg squats, and then she needled me good and left me in the dark for twenty minutes.

She said she was impressed with the way I was walking, could barely believe I lasted an hour and a quarter in the pool, and told me she gave me quite an aggressive acupuncture session.

She said, You shouldn't swim for a bit, with all those holes in your body ....

So, I went straight home, made lunch, and got right back into the world of Jake and Martin Bohm, not to mention Gugu Mbatha-Raw.

Little Jacob never says a word to anybody, but we hear his inner thoughts at the start and end of each show, as he goes on about numbers and connections between people.

He has received no formal training in complex mathematics, he has discovered the Fibonacci sequence on his own simply through observation and his unique perception of the world around him.

He says things like, ""The ratio is always the same: 1 to 1.618 over, and over, and over again. Patterns are hidden in plain sight, you just have to know where to look. Things most people see as chaos actually follow subtle laws of behaviour. Galaxies, plants, seashells... the patterns never lie, but only some of us can see how the pieces fit together. 7,080,360,000 of us live on this tiny planet ....

And, "Human begins are hard wired with the impulses to share our ideas."

And "Numbers are constant until they're not."

Jake is one of the few who can see the "pain of the universe" through numbers. Teller also alludes to the interconnectivity of humanity as envisioned by the Chinese legend of the red string of fate, which says actions, seen and unseen, can change the fate of people across the globe for the better.

The red string of fate, or the red thread of destiny, red thread of fate, etc., is a belief from Chinese and Japanese legend. They say the gods tie an invisible red string around the ankles of people destined to meet in a certain situation or help each other.

In Japanese culture, it is thought to be tied around the little finger. This magical cord may stretch or tangle, but never break. This myth is similar to the Western concept of soulmates.

Other threads, as subplots, go around the world, and get tied together by wild coincidences. They involve terrorism, war, new media, social media, and contemporary communications devices, as well as numbers, plus other members of the lamad vav.

The name Bohm may be an allusion to David Joseph Bohm, who was an American quantum physicist who contributed to theoretical physics, philosophy of mind, and neuropsychology.

After physio and acupuncture, I felt fairly "out of it", so I drove home carefully, went to Netflix to watch Touch, again, and fell asleep until sometime around five o'clock in the afternoon when Edmonton phoned.

She was finished work early and said she was suprised to see it was as dark as night at five o'clock, already.

She was leaving on a jet plane in the morning, heading back to Texas, for American Thanksgiving, so I said, Bon voyage, but she said she would see me online.

While I was on the phone, I checked Facebook, e-mail, and the news, online, of course and discovered the Grey Cup was in Toronto, Alberta doctors had a new deal, and Hostess had gone out of business.

Toronto was hosting the 100th Grey Cup Game. The Argonauts had to beat the Montreal Alouettes to take on the winner of the big game in the West, with the Calgary Stampeders in Vancouver to play the B.C. Lions. My bet: the hometown team from the East would beat the Lions.

A Montreal/Calgary final for Grey Cup 100 in Toronto's old SkyDome? -- I don't think so!

Toronto's mayor was in the news, as usual, as he was the opposite of a media darling. The headline in the Toronto Star said, Rob Ford's libel trial: Mayor acknowledges 'corruption' comment.

There were a couple of new videos out, promoting Toronto, with time-lapsed photography, designed to make Toronto haters change their minds and think about visiting the biggest city in Canada, again. Planet Toronto had some spectacular photography but, to me, it emphasized the things we don't like about The Big Smoke or "T.O.": It had too many cars driving too fast along high-rise canyons of vertical ice-cube trays, with the light pollution of urban sprawl spreading to the horizons. The place we called SkyDome, beside the CN Tower, now had the name of a cellphone company attached to it.

Everybody loved Rogers. Right?

Canadians loved their cellphones, but not the cellphone companies.

The Alberta government was making doctors take a new contract after 20 months of negotiations failed to reach a settlement and the news came as a shock to the AMA president as the docs never had a contract imposed on them before.   The contract included a one-time lump sum payment of 2.5 per cent to each doctor based on their billings in 2011-12, annual increases tied to the Cost of Living Adjustment, and a $12 per-patient increase for Primary Care Networks.

 The changes were valued at about half a billion dollars over the next four years.

 Alberta doctors were already paid almost 30 per cent above the national average.

 In other words, Alberta's doctors, already among the highest paid in the country, were getting another hike in pay, and according to the AMA, the increase was not enough. -- That's the way the Edmonton Sun put it.

Meanwhile, people were lamenting the end of the Twinkie and Wonder Bread and other Hostess products, as Hostess Brands went under. Everybody joked about their Twinkies becoming collectors' items that would still be good a decade and a half from now.

 The Hostess Snowball was gone, too, of course.

 Hostess had gone the way of Zenith, PanAm, Polaroid, and Texaco, to the big brand name graveyard.

 I can honestly say I never tasted a Twinkie, Ding Dong, Sno Ball, Ho Ho, Cupcake, WonderBread, or any other Hostess product, except potato chips, and Hostess potato chips had vanished more than half a decade ago.

Watching the TV series called Touch a second time, while playing Scrabble online, instead of swimming after acupuncture, I liked it even more. That's something I've learned from my students, who like to watch the same shows over and over again. When I was a kid, we couldn't do that, so I'm in the habit of watching something once, and that's it. But sometimes, if I really like a movie, I'll watch it again and again. Watching Touch the second time, I Googled some of the things that come up on the show, like the Amelia sequence: 3185296328795229751188. It's a sequence of numbers that is said to tie everything in the universe together and can be used to unite the past, the present, and the future.

Although I was hoping to go swimming at night, from 8 to 9, I decided to follow the physio's advice: Don't swim on the same day as acupuncture.

 -- There's always tomorrow.

The Ice kept their streak alive with an 8-3 win in Lloyd versus the Bandits.   That win made it 7 straight and improved their record to 11-1.

 Matt Laboucane and Dallas Ansell each scored a Hat Trick!

 Next up: The Killam Wheat Kings Saturday at 8PM @ the Events Centre!

Day 55 of 80: People Like Us

In the morning, I woke up in the right position. -- That was a first.

In physiotherapy, as well as re-learning how to walk, sit, stand, get up, get into and out of a car, and so on, I also re-learned how to sleep. At the Zen Forest, I learned how to sleep, or how to do sleeping meditation, and it was like that, but with your feet tucked up closer to your rear end, your ribs aligned with your pelvis, and a curve in your spine.

Although I go to sleep in that position, I wake up with my legs stretched out, and my back straight, or some other position. But this morning I woke up in the correct position.

And it helped.

First, I got up at 3 in the morning, as usual, but then I went back to bed and got up at 7:11, without an alarm clock.

From 3 to 4, I exchanged e-mails with different time zones and checked out the news. My friend in Nashville told me her back was bugging her, again, with bulging discs after doing a lot of driving, as they were moving from a rented house to one they bought, so she lost a few days to back pain. Also, she was worried about the war in Syria, which was spilling over its borders, and now Israel was calling up reserves and going to war with its neighbour, apparently.

Closer to home, schools were calling on retired principals and vice principals, in Ontario, to help supervise playgrounds, as the teachers were cutting back on administrative duties. The big news was that more and more school districts were in a legal position to strike and they were making announcements about teachers walking out soon.

That's a scary, painful, process. And I've been there. So, I felt for them.

But going to war with a nation that is your neighbour is something else again.

In Alberta, they were talking about importing an army of workers from the USA as there were thousands and thousands of new veterans who were unemployed and a shortage of workers in the oil patch meant problems in productivity as there was infrastructure and pipelines to build as well as oil to be extracted.

Not all of those veterans would want to work in the oil patch up north, of course, as they likely thought Northern Alberta was as cold as the North Pole.

But the truth is that not even the North Pole is as cold as the North Pole, anymore.

My phone app shows that the weather in Fort MacMurray isn't a whole lot different than weather in Cold Lake, and Cold Lake isn't much different than Central Ontario, and last winter here was as warm as winter in southern Ontario, which gets warmer all the time. If I was an American, I would probably think that Northern Alberta would be twice as cold as Minnesota, and I would not want to go there.

Last night I had the strangest dream I never dreamed before: I was in Stratford, at the Shakespeare Festival, and my former favourite English teacher was there, running amuck, because he was freaked out about the state of local parks. None of this dream has any connection to reality, by the way. The last time I was in Stratford, the parks looked fantastic, and my old English teacher was not an environmentalist or a landscaper or an outdoorsy kind of guy who knew how to wield the equipment he was using in this dream. He was single-handedly tackling one park after another, repairing rivers, riverbanks, gardens, the whole landscape, cleaning them up and fixing problems, so they worked better and looked a lot better. In the dream, I was the go-between, preventing the management from going after him and stopping him from attacking them.

The dream came complete with full colour theatrical productions of big plays on outdoor stages, with elaborate special effects created by the technical crew.

Where did that dream come from?

I had a theory.

My research, online, into the regeneration of nerves, led me to St. John's wort. The alternative practitioner who does reflexology told me to wait a week or ten days after finishing the painkillers and anti-inflammatories that doctor gave me before starting any herbal remedies. So, I waited longer than that. But last night, I put 20 drops from a bottle of St. John's wort into a glass of water and drank it down, just before bed.

Sweet dreams are made of this, they say. And who am I to disagree?

It's Saturday and I have few plans for the day but lots of desires, but it depends on how much energy I have. At the top of my list is getting back into the swimming pool for more deep water walking.

Pool time on Saturdays is limited. There's lane swimming in the afternoon, for an hour, so I was planning to do that. -- But what about the rest of the day?

The Cold Lake Ice hockey team was on the road, so there was no Jr. B game in the Energy Centre.

In another dream, the team's name was changed to Energy. It was the Cold Lake Energy, instead of the Cold Lake Ice. Edmonton and Fort MacMurray had team names with the word "oil" in them. There was the Oilers of the NHL and the Oil Barons, the Oil Kings, and so on. -- How about calling a team the "Energy".

It might not strike fear into the hearts of hockey players on other teams, like scary sounding names such as the Phantoms, the Hitmen, or the Thrashers, Predators, Avalanche, Devils, Lightning, Panthers, Sharks, or Raptors. But Dr. Masaru Emoto, the star of What The Bleep Do We Know!?, would approve.

Dr. Emoto was the Japanese author best known for his claims that human consciousness has an effect on the molecular structure of water. He claimed that high-quality water forms beautiful and intricate crystals, while low-quality water forms the opposite sort of crystals.

According to Emoto, an ice crystal of distilled water makes a basic hexagonal structure with no intricate branching but positive changes to water crystals can be achieved through prayer, music, or by attaching written words to a container of water.

Emoto's book, Messages from Water, contains photographs of water crystals next to essays and "words of intent." His ideas appeared in the documentary called What The Bleep Do We Know?!.

Did he have a Facebook page? I checked. It is full of information about crystals, global water ceremonies, and his efforts to create a worldwide wave of peace, love, and harmony.

Please don't tell the Science teacher at our school this, but I have a tee-shirt with words in many languages for peace, love, and harmony printed all over it.

If a word on a bottle of water can change its crystals, imagine what words can do for the human body, which can also be seen as an incredibly specialized container of fluids. -- That's how the theory goes.

Imagine the effect of a team name on the players. Do the Cold Lake Ice play as though they have ice in their veins?

-- Yes!

What if they were called the Energy?

My friend in Alabama posted a few things first thing in the morning, including a picture of rescue workers digging out a woman's body after an earthquake, with a very touching story attached to it. Alabama added this comment: Caution -- may cause tears.

Here's the story:

After the Earthquake had subsided, when the rescuers reached the ruins of a young woman's house, they saw her dead body through the cracks. But her pose was somehow strange that she knelt on her knees like a person was worshiping; her body was leaning forward, and her two hands were supporting by an object. The collapsed house had crashed her back and her head.

With so many difficulties, the leader of the rescuer team put his hand through a narrow gap on the wall to reach the woman's body. He was hoping that this woman could be still alive. However, the cold and stiff body told him that she had passed away for sure.

He and the rest of the team left this house and were going to search the next collapsed building. For some reasons, the team leader was driven by a compelling force to go back to the ruin house of the dead woman. Again, he knelt down and used his hand through the narrow cracks to search the little space under the dead body. Suddenly, he screamed with excitement, "A child ! There is a child !"

The whole team worked together; carefully they removed the piles of ruined objects around the dead woman. There was a 3 months old little boy wrapped in a flowery blanket under his mother's dead body. Obviously, the woman had made an ultimate sacrifice for saving her son. When her house was falling, she used her body to make a cover to protect her son. The little boy was still sleeping peacefully when the team leader picked him up.

The medical doctor came quickly to exam the little boy. After he opened the blanket, he saw a cell phone inside the blanket. There was a text message on the screen. It said, "If you can survive, remember I love you." This cell phone was passing around from one hand to another. Everybody that read the message wept.   "If you can survive, remember I love you." Such is the mother's love for her child!

Share if the story is touching for you.

So, even though I was moved by that story, I Googled "earthquake mother child text message scam" and got a website that said it was, indeed, a scam.

When I told Alabama, she got mad at me!

The word from a scam reporting website called Waffles At Noon was that clicking on their "Reference" link will yield the same story with one additional sentence the Facebook crowd conveniently left off: "Note – This story may not be true but the message in the story is what really matters."

They explained that the photo comes from a news report of a Chinese earthquake in 2008 – not the Japanese earthquake as this story sometimes states – and of rescuers pulling two bodies from the rubble.

Looking closely at the photo, you can clearly see that there are two dead bodies, neither of which is the size of the infant described. Nor is the larger body in any sort of protective pose. The caption in the original story merely reads: Firefighters dig, bare-handed, two bodies out of the debris in a quake-hit county in Sichuan Province in 2008.

Commenters online wondered how someone would have time to text while a house was collapsing around them. The website concluded that to grab an unrelated photo and attach a manufactured story borders on scandalous behavior.

Next I Googled "neuropathy scam" and got a lot of interesting information. There were shysters out there selling snake oil to people who suffered from the pain that goes with neuropathy.

One of the links connected to an Amazon page with testimonials for and against a nerve support formula.

Ads for that formula popped up whenever I checked my e-mail or went on YouTube.

Amazon published lots of positive and negative testimonials. Many people complained about the follow-up phone calls and e-mails from the company's marketing team.

On Facebook, a Buddhist meditation group posted this comment: "Suffering is our call to attention, our call to investigate the truth of our beliefs." -- Tara Brach.

Love Tree posted this one: When you realize how PERFECT everything is...you will tilt your head and laugh at the sky. ~Buddha

Edmonton sent me a message on Facebook chat to say her airplane was stuck on the tarmac for an hour because they had to change a spark plug. However, she was happy because they showed a movie she liked called People Like Us.

She said, You HAVE to see this movie.

And she said it was great.

So I said, Bon voyage, Sparky! And don't forget: Everybody loves you!

-- I often tell her that.

People Like Us turned out to be a Michelle Pfeifer movie released in 2012 and already on DVD about a guy whose father dies and as a result he finds out he has a sister.

Last weekend, the Cold Lake Royals won their quarter-final game against the West Central Rebels by a score of 43 to 20, so this weekend they were in Edmonton for the final game before the championship. The CLHS Royals traveled to Edmonton Friday evening for the Tier 4 provincial semi-final against the Ardrossan Bisons.

It would be one of several games at Foote Field for all tiers of football across Alberta.

While I was watching People Like Us, there was a Cash Mob in Huntsville, the town in Muskoka where we had the novel marathon and where I used to hang out.

It turned out the destination of the Cash Mob, the store they went to, was .... Love Tree.

-- Nice!

People Like Us turned out to be a very moving movie for adults with a big kick in the heart at the end.

Reviews were mixed, so Rotten Tomatoes gave it a 56, but what do they know?!

-- I loved it.

Of course, I'm emotional, these days, due to my journey of the past 80 days, plus acupuncture, and a lot of time to meditate and contemplate.

Is it sentimental and Hollywood?

Well .... Yes.

Should you watch it for an emotional work-out?

-- YES!

The Cold Lake Royals lost their semi-final football game in Edmonton by a score of 49-33. They made a big comeback in the second half, but could not make up the difference, so their season ended just one game away from the provincial finals.

Their Facebook page filled up with tributes from parents and fans, saying congratulations on a great season, and commenting on the growth of the players as young men.

Somebody said they were on Global TV, with another team, and the reporter got them mixed up, but it was a good report anyway. Somebody else posted a video of one of their pre-game rituals. Before games, the boys took the field in uniform and sang the national anthem at the top of their lungs. They did warm-up exercises together, chanting their team name, spelling it out and then yelling it. -- Their fans in the stands always got choked up when they did that.

Swimming, the second time, was not as much fun. It was more like work. It was only half as much fun as the first time.

Why? I don't know.

It was a little more crowded on Saturday. And for some reason, I had a lot less energy.

Odd. -- I'd been looking forward to it for days and had a relaxed day, so I should ha e been ready.

My glutes hurt, on the left side, half the time I was in the pool, but I carried on, trying to work it out.

The hour went by as fast as the guy swimming in the far lane, while I walked forward and backward beside the slow lane.

Swimming a few lengths felt good, but I focussed on water walking, so I could get back to work in just over one week.

Frankly, I could no longer see that happening. While I was walking in the water,, I worried about that and planned to call for a doctor's appointment, and wondered if I'd need a physiotherapist's assessment.

I already knew what she thought I should do: Go back to work half time.

Why didn't that appeal to me?

After swimming, my butt muscles hurt a bit, so I walked slowly, compared to the long, smooth, strides I practiced in the pool.

Steak with mushrooms at Clark's General Store, beside the lake, helped a bit.

What about the hockey game? The Cold Lake Ice versus the Killam Wheat Kings. -- I drove right by the hockey rink at The Energy Centre in Cold Lake without stopping.

Cold Lake looked like the Arctic Ocean as it was a grey day and overcast skies were reflected by the big body of cold water.

It looked as bleak as Iceland in winter and reminded me of a dull day in Grey County, when I lived in Owen Sound, beside Georgian Bay, where the locals, like Icelandic people, took delight in distinguishing between shades of grey, because that's all there was.

Recovery Day 56: A Leg To Stand On

Last night, instead of going to the hockey game, I decided to stay home and watch TED videos. The Cold Lake Ice game against the Killam Wheat Kings went to overtime and then they lost, 3 to 2. That was just the second game the Ice lost all season.

So, the Royals and the Ice both lost on the same day.

It was not a great day for Cold Lake.

The TED videos I saw were great.

Instead of venturing out, I decided to turn up the heat, make some comfort food, and hit the sack, watching Oliver Sacks do a TED video.

TED stands for Technology, Entertainment and Design, and the TED Talks come from a global set of conferences owned by the private non-profit Sapling Foundation, started in Monterey, California, in the Silicon Valley, in 1984, formed to disseminate "ideas worth spreading."

They have proved to be very popular, with over a billion viewers, over the years. The "talks" or lectures have taken place in Europe and Asia, as well, and there are Canadian TED Talks, too. My friend Marty Avery did one in Banff. She was brilliant.

The speakers are given 18 minutes to present their ideas and past presenters include Bill Clinton, Jane Goodall, Malcolm Gladwell, Al Gore, Bill Gates, and many Nobel Prize winners.

Since 2006, the talks have been offered for free viewing online. Sir Ken Robinson's 2006 talk about how schools kill creativity tops the list with 13,409,417 views. It's followed by brain researcher Jill Bolte Taylor's epic story about suffering a stroke — and documenting her body shutting down.

The top 20 list includes Tony Robbins, Steve Jobs, and Stephen Hawking. And then there was Mary Roach's talk on 10 things you didn't know about an orgasm.

Olive Sacks talked about hallucinations.

Sacks is a British born doctor who has lived in the U.S.A. most of his life and is famous for his book called Awakenings, turned into a movie starring Robin Williams and Robert De Niro. His other books have become best-sellers, too. I read The Man Who Mistook His Wife For A Hat years ago and went to see Sacks speak at the University of Toronto. For a guy with major celebrity status, he appeared to be incredibly humble, like a nervous, first time, public speaker.

Sacks writes about his neurological patients in a way that has made him the poet laureate of contemporary medicine. Some of his other bestsellers are titled An Anthropologist On Mars, Seeing Voices, and The Island Of The Colorblind.

He writes about tourettes, visual agnosia, autism, achromatopsia, and other neurological disorders. His fans say he writes with compassion and makes obscure topics fascinating. His critics say he exploits his patients. Tom Shakespeare, an advocate for the rights of people with disabilities, called Sacks "the man who mistook his patients for a literary career".

I think he would love the tv show called Touch.

In his book The Island of the Colorblind, Sacks writes about an island where many people have achromatopsia (total colour blindness, very low visual acuity and high photophobia), and describes the Chamorro people of Guam, who have a high incidence of a neurodegenerative disease known as Lytico-Bodig disease (a devastating combination of ALS, dementia, and parkinsonism). -- That one hit home with me as I'm a little colourblind, like most men, including my brothers, and the neurological specialist at the Edmonton Clinic scared the heck out of me by suggesting the cause of my current problems with walking could be ALS or Parkinson's, as well as other things, like cancer, MS, or AIDS.

-- I forgot to send that guy a 'thank you' note.

Bill Murray played a character like Sacks, or a caricature, in the movie called The Royal Tenenbaums. The film stars Gene Hackman and Anjelica Huston, with Danny Glover, Bill Murray, Gwyneth Paltrow, Ben Stiller, Luke Wilson, and Owen Wilson.

In his TED Talk, Sacks said hallucinations are a lot more commonplace than most people are willing to admit. Have you ever seen something that wasn't really there? Heard someone call your name in an empty house? Sensed someone following you and turned around to find nothing?

His new book, Hallucinations, was published this month, November of 2012. Dr. Sacks weaves together stories of his patients and of his own mind-altering experiences to illuminate what hallucinations tell us about the organization and structure of our brains, how they have influenced every culture's folklore and art, and why the potential for hallucination is present in us all, a vital part of the human condition.

Most people don't want to talk about their hallucinations because, of course, we don't want people to think we are crazy, have dementia, or should be off work for a long time, talking with a team of psychiatrists.

That's a line from Field Of Dreams, the movie based on the novel Shoeless Joe, by Alberta author W.P. Kinsella. "You're seeing a whole team of psychiatrists, aren't you."

Dr. Sacks is a shy guy who lives alone, is celibate, he says, and goes swimming, alone, every day. He refers to his shyness as a disease. Also, he has prosopagnosia, or face blindness: although he has an excellent memory and has no trouble recognizing objects, he has a terrible time remembering faces. He could almost be 'the man who mistook his wife for a hat', except he never had a wife.

In his book called A Leg to Stand On, the doctor becomes a patient, as Dr. Sacks chronicles a mountaineering accident which left him with the uncanny feeling of being "legless," and raises profound questions of the physical basis of identity.

An encounter with a bull on a desolate mountain in Norway left him with a severely damaged leg. But what should be a routine recuperation is actually the beginning of a strange medical journey, when he finds that his leg uncannily no longer feels a part of his body.

Sacks's description of his crisis and eventual recovery is not only an illuminating examination of the experience of patient-hood and the inner nature of illness and health, but also a fascinating exploration of the physical basis of identity.

"Here the roles were reversed and I was the patient myself, bewildered by an experience, a sort of "alienation" of an injured leg, which I could not comprehend or communicated to my doctors," Sacks writes. "My only relief was to write about it."

-- I could relate!

My sister sent me some photos of her trip to a little get-away place in a wine valley in South Australia and a brief note saying 'good luck getting back to work' and she was excited about getting away from work.

With her co-vivant, she went to Lyndoch Hill in the Barossa Valley for the weekend. -- She is the relationship counsellor who does not believe in marriage. She believes they are 'better than married'. They say they don't need a piece of paper from the government or a service in a church to be married. They are committed to each other, and that's that. They have been together for three decades.

My sister has a theory about me. She said, You identify with your job, don't you. -- Your identity is wrapped up in what you do for a living. If you aren't teaching, you don't feel like yourself. Right?

She said she didn't have that problem.

It's not a new theory. However, I had never heard anyone apply it to me, before. Nobody said, Teaching is his life. They said, That guy loves teaching and he loves working with kids and he loves young people and he loves his students, but nobody ever said my identity comes exclusively from my job.

It's true that I love teaching, et cetera, and I love it more and more every year. But I have other loves, too. And I also love doing nothing.

I love writing, reading, hiking, drama, media, meditating, working out, coaching, communicating with people close to me, thinking, and many other things. I miss skating and skiing. But these things do not define me or determine my identity.

My identity has never been an issue. I've always felt like myself! I've taken on lots of challenges and adventures that have changed me and I've had lots of life-changing experiences, but I've always come back to myself.

Shakespeare said, "We know what we are, but not what we may be."

Mahatma Gandhi said, "The best way to find yourself is to lose yourself in the service of others."

The narrator in the magical realist novel called Love In The Time Of Cholera, by Gabriel Garcí-a Márquez, says, "He allowed himself to be swayed by his conviction that human beings are not born once and for all on the day their mothers give birth to them, but that life obliges them over and over again to give birth to themselves."

Zen says that if you drop knowledge - and within knowledge everything is included; your name, your identity, everything, because this has been given to you by others - if you drop all that has been given by others, you will have a totally different quality to your being: innocence. This will be a crucifixion of the persona, the personality, and there will be a resurrection of your innocence.
You will become a child again, reborn. - Osho

Those are my thoughts on Sunday morning.
        Mostly I'm thinking about the pain in my butt that is preventing me from walking.
        First I lost my nerves, apparently, and then my muscles, in my legs and feet, and now I have a pain in my gluteus maximus. Or is it my piriformis?
        The good news is that means I'm getting better. -- Slowly.
        My nerves and muscles are making a comeback, but it's a pain in the butt!

I Googled glute pain and found a long article called Piriformis Syndrome: A Real Pain In The Butt, by Dr. Nick Campos, sports chiropractor in West Hollywood, California. He works at the West Hollywood Healing Arts Center.
        Dr. Nick's long, informative, piece can be summarized in four short points:
1. Piriformis syndrome is painful and can lead to nerve impingement which is never something to take lightly. Nerve damage can be permanent if not addressed soon enough.
2. Piriformis syndrome is usually the result of some other dysfunction whether it's a flexibility issue, a muscular imbalance, or a foot dysfunction.
3. Treating the piriformis muscle alone, without addressing the underlying cause will lead to a chronic problem.
4. Piriformis syndrome is reversible but it takes a little care.

Short, tight muscles can cause a number of problems including joint dysfunction, circulation problems, poor posture, and as in the case of piriformis syndrome, nerve entrapment.
        One solution to piriformis syndrome is to stretch regularly.
        Two stretches in particular are superb for relieving piriformis syndrome—the supine piriformis stretch, and a yoga pose called pigeon.
        To do the supine stretch, you need to lie on the floor on your back with knees bent. Crossing one leg over the other, push the lever side (which is the side not being stretched) actively toward your chest. By default, the bent leg will also move closer toward your chest, stretching the piriformis muscle. This stretch is excellent, especially for people that

have poor flexibility to begin with. For those a bit more flexible, the pigeon stretch will be much more effective.

The pigeon stretch is done on the floor or on a yoga mat. You need to bring your affected leg in front of you with a bent knee and bent hip. The unaffected leg should be stretched straight out behind you. The benefit of this stretch over the supine one is that you will get a much deeper opening of the piriformis muscle—great for those of you who already stretch a little now. Further, the hip flexors of the unaffected side also get stretched in pigeon, making it a great all around hip opener.

On my way to the pool, I heard some of The Vinyl Cafe on CBC Radio One. Dave swallowed a fly and feared he would get a dread disease that would cause neurological problems that would lead Dr. Sacks to seek him out to watch him do complicated math problems on a blackboard at some university while trying to find a cure.

That was funny because I had a dream about Dr. Sacks coming to visit me to find a cure for the ideopathic neuropathy that left me without a leg to stand on.

When I walked into the sports complex on the base, the Argonauts and the Alouettes were tied at 17, but when I walked out, Toronto had won the game over Montreal by a score of 27 to 20.

Next up: Calgary at Vancouver.

Next weekend: Grey Cup #100.

Would Toronto be happy? Or did they just want to get out of the CFL and get an NFL team? They were looking at the Buffalo Bills. Would their new name be the T-Bills?

That's a little joke. Toronto is known as a financial city. It is not a manufacturing centre or primarily a port or an industrial city. It is known for Bay Street, which is the home of the TSE, the Toronto Stock Exchange. T bills are treasury bills -- usually seen as safe, short-term investments, issued by the federal government and some provinces, in Canada.

Toronto! How we all love to hate Toronto and its big league sports teams!

Toronto is so unlike the rest of Canada, or even the rest of Ontario.

Toronto should be a city state, a new province of Canada, even though Torontonians identify more with people in major American cities and might prefer to be an American city state. -- That's just my opinion.

I used to love Toronto, in the Sixties, when the population was a lot smaller, around two million, and you could actually get around the city fairly easily. Now the GTA has six million people and gridlock. In the near future, the GTA will spread from Toronto to Oshawa, to the east, and over to St. Catherine's, in the west, or maybe all the way to Niagara Falls, and up to Barrie, in the north, and it will be called The Greater Golden Horseshoe, with a population of twelve million.

That seems un-Canadian, to me.

Now Edmonton is my favourite city. It has a population of around one million.

Well, actually, Cold Lake is my favourite city, with a population of around 15,000.

The swimming pool in the City of Cold Lake, on the base, was quite un-crowded on Sunday afternoon. There were a couple of other guys in the water, but they left after half an hour. I was the first person in the pool and the last to leave, and I had it to myself for the final half hour.

To make the physiotherapy session in the pool go better than the last time, I paid more attention to my warm-up. First I had some caffeine, then I walked over to the two NHL size hockey rinks in the sports complex, and I did some stretching in the shower before I got into the pool.

It was one hour of water walking, with just a little swimming, and it felt pretty good. The pain in the butt came and went. The more time I spent in the water, the better I felt.

Just for fun, I raced the guy swimming in the lane beside me, a few times. He beat me the first time, but I beat him the next three lengths.

This time, I added the overhand crawl to my repertoire. I beat the guy next to me doing the crawl and then I beat his crawl with my breast stroke.

My breast stroke is pretty fast. My best time was 14.7 for 25 metres. That was a decade ago. When I coached swimming, we had a technical coach who was just off the Olympic qualifying time for the breast stroke, and he showed me a few tricks. My swim team did well, for a small school. I took 4 swimmers to the provincial finals, and they finished in the top ten.

The men's 100 metre breaststroke event at the 2012 Summer Olympics at the London Aquatics Centre was won by South African Cameron van der Burgh in a world-record time of 58.46.

After swimming, I felt okay, so I walked around the superstore, got some swimming goggles, and noticed that my butt muscles were sore, again.

While I was in the pool, water walking, I felt optimistic, and made plans, in my head, to go to work for a short time every day for the next week, to get back into the flow. I wanted to go in at the end of the school day and talk to colleagues. Maybe I could do a drama and basketball practice, too.

While I was walking out of Walmart, I was in so much pain, I wondered if I would ever be able to handle a day at work in a school.

Want to hear a horror story? One winter when I was coaching swimming in Ontario, I took part of the team to a meet down south in Orillia. The boys on the team travelled with me, in my Tracker. There had been freezing rain and the four lane highway had not been plowed. We had to leave early in the morning in order to make it to the meet on time. One lane of the highway was good and all the traffic was in a slow moving convoy, which would not get us there on time. So, I pulled into the other lane, to pass ....

When my little truck hit the ice, we were spun around, so we wound up going backwards, my tires spinning on the ice, the median right beside us, and the three guys who were my passengers were silent as a graveyard in winter.

I took my foot off the gas, of course, turned the wheels to the left, and spun around again, to get back into the right lane.

The car behind me had backed off to give me room. -- It was the car carrying the girls on our team. -- They were all watching us slide backwards on the highway, horrified.

As soon as I could slow down safely and pull off the road, I parked beside the highway, walked around my vehicle, made sure my wheels were all pointing in the right direction, then got in again, and rejoined the convoy.

We got to the meet a little late but my guys said, It's alright, we don't have to warm-up, you got our adrenaline pumping with that awesome display of winter driving!

They each had their personal best times in the pool that day.

I was just happy to be alive!

The big news of the day, internationally, was that U.S. President Obama was at a Buddhist temple in Thailand.

A crowd chanted "Obama, Obama" as he arrived in an armoured Cadillac at Thailand's most famous temple compound and Wat Pho or The Temple Of The Reclining Buddha.

The centrepiece of the temple is a Buddha twice as long as a swimming pool or half the length of a football field, and fifty feet high, covered in gold.

My Calgary buddy in Toronto posted this note in Facebook: There is going to be a horse in a downtown Toronto hotel lobby this week!!! GO STAMPEDERS!!! Yahoo!!! Every Grey Cup that Calgary has played in, a horse has been ridden into the lobby of a downtown hotel of the host city. The tradition started in the Royal York lobby in 1948. Now I'll have to get a ticket to that game without putting myself deeper in debt!

It was true: The Calgary Stampeders upset the B.C. Lions 34 to 29, to advance to the Grey Cup.

For the 100th Grey Cup, the Stampeders would face the hometown Toronto Argonauts.

The Argos would play in their first Grey Cup at home since 1982, when the Edmonton Eskimos edged Toronto 18-17 at Exhibition Stadium. They haven't won the CFL's championship game at home since 1952, when they beat the Eskimos 21-11 at Varsity Stadium.

It sounded like a movie script.

If we lived in Hollywood, the hometown team would win, after a few complications, with a Hail Mary in the final moment of the game, after trailing at half time, and an injury to a key player, and in the end it's the team concept that overcomes all.

But there's more drama: First of all, there was a mass exodus of players and a coach from Calgary to Toronto in the off-season. The Argos got fined by the CFL for tampering.

Toronto beat Calgary the last time they played -- and injured their quarterback's shoulder. The game before that, the Argos won with some illegal hits on the Stamps' star running back.

So, who's cheering for Toronto?

Not me!

Meanwhile, Yahoo! News Canada posted an ABC news report saying climate change was still controversial in the U.S.A., but only in the U.S.A., and it wasn't mentioned during the American presidential election, but the American Broadcasting Corporation was reporting that "global warming is really happening".

I posted it on my Facebook page, and another page I have, called Arctic. But would anyone respond? Or were they obsessed with the upcoming Grey Cup game? Or were they consumed by the #1 interest of FB fans: pictures of cute cats and kittens doing silly things?

Facebook had a shortcut code for a cat-face emoticon :3 .

The cat faced smiley emoticon is meant to represent a cat or kitty's face and was created so you can add it to your messages to act kinda cute and quirky.

-- This is how we fiddle while the Arctic melts.

Recovery Day 57: I Don't Like Mondays

On Monday morning, I got up in time to go swimming, without the alarm clock, but decided not to go, since I had physio later in the day. Besides, I was stiff and sore. It felt as though I had been in a hockey game the night before.

In particular, my thigh was sore, on my right leg.

That upset me for a while, but then I realized my butt muscles were NOT sore, for a change. I went back to bed and put a cold pack on my thigh.

What else? -- I was tired: sleepy and .... not exhausted, but .... well .... tired.

I got up again at 8:00, yawned a lot, decided to have a breakfast bite, and went back to bed. -- What an exciting Monday morning.

What if I had to go to work today? Would I have made it?

-- Unlikely!

Would I be able to do it one week from today?

On Facebook, Love Tree shared Love, Light & Truth's photo, which said, Surround yourself with light. Keep your focus on love and have faith in your abilities.

I liked that, but I wanted to turn off the lights, crawl back into bed, and pull the covers over my head.

As for love and faith ....

I would love to go back to work.

Did I have faith in my physical ability to handle my job as a teacher?

My plans for the day included physiotherapy, swimming, going to work for a little while at the end of the school day, and making an appointment to see the nurse practitioner in Cold Lake, instead of the doctor in Lac La Biche.

Physio wasn't until 1:30. Swimming was from 11:30 to 1:00.

The bed was right beside me ....

Should I see the doctor or the nurse practitioner before I returned to work?

The doc asked for a report from a physiotherapist. He recommended four to six weeks off work. The doc wrote a note for four weeks. I changed physiotherapists and I signed on with the nurse practitioner.

Should I see her before going back to work?

-- Couldn't decide.

After sleeping for another couple of hours -- almost three -- my leg was still bugging me, so I put my Dr. Ho's pain therapy machine on it for a little electrical massage.

Although I missed the start of the noontime swim, I could still make it for part of the session. But .... Maybe I'd give it a miss for the day. -- Or go in the evening, form 8 to 9, if I was feeling better.

Sometimes physiotherapy and acupuncture can take a mickey out of you.

Does anybody ever say that, any more?

"Take a mickey out of you."

I've never said that before. -- Sounds like something my father or grandfather used to say.

The Urban Dictionary says it's a racist term but other online sources say it just means to tease somebody. -- I always thought it meant to take the fight out of somebody by tiring them out.

I feel as though I need a shot of energy.

If I was a coffee drinker, this would be a good time for a cup of jo, some java, some hot liquid sunshine. -- But I'm not.

I like cold drinks more than hot drinks. -- Last year I tried the energy drinks. Coffee flavoured Java or Monster energy drinks. -- Despite the warnings.

Yesterday the Toronto Star had an article about energy drinks being linked to the death of three teens, but the energy drink companies said they sold millions of cans without any ill effects. Fox News did a feature on 5 Hour Energy as it was under investigation by the FDA. Thirteen deaths were being investigated in the U.S.A.. The CBC said there were renewed requests for research in Canada, because the FDA was investigating in the U.S.A..

Eat Right Ontario said most energy drinks have ingredients like caffeine, sugar, taurine, vitamins and herbs and pointed out that energy drinks can be found anywhere you buy beverages, beside the pop, juices, and sports drinks.

Men's Health Magazine was against most energy drinks, too, saying a lot of caffeine and sugar is bad for you, the extra B vitamins did nothing for you, the other ingredients were unproven, and if you took energy drinks in combination with multi-vitamins you could raise your risk for cancer.

They said, For a caffeine-free boost, sip on FRS Healthy Energy drinks. They're free of folic acid and contain reasonable levels of the other B vitamins. What makes FRS effective is quercetin, an antioxidant that can help you fight fatigue during exercise, a 2010 University of South Carolina study found. Like caffeine, quercetin also blocks brain receptors for adenosine—a chemical that makes you sleepy—to make you feel energized, says study author Mark Davis, Ph.D. "Over time, it can also increase the number of mitochondria in your cells," he says, "which provide energy for your muscles."

But it wasn't available in Cold Lake, Alberta.

After ice, rest, electricity, hot water .... my leg feels about the same.

But it's time for physio and acupuncture.

-- At least my glutes are no longer the problem.

-- Don't know about swimming after physio ....

It turned out that physio made me feel better and acupuncture made me feel worse, so swimming was out. All I did was stagger home and sit down.

As for my plan to go to work .... Well .... I put that on hold.

While I updated my notes on physiotherapy, I listened to a Metro Morning podcast from Toronto on CBC Radio One. Matt Galloway spoke with Sam Hammond, president of the Elementary Teachers' Federation of Ontario, about the teachers' plan to 'work to rule'. He said, "Job action will not affect day-to-day, in-class activities because the focus of the action will be on administrative events such as staff meetings".

He also said that the provincial government created this crisis in education and he wished the Teachers' Federation wasn't in this situation, but they have to deal with it, and combat Bill 115.

The Ontario Secondary School Teachers Federation announced it reached an agreement with three or four school boards, but no details of the deals had been released.

The sports new from Ontario was all about the goalie from Gravenhurst playing for the Jr. A Owen Sound Attack as Jordan Billington was named the OHL player of the week, after earning two shut-outs, and setting an O.S. franchise record for wins, with 66.

He had a big game with Team Canada coach Steve Spott on the opposing bench and the Team USA goalie in the opposite net. Binnington has a good shot at being named the #1 goalie for Team Canada in the World Juniors tournament.

That was one of the worst days of the past three months of injury and recovery and relapse.

Would tomorrow be any better?

Day 58: Another Monday

What if one week had two Mondays?

This morning I got up at 5:45 instead of 3:00 or 4:00 and I felt the same as the day before. After a day of rest, except for physio and acupuncture, nothing had shifted. The sensations were the same. I was still in a good deal of pain. And it was not having a good effect on my brain.

Well, maybe it had shifted. Maybe the pain in the butt, on my right side, had gone deeper. -- It had moved up from my thigh to my hip.

When I read a novel I love, I don't want it to end. It's like the long stories of Alice Munro that you wish were full-length novels.

This feels like a novel sequence, a ten book series, that should have been a short short story.

If it was a movie, like those I saw yesterday, I would have clicked it off a long time ago.
-- The Jetsons and Picture Perfect really didn't do it for me!

Today, for acupuncture, I'm supposed to wear shorts, so it's easier to needle my legs. And in physio, I'm supposed to be lifting weights.

That's what I have to look forward to.

I have to work on my attitude, too.

Where is Mr. Positive? What happened to the attitude of gratitude?

Alabama sent me a few notes on FB chat. One said: It is turning out to be a REAL Thanksgiving! One best little Buddy has found out she WILL be a grammie after arduous years long expensive trial and previous disappointments in efforts to make a baby. thank you thank you thank you!

FooFoo (Small PAWS rescue dog) was returned to my son and oddly the 3 legged cat MEEKA was returned to my daughter TODAY after a 2 month incredible journey.

It feels like the family and friends are truly Blessed.

Happy Thanksgiving Y'all!

Edmonton was in Texas for American Thanksgiving. Her plane was delayed on the tarmac as it needed a spark plug, and then it was diverted to a different airport, so they gave her a night in a hotel. That screwed up her rental car reservation. She went to the same company in the new city and they told her they didn't have any cars left, so they rented her a pick-up truck. But she thought it was funny, and was still laughing, so it's all good.

In Canada, stores were following the American tradition of celebrating Black Friday right after American Thanksgiving, even though Canadian Thanksgiving is celebrated earlier. Black Friday is the day stores say they are finally in the black, or out of the red, so they celebrate with sales that kick off the Christmas shopping season. The following Monday is Cyber Monday, they say, when online retailers have big sales. In America, they are afraid the government is about to go over a fiscal cliff. They fear Christmas shopping won't be enough of a stimulus to the economy to prevent the country from going over the cliff.

Canadians who live close to the border with the U.S.A., which is the majority, are getting into the habit of heading south for Black Friday sales, so Canadian retailers now have Black Friday sales, to keep Canadians shopping in Canada.

I'd like to buy .... health!

In Northern Alberta, we are not close to any American cities. Even in southern Alberta, it's a long drive to the U.S.A., and you would have to travel for a long time until you reached a city that's larger than Calgary.

Calgarians drive up to Edmonton and Edmontonians drive down to Calgary, or they fly down to Las Vegas, Nevada, population two million.

Next weekend, a lot of Calgarians and a few Edmontonians will fly to Toronto, for the Grey Cup Game, dressed in Stampeders gear, to watch the best in the West take on the Argonauts.

Andrew, my buddy from Alberta who now lives in Toronto, says he is dressing Calgary all this week, heading into the Grey Cup Weekend. That means cowboy boots and Stampeders gear. And I hear Toronto is getting excited about the big game, as sports fans in T.O. are starving for a championship of some kind to celebrate.

A friend of mine in the big city posted a video on Facebook shot by Argos TV that showed the action in Toronto's dressing room after they won their big game against the Montreal Alouettes and celebrated their Eastern Championship, briefly, while starting to think about the CFL final.

Meanwhile, some NHL players in Montreal created a baseball cap that said Puck Gary on the front, to express their feelings about the hockey lock-out.

It was time to watch Blindside, Alice's Restaurant, Home For The Holidays, and A Charlie Brown Thanksgiving, once again.

-- Wait a minute .... I'm a Canadian! I don't celebrate American Thanksgiving!

It's time to watch the Grey Cup Game!

What I really wanted to do this week and weekend was swim my way from hell to health and then go to work next week.

What time was my physio appointment?

When did I have reflexology again?

I felt as though I had played in a football game and needed more than a week to recover.

The New York Times best seller book list had nothing that interested me.

Alabama posted a cool video on FB. It was made from security cameras capturing the opposite of crimes: It showed people stealing kisses, dancing in the street, returning dropped wallets, hugging old friends, helping motorists in trouble, people protesting against racism and for peace, and heroes doing all sorts of things from pushing stalled cars off railroad tracks, just in time, to just buying somebody a Coke or a coffee, paying it forward.

The little video had a cool soundtrack of the old Supertramp song called "Give A Little Bit" (of your love to me)!

-- That was inspiring!

I went to physio and the therapist wrote up an assessment for the doctor recommending more time for daily physiotherapy and not returning to work on Monday or any time soon. She says I am "significantly deconditioned with residual neurological weakness and sensation deficits".

In short, it looks as though it might be another couple of weeks before I return to work.

And I'm not even upset about that.

Last week, I wanted to fight it, and get back in there, go to work, return to the classroom, get drama and basketball going, and all the rest of it.

Yesterday and the past few days have been bad, though, so it's all changed. I've been stiff and sore and walking less.

The physiotherapist says my recovery is no longer a straight line, it's more like stairs that go up and down, as I've been going up two and then down one, which is not unusual.

She also says I should limit my time in the pool to half an hour and go to her for daily physio for one to four hours for a couple of weeks.

And right now, that sounds good to me!

Before I went to physio this morning, I had an idea for a novel for my Grade 3 kids, and I started writing it. And it felt good to do some creative writing, some fiction, for the first time in about three months.

While I was stuck full of pins for acupuncture, lying face down in the dark, I got an ending for the novel, plus a quote for use as an epigram at the start, from The Answer Man, and felt I could write the whole thing as soon as I had the time.

In The Answer Man, the author goes to a parent-teacher conference and the teacher says the student they are discussing is immature. So he says, He's seven!

She says, He tells jokes to his seat-mates while I'm talking to the whole class.

Google gave me the script. Here's how it goes:

Teacher: And he's also... oh, how do I say this? Well, immature.

Answer Man: He's seven.

Teacher: Yes, but he still jokes with his seatmates during a lesson and sometimes while we're trying to have peaceful time. The other students still have to learn all the same lessons, and Alexander often makes this difficult. I think if Alexander's really going to thrive...

AM: Let me stop you right there. I'm gonna tell you a couple of things. All kids develop in different ways at different times and in different directions.

Teacher: I don't think you're hearing me.

AM: What I am hearing is that Alex is not a good candidate to be a robot in your clone army. In the meantime, why don't you try not making him feel like being who he isis the problem? That's what happened to you and me, remember? Oh, and by the way, his name is Alex. He fucking hates "Alexander."

The new physio assessment letter said: Mr. Avery has been attending pysical therapy since October 20, 2012 for treatment of a severe back injury with significant neurological involvement. He is currently on short term disability and about to transition to long term disability. Treatment has included exercise, education, core stabilization, balance exercises, manual therapy, and traction, as well as acupuncture. We have been successful in improving his sensation in the lower extremities, decreasing his overall pain, and improving his function to a degree. He is now able to walk unassisted, he can climb stairs, is able to tolerate sitting for approximately half and hour, and transition movements. He is now going to the pool regularly to assist with his recovery. It was initially planned that Mr. Avery would try to return to work November 26, 2012. This gentleman is significantly de-conditioned and continues to have residual neurologic weakness and sensation deficits of L4, L5, and S1 nerve roots. It is my opinion that return to work at this stage is premature and may cause regression in his condition.

I recommend a functional abilities assessment followed by a comprehensive rehabilitation program that would more successfully prepare hi for a gradual return to work program. I have had conversations with his employers re: the benefits provider and to discuss with the superintendent.

Reflexology, after acupuncture and physiotherapy, after four days of that pain in the butt, proved to be a very good idea. Micell Pilon the alternative practitioner and reflexologist took away the pain once again.

She was a little surprised about the physiotherapist's report, as she thought I was a little closer to recovery. Angela said I was operating at about 50% but Mecelle said she would have put it at 70%. -- But she hadn't seen me for a four or five days.

I told her about the pool and the pain. She said to stick with swimming, or water walking, as well as whatever Angela wanted me to do every day, for an hour or four, and not to worry, because she would be there to help me with the pain.

So, the next day I planned to present the doctor with the physiotherapist's letter and two forms: one to go back to work and one to for a couple of weeks of LTD.

And I prayed he would send me back for my physiotherapy and acupuncture.

PS My Scrabble score went up to 1650, with 227 bingos, so I felt lucky, again, but then the pain in the butt came back again.

Recovery Day 59: Are You A Belieber?!

Almost sixty days of recovery after 28 days of going downhill after injury and the pain the alternative practitioner said was sciatic got so bad overnight that I could barely walk at midnight or in the morning when I woke up around six o'clock. Also, I was unusually sensitive to cold, I noticed, when I iced the area.

    It was another night of sleep full of vivid dreams. The one I remembered was in a coastal forest with a buddy and a dog, filling in a bay with old logs, discovered historical artifacts, and telling a girlfriend what we had discovered and accomplished. It was unclear where in the world we were, but it was a deciduous forest with no evergreens and a lot of big old birch trees about two feet in circumference. A lot of them had been knocked down and were in various stages of decomposition but none of them were so rotted that they fell apart. We rolled them down a little hill into the water and jockeyed them into place like lumberjacks.

    After that, I had a day dream recollection or memory of doctors from the time I was a kid, starting with Dr. Fisher in The Fisher Clinic in my old hometown, through doctors in university, Halifax, Victoria, Winnipeg, Toronto, Grey County, and Bracebridge, over the decades, going for an annual check-up, and the time I got plantaar fasciitis, coaching soccer, moving the team from indoor training in the winter onto the field in spring, how the principal didn't believe me, even though the VP and everybody else did, including doctors and specialists, and then moving to Durham Region, having no problem finding a doctor, until I moved into a town up north, called Port Perry, where there was a six month waiting list, so you had to go to a walk-in clinic with a waiting time of an hour or two. And then I moved to Alberta, where there were no waiting lists, and after a year or so I figured out you had to look for a doctor in the region, not just the city you lived in. And I reviewed the alternative practitioners I had met, while I had plantaar fasciitis, and how I tried a lot of things, and they all worked, for a while, until I found an acupuncturist, who fixed it forever. That was my only other health adventure, other than annual check-ups, a broken leg when I was a kid, a cut that needed stitches when I was in high school, playing hockey, and teeth knocked out, playing soccer. And then there was this autumn with three months of emergency room visits followed by a specialist and a doctor plus physiotherapy and an alternative practitioner who did reflexology and it all started with a visit to a chiropractor, which was my choice when I had back pain and there was no doctor available.

    And that brought me up to today, the day I saw the doc again, driving to Lac La Biche and back, and he would decided if I was going back to work or back to physiotherapy on LTD.

    After months of thinking I'd be back at work in a week or two, but never getting there, now what I wanted was daily physiotherapy for a couple of weeks to build up the muscles I lost with nerve damage.

Conversations and communication by e-mail with the principal and superintendent of my school were not exactly comforting, yesterday, so I was a little worried about getting my job back, but still confident physiotherapy would get me back in shape and then the doctor would sign the form saying "back to work" and they would get me back in there.

Instead of mid to late November, it looked like early December.

It was 14 below and snowing in Cold Lake, not the most ideal conditions for driving across the southern section of Northern Alberta, but not terrible, either. So, I planned to go, and hoped for the best.

What did Facebook have to say this morning? Not much .... My Alberta buddy in Toronto posted a video that hit me hard, but in a good way, showing the Nashville Predators of the NHL arranging for a "smashmob" at a hockey game for ten and eleven year olds, with a big crowd in the stands, with signs, plus cheerleaders and a band, so the rink was packed, and the kids on the ice were amazed because they usually got only around 40 people out to their games, and the message was that the NHLPA and the owners group represented by Gary Bettman should see this video, notice the looks in the eyes of those boys, and remember what hockey was all about.

Although I didn't feel emotional that morning, or the day before, that video got me going!

The big international news of the morning was about Palestinian terrorists blowing up a bus in Tel Aviv, Israel intercepting rockets from the Gaza strip, and jets from Israel making the ground shake in the Palestinian territory, while Hilary Clinton tried to get a truce and the U.S. restated their commitment to Israel.

Justin Bieber and Justin Trudeau were both in the news a lot, lately. Trudeau was in Alberta, for the federal Liberal Party. And Bieber was in the U.S.A., winning awards, and breaking up with Selena Gomez, or getting together with her again, or something.

"Justin Bieber may be Canadian, but he was the all-American boy at Sunday night's American Music Awards," AP said. "The pop singer dominated the awards show, winning three trophies, including artist of the year. His mom joined him onstage as he collected the award, beating out Rihanna, Maroon 5, Katy Perry and Drake."

Bieber won favourite pop/rock male artist and he also won favourite pop/rock album for his platinum-selling third album, "Believe."

-- Way to go, Stratford boy!

In the mail, I got a package of forms to fill out for LTD, with one for a GP, one for a specialist, and so on, plus some information about the program. It looked like I would get more than fifty percent of my pay.

Love Tree

Let us take our challenging moments today and look at them as opportunities. Opportunity to surrender, forgive and embrace unconditional love. If we don't we simply continue on the wheel of the same reoccurring patterns. It's time to transcend this negativity and see our light everywhere and in everyone. Be the Light.

Love tree also posted a pretty picture of a sunrise with the slogan Heaven is a state of consciousness, plus Think globally, act homely, socially, religiously, spiritually. That was an update from Think globally, act locally.

Day 59, Part 2: Shades Of White

The drive from Cold Lake to Lac La Biche is so beautiful it should be a famous tourist attraction. It's beautiful in summer, with a green forest, blue rivers, and a big blue sky. It's beautiful in autumn when the aspen forest and tamarack trees turn yellow, wheat fields turn golden, and the black spruce forest remains dark green, while the rivers start to freeze over. In winter, it is all white, with different shades of white for highways, the shoulders of the roads, fields, forest, and sky.

The forest, in places, is bright white, with new snow stuck to trees by freezing rain.

It was foggy and when the winter fog lifted, it was replaced by snow falling lightly over the landscape.

When a truck goes by, there's a whiteout for a few seconds.

There is a section in Lac La Biche County without telephone poles or towers and radio reception is replaced by static, or white noise.

As I entered the white noise, in a white-out, I ws listening to a song on CBC Radio One called "Snowy Soul".

It's a long and winding road with several long, flat, straight stretches, and low rolling hills where prairie meets boreal forest in the Lakelands. And I love this landscape. It reminds me of home, in Muskoka.

On the way over to Lac La Biche from Cold Lake, I saw something else that reminded me of home: A snow plow.

In Muskoka, and all over Ontario, snow plows are everywhere, it seems, and you see them all the time. In Northern Alberta, in Edmonton and Cold Lake, at least, they plow less and do more to maintain the ice pack, or snow pack, putting sand on top to improve traction, rather than scraping the snow and ice off the roads and highways to reveal bare asphalt.

But the snow plows were out, on Highway 55, turning white highways to black.

Alberta Provincial Highway No. 55 was an east–west highway in northeast Alberta, Canada, going from the Saskatchewan border through the City of Cold Lake and the Hamlet of Lac La Biche.

Highway 55 is the eastern segment of Alberta's portion of the The Northern Woods and Water Route -- a 2,400-kilometre (1,500 mi) route through northern British Columbia, Alberta, Saskatchewan and Manitoba.

It continues as Highway 55 in Saskatchewan, where it is the most northerly transportation route in that province.

The highway goes over to The Pas, in Northern Manitoba, and then down to Winnipeg.

Cold Lake, Alberta, is a little further north than The Pas.

The NW&WR starts at Dawson Creek at the British Columbia Spirit River Highway 49 and ends at Winnipeg, Manitoba. It was built in the 1970s.

In the old days, fur traders and early settlers used the rivers and Red River cart roads such as Long Trail in this part of the country.

In the early 1900s, the railroad and bush planes supplemented travel to this northern boreal transition area.

Corduroy roads provided a means for early land vehicles to cross over muskeg and swamp.

Horse drawn ploughs filled low areas, settlers hauled gravel and cleared bush for the roadways surveyed along high elevations following lake and river shore lines.

Municipalities graded gravel roads, providing transportation between trading centres.

Now it's a beautifully engineered two lane paved highway with a speed limit of 100 khm.

In the doctor's office, I had a lot of time to stare at the door, which was made of wood with a pronounced grain. At first, it looked as though the door contained a lot of scary faces. But after the doctor said, Alright, those faces changed, turned into smily faces of joy and celebration.

A few minutes later, I could see both faces at the same time.

Is this a comedy or a tragedy, one of my students used to ask all the time.

-- It's a dramedy.

What would daily physiotherapy be like? I wondered. How many hours would it take?

Did I care?

The more the merrier!

I just wanted to get healthy again, go back to work, and get my life back.

The doc left to make a copy of the physio report and to mail it along with the LTD form he filled out and he was gone so long that somebody from reception came in to ask me who I was so s she could remind the doctor.

While I was waiting for the doctor, in the waiting area and in the examination room, I wrote several pages for my new novel called The CLAW!

Day 59, Part 3: Ice Days And Lice Days

After the appointment with the doctor, I made a phone call to work, to let them know what was happening, so the principal could book the supply teacher, and I said I would drop in, on my way home, to deliver the doctor's note.

The drive home was completely different. How long was I in the clinic? First I had to wait in the designated area, and then in an examination room. The doctor's visit did not take long. He read the letter or report from the physiotherapist, said, Okay, filled in the LTD form, went away to make a copy of the report and mail it off with the form, and I had to wait for him to return, to get a doctor's note for work.

It's a good thing I take my journal with me wherever I go, as I had time to write several pages while I was waiting. -- Eventually a nurse came in to say, Who are you? She wrote a note for the doctor and posted it on the door, in case he forgot to come back, she said.

The weather changed over the noon hour, while I was inside, so the drive home looked like a journey on a different day. The winter fog had lifted, the snow stopped, and the overcast skies cleared up, so it was a clear day with blue skies.

Snow plows had cleared the roads down to black pavement.

Because I could see further, I noticed birds and animals. There were brown cows in the fields and big black crows by the side of the road as well as the odd magpie.

In one area, I saw a dozen crows with one magpie at the side of the road, so I guessed there was some roadkill there, that they dragged off the highway, and in the middle of that murder of crows there was a bigger bird that left them behind as I drove by. An American bald eagle, with bright white tail feathers and a wingspan of about ten feet, flew beside my car for a quick minute and then climbed up the sky to my left, heading north.

What a sight!

In my whole life, I have seen just a few bald eagles, and all of them have been Alberta. This one was the biggest, healthiest looking, and most impressive specimen of the species.

Some people believe that seeing a bald eagle, like that, is a good omen. Others just enjoy the show. Some consider the bald eagle to be one of the most sacred animals in existence. The tail feathers of the bald eagle are prized for use in prayer ceremonies. The one I saw that afternoon had huge, bright white, tail feathers. -- Just seeing this bird flying beside me for a bit made me feel stronger and happy.

The rest of the trip along Highway 55 was easy but uneventful, and I was happy about that, too. There were only a few cars and pick-up trucks, plus a few big rigs with tall and wide loads around La Corey, the little town, or hamlet, in the heart of the local oil patch.

It took the usual amount of time to get to Cold Lake, about an hour and a half, so I thought the school day would be well over and the only people left at the workplace I was missing would be the principal and the head secretary. But the parking lot was full of snow-covered cars.

As I walked in from the handicapped parking spot by the doors, I saw a few people I didn't know walking out, so I guessed there was a special event that kept the teachers in the building. It turned out to be the lice patrol.

All elementary schools go through this ordeal from time to time: One kid gets head lice and they spread from student to student until a lot of them need to have their heads washed with a special solution. In bad cases, head shaving may be required, plus a day off school, to make sure they are gone.

Our school combines secondary and elementary, and some of the high school kids work with the younger students as tutors or peer helpers, so a few of the teenage girls with long hair found out they had head lice, too.

-- I could imagine there would be a lot of drama around that little discovery!

The teachers were scratching their heads and shuddering at the thought of head lice and their experiences dealing with little kids and lice all day.

First they had an ice day and then they had a lice day.

The next day was parent interview day and report cards were due.

My colleagues looked and acted exhausted.

After talking with my supply teacher for half an hour, my legs felt shaky, so I sat down on the pew in the front hallway, and a few of the teachers came over to sit beside me. One gave me a hug, one put her head on my shoulder, and a few others gathered around to ask me how I was doing and when I was coming back.

Our small school is like a big family and it was a little bit like a family reunion, rather than just running into the gang at work.

There were only a few students in the building at that hour. -- We had even bigger reunions.

I realized I missed that place even more than I thought. -- Even so, there was a part of me that was not unhappy to be missing the stress of head lice, ice days, report cards, parent conferences, and the rest of the long list of demands on the days of teachers.

While I was there, I recruited a new teacher to supervise my boys basketball team while I was away and a couple of students took over training the team.

And I found out the junior drama group had folded but the senior group was still good.

From the school, I drove over to the physiotherapy office, to let them know the doctor approved the plan, so they could book me for daily sessions.

I already had an appointment for the next morning.

By the time I got home, I was hungry and tired and my right leg, in the thigh and butt area, was so sore that walking was a painful chore.

It was like that when I woke up at three in the morning, too, so I put the Dr. Ho EMS or TENS electrical pulse massager on it.

Recovery Day 60: American Thanksgiving

On the day we call American Thanksgiving, there was a cartoon on Facebook with a photograph of three big turkeys facing each other as though they were having a conversation and the caption said, Something's up! The farmer just unfriended me on Facebook!

My status update: "saw a big beautiful bald eagle flying beside me as i drove highway 55 from Lac La Biche yesterday afternoon ... it left a murder of crows the size of ravens -- and whatever they were snacking on -- to show me its bright white tail feathers .... what a sight!"

Google told me bald eagles summer here and winter much further south. Google also said the American bald eagle has a wingspan of up to eight feet.

So, what I saw must have been a new breed: the Canadian bald eagle, which winters in Northern Alberta and has a wingspan of ten to twelve feet.

Remembering the band I heard on CBC Radio as I drove across Alberta yesterday morning, I Googled "Trent Severn", the new band named after the water system that cuts through the place I used to live and links Lake Ontario to Georgian Bay, and I discovered they have an elaborate website loaded with info about the group, their gigs, their music, their album, and even the words to one of their songs: Snowy Soul.

They had a new post called Wow & Thank You that said, We just wanted to post that we are so grateful for all of the email we received today after our appearance on CBC's Q. We've had an exhilarating day of performances and rehearsals, and we are certainly looking forward to replying to you over our coffee in the morning. Night – TS.

A few rave reviews were posted, too. Rootsmusic.ca said, "Trent Severn has the potential to be a standing-ovation, crowd-pleasing staple at folk festivals for as long it chooses to be."

Exclaim! said, "Joyful, musically flawless and warmly patriotic, Trent Severn's debut is a doozy."

They had a link to Q on CBC Radio, but it didn't work, but I found it on the CBC website and listened to it again. -- I wanted to hear their song called "Muskoka Bound".

It started like this: ".., Muskoka bound / someone said in Gravenhurst, he'd been hangin' around ...." and went on to mention the Toronto Maple Leafs and more Canadiana.

I loved it!

Emm Gryner, Dayna Manning and Laura C. Bates, the trio in Trent Severn, were going to be big stars!

Did they have a Facebook page? -- Of course they did! And I 'liked' it.

If there was one thing I didn't like about Facebook and the internet, it was that some people were not on it. A billion people had a Facebook page, including millions of

Canadians, but from time to time I thought of this person or that person from the past and couldn't find them on the internet.

Where were they? Living off the grid someplace without telephones and computers?!

Last night I had a dream about the girl I liked in high school. We never dated or went out together in high school as she was planning to be a nun, or something, and she didn't go out with anyone. But we became friends and we stayed friends for decades after high school. We got together now and then as we reached different stages and ages of our lives and it always felt as though we had never been apart for a day or a year or half a decade or whatever. We picked up the conversation where it left off and it felt seamless.

· Where was she?

Last night, she appeared in a dream, and then I found her online. Her name appeared in her mother's obituary, which was published a couple of years ago, and it said she was in Nanaimo, B.C..

That was odd because she was working as a lawyer in Toronto, the last time we talked, and she had never been to Vancouver Island in her life, before then.

She told me she liked Jamaica. But she was living in Nanaimo?

Using that new bit of information, I tried to find her again, but did not get lucky.

Our story is a little bit more complicated than that, but I will save that for another day. Or after physiotherapy.

I'm looking forward to physio and acupuncture. I hope it gets rid of the intense pain I feel in my right thigh. And I'm looking forward to working on my new novel. And you should be looking forward to hearing the story about my high school sweet-heart, because .... it's a good one!

Walking to physiotherapy was a painful challenge but my routine there made my leg and butt feel better. Ten minutes on the treadmill at 2.5, leg lifts while standing up, no balance board, back extensions lying on a mat, and leg lifts lying down, one at a time, legs straight then bent, went well enough. The first back extension hurt a lot but the more I did the better I felt. Leg lifts, with my right leg, hurt a lot.

Acupuncture was aggressive and aimed at my right leg, with pins in my butt, thighs, ankles, and, for the first time, my knees.

It felt good until I got up and the pain returned, multiplied times two.

It was a painful walk home and I moved at a slow pace.

The physiotherapist was pleased the doctor had approved her recommendations but she said the next step depended on the benefits company as they usually said 'just go back to work' so they didn't have to pay for anything. However, the physiotherapist said she would phone my employer in advance and that might make it happen sooner.

That was news to me. -- I thought we would start right away.

My physiotherapist said it might be faster to have the initial assessment in Lloydminister as they had a bigger practice with more physiotherapists and did assessments every day.

At first, I said, Okay, but later I changed my mind and said I would rather have it done right here in town, if possible, to avoid the long drive down to Lloyd in winter, and back, after what sounded like a demanding four hour assessment.

We left it open, to be determined later, after the benefits claim came through.

It looked like I would not be back at work in two weeks. It looked like daily physio would not start for two weeks or more.

That was a depressing shock, but I bounced back after about half an hour of contemplation.

If I was off for another month, or longer, until the end of the semester, maybe I could do a TEFL course online, start an online M.Ed. course, and write a novel.

It didn't look like I'd be going home for Christmas. -- It looked like I would have to make Cold Lake my home for Christmas.

My FB post about the murder of gigantic crows and the supersized bald eagle got several responses.

Andrew, from Calgary, now in Toronto, said they were probably ravens.

A former student from Owen Sound, now living in Fort Mac said that, this far north, they were definitely ravens.

My friend Zenda, in Vacouver, said, The bald eagle is a very good sign. It looks like you are going in the right direction!

Andrew, in Toronto, also wanted me to know that Stampeders' fans had somehow got a horse into First Canadian Place, instead of the Royal York Hotel, so the tradition lived on.

Love Tree posted a quote from Black Elk, a Sioux Holy Man: The first peace, which is the most important, is that which comes within the souls of people when they realize their relationship, their oneness with the universe and all its powers, and when they realize that at the centre of the universe dwells the Great Spirit, and that this centre is really everywhere, it is within each one of us.

First Canadian Place was a skyscraper in the financial district of Old Toronto, beside The Eaton's Centre, built by Olympia and York, which we called OY. At almost 1,000 feet, it was Canada's tallest skyscraper, the tallest building in North America outside of New York and Chicago, the 9th tallest building in North America and the 62nd tallest in the world.

It was the third tallest free-standing structure in Canada, after the CN Tower, also in Toronto, and the Inco Superstack, up in Sudbury. When it was built, it was the tallest building in the Commonwealth.

The Royal York, now called the Fairmont Royal York, was in the heart of downtown Toronto, across from Union Station, and it was the tallest building in the Commonwealth, for a while.

The Canadian Bank of Commerce was the tallest building in the British Empire or the British Commonwealth, as it was called, for three decades. The big building in downtown Toronto was built by my grandfather.

Well, he was one of the carpenters who worked on that building.

First he built the mines and the miners' houses in Sudbury, after emigrating from Scotland. He moved to Toronto to work on the skyscraper. After that, he built big houses in Forest Hill. And then he retired to the place in Canada that reminded him the most of Scotland, which was where his daughter, my mother lived, in Gravenhurst, Ontario.

In my twenties, I traveled all around Scotland, Ireland, and England, and went to Golspie, my grandfather's old hometown, up north, past Loch Ness and Inverness, on the North Sea, and was amazed at how 'at home' I felt, with a landscape like Muskoka and a lot of old guys who looked like my grandfather.

Fletcher Armstrong, chairman of the Calgary Grey Cup committee, and Marty, their mascot horse, tried to get in the Royal York Hotel. However, the Calgary Stampeders were denied the chance to recreate the magic of the team's first Grey Cup victory.

Staff at the Fairmont Royal York Hotel refused to allow the Calgary players to parade the horse through the hotel, as they had done in 1948.

Despite chants of "Let us in!" from the Calgarians, with Stampeders' cheerleaders and sympathetic locals, the posh hotel called The Royal York kept them out.

Marty was allowed into a branch of the Bank of Montreal, at First Canadian Place, and a bank executive took a turn riding him inside the bank.

Which place got better publicity out of this event, would you say?

Football fans remember 1948 fondly as the year Calgary, Alberta, representing Western Canada, won its first Grey Cup. Every time the Stampeders are in the Grey Cup, fans take a horse through a hotel lobby in the host city. The tradition started in Toronto at the Royal York in 1948. Marty's owner is the son of the man who gave Calgary fans the horse back in '48.

The 100th Grey Cup game would be held Sunday at Rogers Centre, also known as The Rogers Bundle, and formerly known as SkyDome.

Update: The Royal York changed their mind and put out a press release saying Marty the horse and the Stampeders' fans were welcome, after all.

Marty got into CITY TV. The Globe And Mail called Marty a hipster after dropping in on the downtown TV station in the cool Queen Street West part of the city.

An hour or so later, my Calgary/Toronto buddy posted this: A horse in the Royal York Hotel Lobby during Grey Cup week!! Yahoo!! Go Stamps!!

Maclean's magazine reported that Calgary Stampeders fans finally got their chance this afternoon to recreate a Grey Cup tradition by marching the 15-year-old dark brown stallion through the lobby of the upscale Toronto hotel.

Excited fans chanted "Go Stamps go!" as they surrounded Marty.

The Vancouver Sun story was more colourful. They said, From "neigh" to ... "yay?" Well, not exactly, but after a morning of frustration trying to get into a downtown hotel, it's Marty's Party at the Royal York.

The international news was positive and that was a nice change. As well as news of the truce between Israel and Gaza, there was news about the U.S.A. in Burma for the first time in a long time. There was a picture of President Obama with Aung San Suu Kyi and a picture of Hilary Clinton embracing her.

Clinton was in the Middle East, brokering peace, one minute, and in Burma, the next.

Obama became the first sitting U.S. president to visit Myanmar. He praised the courage of Nobel Peace Prize winner Aung San Suu Kyi and gave a nod to the initial reforms in a nation once notorious for political repression.

Obama spent several hours in Myanmar, the formerly secretive country also known as Burma, and met with Suu Kyi at the lakeside villa where she spent years under house arrest for her pro-democracy activism.

Good news is great, isn't it?

-- Happy American Thanksgiving!

Wally Lamb said, Blessings to everyone on this Turkey Day--vegetarians included. Thankful for you all and repeating my mantra: I WON'T pig out, I WON'T pig out, I WON'T . . . . Happy Thanksgiving 2012.

In contradistinction to the previous and the foregoing, a female Canadian Facebook friend posted this: The holidays are upon us and it can be a very joyful time of year. Some of us have problems during the holidays and sometimes are overcome with great sadness when we remember the loved ones who are not with us. And, many people have no one to spend these times with and are besieged by loneliness. We all need caring thoughts and loving prayer right now. If I don't see your name, I'll understand. May I ask my friends wherever you might be, to kindly copy, paste, and share this status for one hour to give a moment of support to all those who have family problems, health struggles, job issues, worries of any kind and just need to know that someone cares. Do it for all of us, for nobody is immune. I hope to see this on the walls of all my friends just for moral support. I know some will!! I did it for a friend and you can too. (You have to copy & paste this one, no sharing)

After physio and acupuncture, I slept, missed the noon swim at the pool, and didn't get up for about five hours. Around three in the afternoon, I went to the book store and the drug store, to stretch and test my legs, but they hurt a lot, so I headed home again.

At the bookstore, the book I ordered weeks earlier finally arrived: No Laughing Matter, by Joseph Heller, about his bout with GBS.

At the post office in the drug store, I got mail from the U.S.A. Months earlier, I bought a chair at Walmart, and it came in a box, so I had to assemble it, but there were no nuts or bolts or screws. When I phoned the manufacturer, in the States, they said they would mail me the missing parts right away. Two months later .... Here they are!

The Heller book looked good. When I walked into the bookstore the other day, looking for something to read, while I was off work, absolutely nothing appealed to me, which was odd, because it's usually easy for me to find something I like, as I have broad tastes in books.

I ain't no literary snob.

Okay, that's a lie, I hate to admit. But I do read around.

No Laughing Matter made it clear that what I had was not GBS, as a couple of people had suggested. The only connection between what hit me and GBS was that my condition developed rapidly and I lost the use of my muscles quickly.

Heller went downhill even faster, and he lost the use of just about every muscle in his body.

He never lost the ability to talk, but that was about it.

In even more severe cases of GBS, you lose the ability to talk, too.

All I lost was the ability to walk.

Maybe I had a mild case of GBS.

The mildest cases go undiagnosed and clear up after a month or so.

This morning I asked my physio why my muscles hurt so much that I could barely walk, and why it was one area one day and another area the next day.

She said there were two principal causes: referred pain and sore muscles.

Is it normal, or typical? I asked her.

Absolutely, she said. It's probably going to be a roller-coaster.

And she added, I sure wouldn't want to be you or go through what you're going through.

That was the first time she ever said anything like that. She has always treated me as though I have something that's not uncommon, something she sees all the time, and something she knows how to fix, no problem.

I asked her if my muscle pain was the result of the fact that I had lost the use of most of my muscles, from my lower back down, and rebuilding them led to strains and pains.

She said building muscle can be a little painful at the best of times. And this was, in a sense, the worst of times, as I was building up, again, after going down to zero.

Oh, I said. -- I used to like roller-coasters.

I'm on a roller coaster of recovery. Two steps forward, one step down. And it gets longer instead of shorter. -- Frustrating!

Instead of going swimming at 8:00, I got involved with birding. First I changed my FB status to claim that I saw a new species, the Canadian Black And White Hawk, or the Canuck Hawk, and then I changed my profile picture to a stock photo of an American eagle posing in front of a Canadian flag.

Next, I watched the best birding movie ever, The Big Year, with Jack Black, Steve Martin, and Owen Wilson. It showed a raven and a crow, as well as a magpie and almost five hundred other North American birds. The birder of the year had 455.

My brother is a birder and his life list total number was over 500.

That number put him in the top five in Canada.

The current leader was Roger Foxall, with 541, including a fish crow, a cave swallow, and a citrine wagtail.

Last summer I added several birds to my life list, by traveling to good birding places in Bonnyville, Lac La Biche, Saskatchewan, Lloydminister, and Cold Lake Provincial Park. In Cold Lake, I saw a pileated woodpecker. In the lake beside Bonnyville I saw so many birds I could not identify that I felt disoriented for a while. On the surprisingly beautiful beaches of Lac La Biche, I saw pelicans, which amazed me, as I thought their northern range was much further south, like the Gulf of Mexico.

Apparently pelicans follow the pipelines, more or less, from Texas to Alberta and back again.

I thought I saw a whooping crane flying over Cold Lake but I wasn't 100% certain. I saw one in southern Ontario, once, where they shot that movie about the guy who flew with cranes and geese over Lake Ontario.

It's possible that there was a whooping crane here, as they summer north west of here. Global Forest Watch Canada, an environmental group, says the world's only wild flock of whooping cranes flies over the oil sands as it migrates annually between breeding grounds in Wood Buffalo National Park, located on the Alberta-Northwest Territories border, and winter grounds along the Texas coast. Global Forest Watch Canada says the oil sands may be threatening the survival of the big bird, which is an endangered species, close to extinction. If a whooping crane lands in a tailings pond with the toxic waste byproduct of bitumen processing, it will be game over for the bird and incredibly bad publicity for the oil sands.

There are about 10,000 species of birds in the world, and there's an Englishman named Tom Gullick who claims he has seen 9,000, and everybody believes him, apparently.

My life list is around 1,000, so I'm out of the competition.

These days, global warming is bringing new species to Canada, and they say new bird species are evolving faster than ever, especially in the Americas.

There is at least one new bird species identified every 700 years or so.

That's why I was so excited about spotting that big bird that looked like a bald eagle but was out of its range for this time of year, or the season, and was bigger than the famous American bird.

If you spot it, you get to name it, I think, so I made a list: Canadian Eagle, Canuck Eagle, Black And White Canadian Eagle, Great White North Eagle, Oreo Eagle, Great Northern Eagle, The Regal Canadian Eagle, Canadian Spirit Eagle, The Black And White Alberta Eagle .... How about The Eagle Of The Tar Sands.

The fact that I changed my profile picture to a photo of a Canadian Eagle posed in front of the red maple leaf of our flag goes to show one thing: I have definitely spent too much time on my own, away from the job, because of the back injury that caused severe nerve and muscle damage to my legs, and I have gone completely crazy. Or, as McMurphy said in One Flew Over The Cuckoo's Nest, "the bull goose looney".

With that thought in mind, I spent a quiet night reading No Laughing Matter, by Joseph Heller, getting lots of confirmation that I do not have GBS, which made me feel glad all over that I avoided EMG, which is described as quite painful, and lumbar puncture or a spinal tap.

After reading that book, I had a hard time falling asleep, so I watched The Answer Man on Netflix one more time.

Recovery Day 61: Rock And Roll!

Sometime in the middle of the night, I started feeling hopeful, again, and making plans for Friday. My colleagues had the day off, after working late, for parent interviews, so they would all be sleeping in, no doubt, except for the teachers who had little kids. If my back was okay and the pain in my butt was gone and my legs worked, I wanted to get steak and eggs for breakfast, swim at lunch, do my physio exercises, and try hard not to "overdo it".

What about the weekend? What about the next four weeks? What about the Christmas Break? -- It was hard to make plans for the future since I didn't know how well I would be walking or what kind of pain I'd be feeling.

The suspense was not killing me.

Would I live? -- No doubt!

Would I become that guy with the bad back, who can't do much, physically, and is often in pain?

-- I wanted to be the guy who had that mysterious experience with nerve damage and muscle loss who built himself back up again and was stronger than ever.

All I had to figure out was how to do that without "overdoing it".

There were conflicting reports about the teachers' federations in Ontario. Some boards had agreements with the government, some teachers' federations were taking job action leading to strikes, and some boards were talking about locking teachers out.

Matt Galloway of CBC Radio conducted a hard-hitting interview with Education Minister Laurel Broten and then teachers' federation leaders. He said teachers were demoralized or furious.

About 5,000 teachers at York Region were the first to take strike action, and were joined by about 134 teachers at the Rainy River District School Board. About 1,111 teachers at the Trillium Lakelands District School Board and 1,340 at Kawartha Pine Ridge were poised to start work-to-rule.

OSSTF, which represents 60,000 members, and ETFO, with 76,000 members, were protesting the Liberal government's anti-strike law, which also cuts benefits and freezes wages for most teachers.

Four unions were taking the government to court, arguing the law is unconstitutional and violates collective bargaining rights.

Bill 115, which passed with support from the Progressive Conservatives, gave the government the power to stop strikes and lockouts.

The province's education minister Laurel Broten is a "liar, pure and simple. This is about union busting," the president of the Hamilton bargaining committee for public secondary school teachers said.

Chantal Mancini, the president of OSSTF district 21 made the comment on Twitter following the CBC Radio interview.

Leading up to the Grey Cup game, Jon Cornish of the Calgary Stampeders won the CFL's award for Best Canadian and was the West Division's nominee for the CFL's award for Outstanding Player.

Chad Owens of the Toronto Argonauts, known as The Flyin' Hawaiin, won the Outstanding Player award over Cornish.

Owens was the CFL leader in receiving and return yards and accumulated a league-record for all-purpose yards.

Cornish led the CFL in rushing, breaking the 56-year-old record for most rushing yards in a season by a Canuck.

After winning the award, Cornish talked to the media about his mother, a teacher in B.C., where he grew up without a father, and the fact she had a female partner. He spoke about his lesbian mother with pride. She became an Anglican minister. He said she had just returned from a vacation in Israel to see the Grey Cup game in Toronto.

Pictures of Marty, the Stampeders' horse, in Toronto showed that southern Ontario had mild weather, but temperatures were expected to drop to zero by the time of the Grey Cup Game.

It was twenty below in Cold Lake on Black Friday.

Down in Texas, the weather was news, as fog was blamed for a chain-reaction crash of 140 cars.

Canadians shopping in the U.S.A. were expected to drop about 5 billion bucks, which other Canucks would rather see in our own economy.

Even Apple had some discounts on Black Friday. They were famous for not having sales. Other retailers had huge sales on B.F. and Apple dropped their prices by about ten percent.

Was the Apple Thunderbolt computer with the 27 inch screen on sale anywhere? Did I need one? I saw it in my dreams before it was created. -- Maybe I could get by with my MacBook attached to an ASUS monitor. -- Sigh!

My Friday started at six thirty in the morning when I staggered to the kitchen, feeling a little pain in my back and a lot of pain in my rear end.

What about my plan from the night before about getting steak and eggs for breakfast and then doing physio in the pool?

The sun was starting to rise at seven thirty and the temperature had climbed up a little bit. It was 20 below at six thirty and only 18 below an hour later but the forecasted high was for 7 below, around noon, which we considered balmy.

In Ontario, environmentalists were celebrating, for a change. The Mega Quarry in southern Ontario was killed, cancelled, called off.

An American company called Highland, operating out of Boston, bought 8,000 acres between Orangeville and Collingwood, north of Toronto, in Dufferin County, and said they were going to turn it into the biggest potato farm in Ontario. But then they revealed that what they really wanted to do was use the land for a huge limestone quarry.

They had 2,000 acres of prime farmland in an area famous for potatoes but they planned to dig up all that land and more to a depth of around 200 feet, going below the water table, and destroying and area that supplied drinking water for over a million people.

A coalition of farmers and foodies from the city killed the mega quarry with petitions, according to The Globe And Mail.

The National Post said it was big protests that stopped the project.

The company said they withdrew their application for a big quarry because the regulatory process changed and the community did not support it, even though they promised their activity would not disrupt the water and they would return the land to agricultural use after taking out the limestone to build houses and roads.

The president of Highland resigned.

It would have been the largest quarry in Canada and the second largest in North America and would have provided one billion tonse of limestone.

Highland also bought the old rail line that went north to Owen Sound, so they could use O.S. as a port for shipping their product to the U.S.A.

-- I used to live in a condo on the harbour right beside that old rail line. -- The people who lived there would just love to see and hear big box cars rumbling by, full of limestone, to be loaded on Great Lakes freighters.

The mega quarry was dead, but a smaller quarry was still a possibility, down the road.

Alabama posted this: Life is full of meaning when one's values are based upon Nature. The deer is content to greet the sunrise and have food in its belly. The wolf is happy knowing its clan and cubs are safe and cared for. When you realize that human life is consumed by things that lack meaning--the bigger house, the glittery gem, the sleeker car--and refocus to Nature's priorities, life becomes richer and deeper and easier.

A Cold Lake woman told me her husband was excited because he just took down a big buck he had been stalking for about five years, using cameras and deer blinds set up in trees, and the animal was so big it had to be in the top five in Canada this year. He had been watching it, using webcams and motion detection sensor cameras, and knew its every move, all its habits, where it went and when it went there. Soon it would be in their refrigerator.

It was twenty below outside, in Cold Lake, like a big, walk-in, freezer.

Speaking of potatoes .... Wally Lamb posted a link to a video made of a bunch of Sixties TV shows with Rob and Laura and Jeanie and The Addams Family, including Thing, dancing to music from the Sixties -- namely, Let's Dance:
Hey baby won't you take a chance?
Say that you'll let me have this dance?
Well let's dance, well let's dance.
We'll do the twist, the stomp, the mashed potato too.
Any old dance that you wanna do.

There were lots of songs called Let's Dance. This one was the 1962 hit-single by Chris Montez, written and produced by Jim Lee that went to #4 on the Billboard Hot 100 chart in the U.S., and to #2 in the UK Singles Chart.

It was such a great song, it inspired me to dance, sitting down, and then do some weight-lifting. Twenty arm curls with a twenty pound weight was a good start, I thought.
There's nothing like a little music to inspire you for physical and emotional recovery. But, as John Lennon said, ... It's gotta be rock n roll music if you wanna dance with me!

Just let me hear some of that rock and roll music
Any old way you choose it
It's got a back beat, you can't blues it
Any old time you use it
Gotta be rock 'n' roll music
If you wanna dance with me

After getting ready to go to the pool, despite the fact that my butt hurt, I had a funny idea: My new note from the doctor would take me from November 23rd to December 21st, and that was the day they said the world as we know it is in for some dramatic and devastating changes. It would be 21/12/2012!

As I headed for the pool, I caught some of Q, the CBC Radio show, starring Jian Ghomeshi, and this one was a special Q from Edmonton.
I'd been looking forward to it.
The Q Special from Edmonton featured electro-punks Shout Out Out Out Out, the cast of CBC comedy The Irrelevant Show, which I loved, an Edmonton media panel, plus folk-pop musician Colleen Brown, and writer Todd Babiak on rebranding Edmonton with the Make Something Edmonton campaign.
It was such a good show, I listened to it again, online, when I got home after swimming.

Before swimming, I went to the grocery store, as I was running out of things, and I used the walk through the big store, pushing a grocery cart, as my warm-up for swimming. My plan was to get to the pool late, so I would not be tempted to over-do it, again. The physiotherapist said to do half an hour only.

For half an hour, I practiced water walking, in shallow water and deep water, going up and down the same lane. It felt great to get into the water and experience the buoyancy, again, but that pain in the butt, on the right side, never went away. Getting out of the pool was painful as I felt the effects of gravity without the water to hold me up.

After swimming, I went to the superstore to get some things for physiotherapy, and then I was hungry so I went to Clark's General Store Eatery for fish and chips. -- My physiotherapist told me their fish and chips were great, and she was right.

By the time I got home, I was pooped and sore, and I just wanted to climb into bed and listen to the radio for a while. My buttock muscles were very sore and so were the muscles of my upper thighs, at the front.

But first I checked the news. Trudeau was all over the place.

Justin Trudeau said he was only bad-mouthing one Albertan, not all Albertans, in an old rant that has come back to haunt him. A French interview he gave two years ago surfaced with him bemoaning western influence on national affairs and arguing that Canada does better under prime ministers from Quebec.

Trudeau said he didn't mean to smear the whole province - just one Albertan: Prime Minister Stephen Harper.

The best part of the radio show Q in Edmonton, aside from a marketer talking about rebranding the city, was a sketch by The Irrelevant Show, inspired by the hockey lock-out, with a guide to recreating the NHL hockey arena experience at home. Essentially, they said to take a lot of money and burn it, to represent tickets, parking, food and beer, and they made fun of the food and the beer as well as the washrooms, and just about every other aspect of attending the event.   It started with getting the worst chairs in your house, pushing them together, sitting in the kitchen, with a hockey game on your TV in the next room, so you could barely make out who was playing.

They said you should get somebody to pour beer down your neck every ten minutes or so.

And in the end they said you should get some friend to park their cars around yours, in your driveway, so you couldn't move for about an hour.

The part about bathrooms was pretty funny, too.

That evening, the pain in the butt, on my right side, was so bad that I considered going to the emergency department of the hospital. It reminded me of the time, a few months ago, when my legs were giving out, and I wondered if I should go to the hospital, and everybody I asked said, GO!

Walking from the bed to the kitchen or anywhere, even taking a few steps, hurts so much that I hobble and lurch. I'm using my chair like a walker, as it has wheels, and I'm feeling scared, not to mention demoralized.

For distraction, I watched a documentary on Netflix called How To Start A Revolution. It was a fascinating look at the roots of the Arab Spring and the movements which overthrew repressive regimes in Moscow, the former Iron Curtain countries, Burma, and many other places.

A day that started in hope ended in despair.

Recovery Day 62: Argo!

At six o'clock on Saturday morning, the pain in the butt was so bad I could barely get out of bed or walk across the room. Despite the fact that I got a lot of rest and sleep, I felt exhausted. I guess pain does that to you. -- I really wanted it to be over.

They say that pain becomes overactive in sending messages through the nervous system and that energy and the brain's interpretation and response to those messages consumes a lot of energy.

Put more simply: pain is exhausting.

The Calgary Stampeders were not the underdogs, apparently, despite the fact they were playing the Grey Cup game against the Argos in Toronto on their home field surrounded by their fans. The Stamps were the favourites of the betters and the odds-makers.

There would be lots of Stampeders fans at the game. And there would be CFL football fans there who were not hardcore Stamps or Argos fans but big fans of the game and the Grey Cup.

Imagine going to the Grey Cup every year, as a fan, and enjoying the enormous celebration first hand.

The Winnipeg Free Press said the 100th Grey Cup was a 50-50 proposition. The Globe and Mail said it would be like a chess match between old foes.

The Edmonton Journal pointed out that Toronto beat Calgary both times they played this season, and the Argos held the Stamps' star, Cornish, to just 50 yards.

The CBC website said Ricky Ray was just too hot for the Stamps to be able to handle and called him the best gift Edmonton gave Toronto until they get that oil pipeline reversed.

The Montreal Gazette said it was a welcome distraction form the NHL lockout and the feud between billionaire owners and millionaire players.

CTV News announced that Gordon Lightfoot and Justin Beiber would be the two acts that bookend the Grey Cup halftime show.

The halftime show will also feature the band Marianas Trench and Carly Rae Jepsen.

What about Walk Off The World and Trent Severn? -- Those would have been my choices.

The Life Of Pi was in movie theatres, along with Rise Of The Guardians.

They Grey Cup game, for me, came down to Ricky Ray, formerly the QB of the Edmonton Eskimos, now leading the Argos, versus Jon Cornish, the Stampeders' running back who was named the Outstanding Canadian in the CFL.

The big question, for me, was this: Would Cornish moon the huge crowd at the Rogers Centre, dropping his pants in front of the Argos and millions of people watching on TV, the way he did earlier in the year in Saskatchewan in front of Roughriders' fans?

Since I couldn't walk, I went back to bed. With the Hellen Hunt / Jack Nicholson movie, As Good As It Gets, playing in the background, I fell asleep and didn't get up until he said "You make me want to be a better man" and then they kissed.
    -- Great movie.

Love Tree says, "Have patience with all things, but chiefly have patience with yourself. Do not lose courage in considering your own imperfections but instantly set about remedying them - every day begin the task anew."
~ Saint Francis De Sales

Mr Positive says, When it seems things aren't going the way they should, or even the way you want them to, STOP, GO BACK, TRY AGAIN. - Jerry

-- Good advice.
-- I'll try it.
-- Fall down nine times, get up ten.

So, I got up, went back, and tried again. This time, I strapped on my Dr. Ho belt, pumped it up, and sat up to watch Argo, the movie, which had nothing to do with the Toronto Argonauts or the Grey Cup Game.
    Argo is a great Canadian movie about the 1980 joint CIA-Canadian secret operation to get six fugitive American diplomatic personnel out of revolutionary Iran. It's about Canada's ambassador to Iran, Ken Taylor, and his covert operation, with the cooperation of the American Central Intelligence Agency, to get half a dozen Americans out of Iran by getting them Canadian passports so they could get past the Iranian Revolutionary guard by posing as a Canadian film crew scouting locations for their film called Argo.
    The film stars Affleck as Mendez with Alan Arkin and John Goodman, produced by George Clooney.
    I've always been a big fan of Alan Arkin's.
    Another way of looking at it: Argo was a 2012 American thriller film directed by Ben Affleck, based loosely on an account published in 2007 of the "Canadian Caper" when Jimmy Carter was president of the United States.
    The movie makes it look as though the CIA and Hollywood were the stars and Canada played a small role but, in reality, Canada was responsible for the six and the CIA was a junior partner.

The movie got a 95 from Rotten Tomatoes, so critics loved it, but some Canucks criticized it for glorifying the Americans and making the Canadian ambassador look like a concierge.

Jian Ghomeshi, the host of Q on CBC Radio, said he thought the film had a "deeply troubling portrayal of the Iranian people" with an unbalanced depiction of an entire ethnic national group.

Somehow, watching that movie, my pain faded away.

Inspired by those quoteable quotes about trying again, I thought about the pain in my butt in a new way. The physiotherapist said it could be muscle pain or referred pain. I'd been thinking it was muscle pain, caused by me overworking muscles that had declined because of nerve damage, and had to be built up again. -- But what if it was referred pain, from my back?

The Dr. Ho belt gives relief from back pain, with a little traction. And the pain in my butt and legs went away.

Woo-hoo!

After lunch, when my pain went away, I thought I'd try to get some sleep that was more restful and rejuvenating, and then get up so I could go to the pool.

However, I slept right through swimming time.

And I wasn't even upset about that.

-- I must have been wiped out!

When I woke up, I got the bad news about the NHL season being wiped out. One third of the season was cancelled on Black Friday, including the All-Star game.

At midnight I Googled 'referred pain from back' and found a very informative article at Spine Health's website. It said, Low back pain with referred pain can vary widely with regards to severity and quality. It tends to be achy, dull and migratory (moves around). It tends to come and go and often varies in intensity. It can result from the identical injury or problem that causes simple axial back pain and is often no more serious.

Referred pain is usually felt in the low back area and tends to radiate into the groin, buttock and upper thigh. The pain often moves around, and rarely radiates below the knee. This type of low back pain is not as common as axial low back pain or radicular pain (sciatica).

Referred pain is analogous to the pain that radiates down the left arm during a heart attack. It is the result of the extensive network of interconnecting sensory nerves that supply many of the tissues of the low back, pelvis and thigh.

An injury to any of these structures can cause pain to radiate – or be "referred" - to any of the other structures. It is important to understand that this type of pain is not due to "pinched nerves".

Unfortunately, the brain cannot determine the specific source of the pain.   In general, referred pain is treated with the same types of nonsurgical care as axial back pain and will frequently diminish as the low back problem resolves. Once the possibility of a serious underlying medical condition as the cause of a patient's low back pain is ruled out, treatment of referred low back pain is non-surgical and may include one or a combination of the following:

A short period of rest (e.g. one or two days)

Physical therapy, active exercise and stretching

Ice packs and/or hot pads

Appropriate medications for pain relief

After going to bed early, I woke up after a couple of hours, and I couldn't get back to sleep, so I watched a movie called Mercury Rising. a 1998 American action thriller film starring Bruce Willis and Alec Baldwin, based on a novel Simple Simon.

Willis plays, a failing FBI agent who protects a nine-year-old boy with autism, targeted by government assassins after he cracks a top secret government code.

The reviews were brutal, but I liked it. Rotten Tomatoes gave it 17.

Willis got nominated for a Golden Raspberry Award for this one, along with Armageddon and Seige, and he won. -- That's the award for bad acting.

Day 62: Grey Cup Day

Alabama posted a note that said: Don't overthink Enjoy breathing Sleep is worth it Go outside Make something Have a thoughtful conversation Play with a dog Smell the roses Laugh at yourself Ask questions
    -- There were no punctuation marks, but it sounded like good advice.
    -- Spellcheck said overthink was not a word.
    -- I'd have to think about that!

The phone woke me up at 9:00 and I'm still wondering who that was. Although I got up to answer the phone, I never made it, as it was on the kitchen counter, and the pain had migrated down my right thigh, so I moved very slowly, and the cricket sounds of my iPhone stopped before I got there.
    Since it worked well the day before, I put the Dr. Ho belt on right away, and sat down, to give the referred pain an opportunity to fade away and stop radiating.

Is referred pain 'real' pain? I wondered. If referred pain was caused by the brain interpreting signals from another area with pain, which part of your body should you ice or zap? If the pain in the thigh was caused by pain in the back, which area do you ice? -- I hoped traction would help both pains.

Over-thinking is for tanks. I loved breathing, counting breaths, re-learning how to breathe, practicing umbrella breath. Sleeping had never been a problem, despite back pain, thank goodness, and I always enjoyed dreamtime. Although I didn't go outside yesterday, I loved getting out there, especially if I could go for a walk in the woods, or the forest, or anywhere. As soon as I had some energy, I wanted to get back to the new book project. I missed talking to Edmonton. When was she getting back from Texas? Maybe I would drop by the pet store and the flower store to play with dogs and smell roses. I wanted to go to the pool to do the dog paddle, or exercise with deep water walking, while the Grey Cup was on. Is there any question I had cabin fever?

It was twenty below at ten in the morning and overcast with a little snow in the forecast, so it was a good day for staying indoors and watching football on TV, but that was not for me. Everybody else in Canada was chanting, Go Argos! or Go Stamps! but I was chanting Go Outside!

Fran Leibowitz was in Edmonton, at the University of Alberta, for the Festival Of Ideas, and the event would be on CBC Radio One at 3:00 in the afternoon. I was looking forward to swimming and Liebowitz.
    Fran did not have writer's block, she said, she had a writer's blockade.

She was the author of Metropolitan Life and Social Studies, two terrific books that were loaded with wit and insight. She was called "the funniest woman in America". And she reminded me of a girl I liked in high school.

Anne Wachtel interviewed Fran Leibowitz privately as well as publicly and both interviews were on the radio station's website.

She said she had 9,500 books in her personal library. She said she doesn't like plot or puzzles, she used reading as an escape.

She said there is no art form like writing because we all have so many words in our heads every day.

She said she is not depressed, although she has been depressed, but she says she is morose and angry. And she has a great memory.

The weather, on Grey Cup day, in Cold lake, called for snow grains, a form of precipitation I had never heard of. Snow grains are characterized as small, white, grains of ice. They are called grains because they are fairly flat and elongated.

Unlike snow pellets, snow grains do not bounce or break up on impact.

Usually, very small amounts fall, mostly from stratus clouds or fog, and they never fall in the form of a shower.

We have many words for snow, in Canada, as Margaret Atwood famously noted in a poem.

What she actually said was this: "The Eskimo has fifty-names for snow because it is important to them; there ought to be as many for love."

There is snow in Central Ontario, now, I noticed. The webcams on the main street of Bracebridge, called Manitoba Street, show a dusting of the white stuff on the grass in front of the church across from the funeral home.

Toronto had no snow and the temperature was a couple of degrees above zero, for the Grey Cup.

Of course, it will be warmer in the skydome and there will be no snow because the roof will be closed. -- There won't be a grain of snow under the skydome.

Somebody on Facebook posted a little video called The Amazing Transformation of a Guy Who Didn't Give Up! He was an American paratrooper, a veteran of the Gulf War, who hurt his back and legs by jumping out of airplanes with a parachute. He was so injured, the doctors said he would never walk again.

Doctors gave him the same message for fifteen years, it says in the video.

He went home, gained weight, and wound up with braces on his knees and hips and back, and could only walk with a couple of canes.

Yoga instructors turned him around.

Finally he found one yoga instructor who helped him.

He lost 140 pounds in 10 months, got his balance back, after falling a lot, did a lot of yoga, eventually walked with one cane, and then no canes. The climax came at the end of the video, with a few seconds of the guy running.

-- Very inspirational!

Swimming was great, once again.

One of the lifeguards was friendly. He came over and asked me how my back was doing. I asked him about bouancy belts, and he got me one to use in the pool.

Buoyancy belts are recommended for deep water walking and other forms of physiotherapy in the swimming pool. And I discovered it helps a lot.

Instead of using energy and motion to stay afloat it frees you to focus on the movements you need to practice for walking.

The blue belt made physio feel like it was fun, again.

I stayed a little longer than the prescribed half hour, but not much. I was in the water for about 45 minutes.

When I walked out of the rec. centre on the base, the Grey Cup pre-game show was on the big screen in the lobby, but nobody was watching it.

After swimming I was moving well and feeling good so I walked all over the superstore, to get milk and a few other things, then I went outside and made my way through the snow from the Wok Box to the Dollar Store and back, and then I was done.

How many steps was that? Two thousand?

My back hurt, a bit, and my knees were shaky, but I felt better as soo as I sat down to Mongollian beef and broccoli.

Walmart was busy, despite the fact the big game was on TV, and that's usually a good time to get into the stores in certain places.

Edmonton stores are quiet during Oilers games. Saskatchewan is silent during Roughrider games. Canada goes dark during Stanley Cup games. Cold Lake did not slow down for the Grey Cup game.

While I was in the pool, I set myself a new goal: Return to work by December 11th. -- That was the day STD stopped and LTD started.

If that worked out, I could work for a couple of weeks, take a break for Christmas and New Year's, then work from January to June like a normal person.

After walking Walmart and its parking lot, I heard the Zen master say to me, "We'll see!"

In the book section of the big store, I found a new trade paperback about a North American's experience at the Buddhist place Thay Thich Nhat Hanh ran in France, and I picked it up to read while I was stuck in the long check-out line.

Instead of reading, I did leg stretches, using the shopping cart, and that made my back feel better. After walking all the way arround the store, I had a little back pain.

-- Not much. -- A 2 on the scale from 1 to 10, with 5 as a median, and 10 as the most pain I've ever experienced.

In the pool, I was thinking about pain. The most pain I've ever felt, aside from plantaar faciitis first thing in the morning, was in romantic relationships.

Lucky for me, I'd had several. Unluckily, they all ended badly.

While I was deep water walking in the pool, I thought about one of them, in particular. But I won't bore you with that sad story.

The West was expected to win the big game by a small margin in a high-scoring game and the hometown team was seen as the underdog. Could T.O. win the milestone CFL game at home, on their own field, in the biggest city in Canada?   The Toronto Argonauts pumped up the hometown crowd by opening the scoring at the 100th Grey Cup, taking an early 7-0 lead against the Calgary Stampeders.

Chad Owens -- who else? -- scored the first touchdown of the game, but the Argos' lead was cut shortly after by a Stamps field goal. Ricky Ray found Owens for a five yard run into the end zone for the first points of the game.

At the end of the first quarter, the Argos were up 7-3.

More than 52,000 CFL fans from across Canada packed Toronto's Rogers Centre for the game.

Burton Cummings sang the national anthem.

The second quarter was a nightmare for Calgary as Stampeders quarterback Kevin Glenn, in the shadow of his own goalposts, threw the ball right into the arms of Argos defensive back Pacino Horne, who returned it 25 yards for another touchdown. The Argos were up 14-3.

Both Argos touchdowns came off Stampeders' turnovers.

Most people picked the Argos as the underdogs and expected the Stamps to win, but the Double-Blue dominated everything except punting and kick-off returns in the first half.

Was everyone looking forward to the half-time show with Gordon Lightfoot and Justin Beiber?

Would they do a duet, singing the Canadian version of "This Land Is Your Land, This Land Is My Land?"

Did anybody want to see that?

-- No!

Calgary staged a comeback with another field goal and a drive that took them deep into the Argos' zone, but Toronto took the ball right back up the field and scored with just 20 seconds left in the quarter.

The first half ended with the Argos way out front, 24 to 6.

In the second half, fans in the stands started leaving, believing the outcome of the game was clear and Calgary was not going to stampede back to score 19 more points than the Argos.

The Stamps marched 81 yards on 9 plays but settled for a field goal with 6:08 to go in the 3rd. third quarter, but the Argos responded with a field goal of their own.

Calgary picked up a couple of points by pushing the Argos into their own end zone. Toronto conceded a safety. -- Why not? They were still ahead by a score of 27 to 11.

A friend in T.O. posted a picture of a sculpture of an Argo guy holding a football beside a mock-up of the Grey Cup in front of a picture of the stadium, so I reposted it when it was clear who was going to win.

Andrew from Calgary said, The Argos played like they wanted it and smacked the Stampeders in the mouth from the opening kick off. Football's a simple game to win if you hit, hold onto the ball and stay disciplined. Congratulations Argos. You came out and played from the opening kickoff like you wanted it. Well done. To my boys: You can't make mistakes like that, get out-hit, and take stupid penalties, and expect to win.

Final score: 35 to 22.

Woo-hoo!

As an Edmonton Eskimos fan, I was cheering for Ricky Ray and the Argonauts.

We wanted the West to win, but we couldn't stand to see the Stampeders win, and listen to Calgarians brag about it for the next year, and we liked Ricky Ray, the former Q.B. for the Esks.

Besides, a Toronto win would be good for the CFL. It might get Torontonians back in the game and make them forget about the NFL.

The 100th Grey Cup game was a brilliant celebration of Canadiana, with nothing but Canadian music played in the stadium during the game, and all kinds of historic photos and videos on the jumbotron, celebrating the game and the country.

After the game, we learned that Mr. Dressup had died.

Ernie Coombs had a stroke a week earlier and then passed away.

With Casey and Finnigan, Mr. Dressup was big star on kids' TV in Canada, always encouraging young people to use their imaginations. He was a Canadian icon.

The Cold Lake Ice had a big win on the road: 5-4 in double overtime in Vegreville versus the Rangers.

Matt Laboucane was the hero with the game winner.

Next up: the St. Paul Canadiens, Wednesday in Cold Lake.

Recovery Day 63: Limbo

It was the Monday morning after Grey Cup 100 and a lot of Canada slept in with a hangover.
        -- Not me!
        After chatting online to Australia until midnight, I still got up in time for the Early Bird swim, but my right leg hurt so much I could barely make it across the room.
        My Scrabble score was stuck at 1650, but I had 230 bingos.
        That's the kind of Monday morning it was.
        It was very dark outside until 7:30 and sunrise.
        After the Grey Cup game, a lot of people in Cold Lake turned on their Christmas lights, so the little city was transformed into a big wintry village awaiting a visitor from the North Pole. We have had snow since sometime before Hallowe'en.

There were no reports of post-game riots in Toronto.
        That was good news.
        There was a good story in the Toronto Star about Elton John in Beijing and how he stunned a Chinese concert crowd by dedicating his show to the dissident artist Ai Weiwei.
        It was the biggest thing since Bjork sang her song about independance and then shouted Tibet! Tibet! in China.
        -- Keep on rockin' in the free world!

My morning e-mail included a note from a former colleague who was on LTD for stress. He asked me if GBS was like Landry's, so I looked it up.
        The syndrome was named after the French physicians Guillain, Barré and Strohl, who were the first to describe it in 1916. It is sometimes called Landry's paralysis, after the French physician who first described a variant of it in 1859. It is now known by many names. Here are some of the Latin and English names, with a short explanation:
    Guillain-Barré Syndrome (GBS), pronounced 'ghee-yan bah-ray'.
    Landry-Guillain-Barré syndrome (LGBS).
    Guillain-Barré-Strohl syndrome.

The long Latin names are packed with meaning: Inflammations end in -itis". As GBS is an inflammation of the peripheral nerves (neuro-), it is often called 'neuritis'.
        Many nerves are involved, so the term becomes 'polyneuritis'.
        More specifically, the nerve roots (the points of attachment of the peripheral nerves to the spinal cord) are attacked by GBS, which leads to the term radiculo-'.
        Diseases are -pathy, making polyradiculoneuropathy an inflammation of many peripheral nerves and nerve roots.

"Idiopathic" signifies that the cause of the disease is unknown. Acute means that it is rapid, while inflammatory means irritating, and finally demyelinating indicates that myelin is destroyed.

GBS is not one disease - the syndrome has several variants differentiated by their symptoms, the preceding infection, the duration of the inflammatory phase, severity, etc. These forms are sometimes called subgroups of the Guillain-Barré syndrome family of nerve disorders.

That's what a physiotherapist thought I had. But I had no paralysis. Over a 24 day period, I experienced the progressive loss of the muscles in my legs.

Was it descending, or did it happen simultaneously?

The recovery happened in a descending order. First my knees came back, then my ankles, and finally my toes.

Right now, I still have tight muscles in my arches, a little numbness and tingling in the balls of my feet and toes, and a lot of pain in my right leg, in the buttocks and thigh.

That doesn't sound like GBS or Landry's, does it.

What it sounds like is a compacted spine with hips out of alignment that got a bad hit by a chiropractor, causing severe nerve damage leading to muscle loss from legs to toes.

My old buddy said he was in limbo, on LTD, so I wished him bon voyage and bon chance, and told him that I wished I could do the limbo!

PS: My Scrabble score finally changed. It jumped up to 1662!

After getting up early, I went back to bed to rest my leg and make the pain go away, but it didn't work. It was still very sore when it was time for the Noon Lane Swim.

While I was in bed, I read the new book I got at Walmart, called

It turned out to be quite a good book by a Canadian woman, a yoga teacher, a Catholic, who spent 40 days at Plum Village in France, at the famous Buddhist retreat, working with Thich Nhat Hanh, called Thay, and reporting journal-style on her experience.

After reading a few pages, I fell deeply asleep.

But not in a bad way!

-- I liked the book and looked forward to reading the rest of it.

Swimming was good, again. As well as deep water walking, I did some shallow water running.

First I tried bounding, which is recommended, and then I added a hop to it, to go with the music. And that led to running in the water.

-- How I would love to run again!

They played a song I liked called What If God Was One Of Us? and I sang along with it as I walked and bounded and ran in the pool.

The lifeguard asked me if I was okay.

After swimming, I went to Humpty's for lunch, as I hadn't had eggs for a while, and I read the Edmonton Sun while I was there. The front page story was a picture of a Calgary Stampeder looking morose with the big bold headline that said BLUE IT!

It should have said Double Blue It.

And after that I went to Canadian Tire to look for something called the Ove Globe. The guy in Canadian Tire called it the Ug Glove. My sister told me she wanted one for Christmas and they didn't sell them in Australia.

The Ove Glove was twenty-four bucks. And that was just for one!

For twenty-four bucks, I told the guy in the store, I'd want to be able to put my hand in an open fire, wearing this thing.

Oh, you can, he said.

Before I bought it, I went outside and phoned my brother, to see if he wanted to get it for our sister, and we talked for fifteen or twenty minutes about health and the Grey Cup and hockey and so on.

He was playing seniors hockey, again, or joining the group of guys that puts on their hockey sweaters and skates around with a hockey stick, firing at a metal target suspended in the net at each end of the ice.

Most of the guys played junior hockey, in their youth, and there's a former NHLer or two. They don't just aim for the target, they aim for a selected part of the target.

I've skated with those guys, so I know they are sharpshooters.

My brother has the coolest hockey jersey. He loves pirates and we love the story of Avery The Pirate, so I got him a Portland Pirates hockey jersey for Christmas a couple of years ago. The logo is a pirate playing hockey.

We talked for a long time and I was outside in the cold. It was about fifteen below. And I loved it.

After walking around Canadian Tire and its parking lot, I went home, but I felt like getting some more exercise.

While I was making dinner, I walked around the kitchen on my toes -- something I couldn't do before -- and then I tried balancing on one foot.

My new record is one full minute for standing in the yoga position called tree pose. -- That's not so great, but it's a lot better than zero, which was all I could do for the past three months or so.

Recovery Day 64: K-Day

The Early Bird swim at the pool on the base, from 6:15 to 7:30, was only held on Monday, Wednesday, and Friday, I discovered this morning, at 6:40, after I put my coat in the locker at the rec. centre.

Was I disappointed? -- Not so much. Frankly, I felt proud of the fact that I was able to get up with the alarm, get going in the morning, and make it there on time, or 24 hours early, as I have not been able to do that for over three months.

After discovering how quiet but friendly the base is at that hour, with everybody saying "Good morning", instead of just going about their business without greeting anybody, I drove around the city to see what it was like.

In the dark, at twenty below, with a little bit of the white stuff falling, decorated with Christmas lights, mostly snowflakes, on city streets, Cold Lake looked like the city at the North Pole in the movie called The Polar Express.

My school was dark but the French teacher was there, doing her marking and lesson planning, and one parent was there with a student, for some reason.

Before I drove home, I got gas at the PetroCanada station, and read the morning newspaper headlines. Like CBC Radio One, the front pages were all about Paul McCartney's upcoming visit to Edmonton. The mayor was declaring it McCartney week and there were pre-concert events for a few days before two sold-out mid-week concerts by the old rocker. The concerts sold out in minutes, apparently, in close to record time.

Edmonton was also having an image make-over, according to the morning news, and they were looking for a new meaning for their old Klondike Festival, which was rebranded as K Days, so they were asking people what the K could possibly stand for.

As a Scrabble player, I know there's close to two thousand words in English that start with the letter k, and most of them sound German or Yiddish, and I like the word kayak, worth a lot of points, but it's rare to get two k's at the same time in the word game.

K, for me, means "Okay", it's the short form for "O.K.", the way we say it while chatting online, as in, "I'll meet you in Edmonton this summer for the big festival," your friend says, and you respond by saying, "K!", or somebody says "See you at that place by the thing" and you say "K".

K also stands for kicky, which means awesome, unless it's used sarcastically. Kicky means colourful, festive, and cool, but it can also mean the opposite, if it's used ironically.

K?

Okay, so I went back to bed. Is that so bad? And I got another hour of sleep before the alarm went off to get me up for physio. Was that a sin?

It felt great.

My leg did not feel so great. So, what can you do? I cheated by putting on my Dr. Ho belt before going to physio. My plan was to wear it during physio, for the first time.

The bad thing about a support belt, they say, is that it slows the development of the muscles you want to build.

But the good thing, I say, is that it allows me to walk while experiencing a lot less pain.

K?

Physio with the substitute physiotherapist went well. In fact, it went better than ever.

Instead of doing 10 minutes on the treadmill at 2.5 mph, I did ten at 3. Instead of the square balance board, I used the round one. Instead of doing the back stretch with my elbows, I used my hands, so I went higher and arched more. I even did the bridge, badly, but without a lot of pain, for a change.

During acupuncture, I had an odd thought, While I paid attention to the pins in my legs and back, and felt the buzz, I was thinking about old relationships, and I was thinking about what I was planning to write about, and I was feeling my feelings as well as my muscles and pains.

So here's the odd thought I had: I was multi-tasking and feeling emotional while contemplating relationships. In short, I was turning into a woman!

The truth is, I've always been a multi-tasked in touch with feelings.

The noon swim was KK. Double plus good, as Alex says in A Clockwork Orange. -- Although the physiotherapist said to limit my time in the pool to half an hour, so I don't overdo things, I did more than double that, at more than an hour, and it felt very good.

As well as the usual deep water walking, forwards, and shallow water walking, backwards, with the buoyancy belt on, I tried something new: I did some cross-overs, football style, and a little bit of the Scottish traveling step from country dancing. My goal was to put one foot behind the other, back and forth, to stretch my legs.

The music was good: Aqua, The Beach Boys, and Walk Off The Earth.

As well as counting breaths, seconds per length of the pool, and laps, and paying attention to my form, making sure my ribs were over my pelvis and I was using the umbrella breath, I monitored the pain in my leg, noticed that it came and went, and I contemplated the course of my injury and recovery, while planning out an essay unit for my Grade 12 class.

My right leg was a little fatigued by the end of the hour, or seventy minutes, I noticed in the shower, but after sitting for a few minutes, getting changed, it felt fine, and after I drove up to the lake and back, it felt great.

Getting into and out of the car has become a problem. My right leg is a little gimpy.

Aside from that, I felt pretty good after getting up early, going to physiotherapy, getting acupuncture, and swimming for over an hour. I got up at six and was still rockin' on at two in the afternoon.

My Scrabble score went down to 1637.

In the pool, I was reflecting on my seven years at the Zen Forest and half a lifetime of being a vegetarian, and wondering about how good that was for me. The Zen Forest diet was su vegeterian, which means no garlic or onions, or anything in that family, including leeks, but I rarely had leeks, anyway, and the goal was to avoid the fetid, or smelly, vegetables that heated you up.

     However, garlic and onions are good for you, they say, and some say they are good for your nerves and the sheath that coats your nerves.

     Also, sitting cross-legged in meditation, on a zafu, or meditation cushion, or two, back straight, was supposed to be good for you, but what did it do for my back?

     And how about all the heavy lifting I did at the Zen Forest, working on forest management and landscaping projects, building roads and cabins, mixing cement and making concrete foundations, and wrestling big marble statues into place? Was that good for the back?

My brother, 72, told me they had snow in Ontario, finally, and he had been up on his roof, shovelling it off, so I told him to be careful and to watch his back.

In the news, there were multiple reports of one of the worst mass shootings in Toronto history, in Scarborough, involving two gangs, plus kids from Pickering and Ajax, where I used to live and work, and the kind of kids I used to work with, when I was a teacher in a school inside a maximum security prison, for kids who kill kids, in Ajax. One of the kids was known as Bam Bam Mosquito. He was over 18, now, so he wasn't a kid, anymore.

     It was the Galloway Gang versus the Malvern gang in Scarberia where, my high school students told me, 'Everybody's gotta gun'.

Sometimes people ask me why I moved from the GTA to Northern Alberta.

When I went to the Wellness Centre for reflexology, I was feeling grouchy, going in, as I was fed up with the slow pace of recovery.

Mecell the magic alternative practitioner did her thing for two hours, instead of one, and took away all the pain in my body. When I walked out of the Wellness Centre, I walked around the snowy parking lot, looking up at the full moon, feeling as though I was in somebody else's body.

     -- How does she do that?!

     I felt kicky!

The weather forecast for the next day was not good: Environment Canada's Official Weather Warning: Heavy snowfall to start early Wednesday with 15 to 25 cm of snowfall expected by Thursday.

A deep Arctic ridge is building down from the Yukon and Northwest Territories bringing below normal temperatures to most of Alberta. Strong winds combined with the cold temperatures will produce extreme wind chill values near -40 by Wednesday morning. At these extreme wind chill values frostbite on exposed skin may occur in less than 10 minutes.

Overnight an intense upper disturbance from the Pacific will cross the Rockies generating a band of heavy snow that will extend into the city of Edmonton by Wednesday afternoon. The heavy snowfall is expected to taper off by Thursday with snowfall amounts in the 25 cm range.

That was for Edmonton.

Would it hit Cold Lake?

Recovery Day 65:

Although I got up in time for the Early Bird Swim on Wednesday, I decided to skip it, go to physio at 9:00, and see how I felt about going to the pool at noon.

There was a bit of referred pain in my hip and I felt sleepy, even though I got around nine hours of sack time and slept like a character in a Washington Irving story. -- I looked like Ripped Van Wrinkle.

It was fifteen below out, but it felt like minus twenty-five. Environment Canada published a forecast for the whole country for the winter of 2012/2013 showing a big blue patch from Edmonton down to Winnipeg and up to Yellowknife and beyond, including Cold Lake, indicating below normal temperatures. Regina set a new record for snowfall in November. The rest of Canada would have normal temperatures for the winter, except for the Maritimes, which would be a bit warmer.

There was a piece in the Huffington Post that said David Suzuki was claiming success for stopping the mega-quarry in Ontario, as his foundation had produced Soupstock after another group created Foodstock, as protests to spread the word about the plan to blast a billion tonnes of limestone from beneath some of the finest farmland in North America.

The article in the Huff Post, from the Suzuki Foundation's communication department, says the project by an American company initially drew the ire of a handful of local farmers and residents who faced overwhelming odds to stop it.

They quoted Margaret Mead saying, "Never doubt the ability of a small group of thoughtful, committed citizens can change the world."

People power won, they said, and it was a victory against a company called Highland that was backed by a Boston hedge fund, the Baupost Group, with assets of more than $25 billion, and represented by Hill and Knowlton, the high-priced PR firm that infamously worked with Big Tobacco to convince smokers that cigarettes don't cause cancer.

Margaret Atwood lived on a farm near Alliston, for a while, and she wrote a famous poem about it, called The Bus To Alliston.

"Snow packs the roadsides, sends dunes
onto the pavement, moves
through vision like a wave or sandstorm.
The bus charges this winter,
a whale or blunt gray
tank ...."

When I lived in Owen Sound, I used to travel through the mega quarry area every Saturday, going down to Toronto to teach a writing course at York University. My soccer and basketball teams used to travel down to Shelburne, Orangeville, and Base Borden, for games. There was one big potato farm after another.

Highland bought a lot of those farms and then hired the farmers back to keep the potato lands in production, claiming they would take out the limestone and restore the land so it could be farmed again. But, apparently, nobody believed them, or thought that was possible.

The anti-quarry group had a Facebook page that showed they were celebrating the fact the Highland Companies withdrew their application for the quarry.

That Facebook group had links to other groups that were protesting oil pipelines, especially the route from the oil patch in Alberta to the West Coast of Canada, through B.C., trapping Canadian oil in Alberta, so it could only be sold to the U.S.A.

Physiotherapy first thing in the morning was good. The physiotherapist was back. Angela Plaquin said I could swim every day, instead of just the days I didn't have physiotherapy, and she recommended I keep a journal, a fitness log, recording the exercises I did and how I felt after doing them.

You like writing, she said. This should be easy for you.

This morning I did ten minutes on the treadmill at 2.5 mph, ten back extensions, ten leg stretches, used the balance board ten times in each direction, and, something new, I sat on a big exercise ball and bounced the smallest bounce while breathing the umbrella breath and maintaining the optimal posture, ribs over pelvis, back arched, shoulders back, blah blah blah.

Acupuncture was less aggressive.

Today I walked in feeling somewhat jolly and walked out the same way, but after standing and walking in the bookstore, my butt hurt a bit, so I went home.

In physio, I told Angela that, the day before, I had done all my exercises bigger, faster, stronger, and longer. She said, That's not what it's all about.

She said to forget the old idea about "no pain, no gain" and focus on doing the exercises well without any pain.

After contemplating that for a while, I told her I had three questions for her.

1. Am I building muscle a lot faster than normal? It sure feels that way.

2. What about a lumbar support belt? It takes away pain.

3. How about swimming more?

She said yes, once in a while, and yes.

She also went over the best way to work out in the water, doing deep water walking, with the buoyancy belt, maintaining the best form, as though walking.

So I told her I was doing that yesterday, while listening to the music, counting laps, breaths, seconds, planning an essay unit for my Grade 12 class, paying attention to pain levels, and thinking about old relationships, and I came to the conclusion that I was multi-tasking with feelings so I was afraid I was turning into a woman. -- That made her laugh!

PS That belief that only women are good at multi-tasking, or all women are better at multi-tasking, is a myth. Psychological surveys and studies clearly indicate that certain people of both genders multi-task better than others, and that's it. The myth is pretty pervasive these days. I've heard lots of women brag about their abilities and put down men for having a deficit in this area.

Speaking of genders .... The new Ontario curriculum teaches six genders: male, female, transgendered, transsexual, two-spirited, and inter-sexed. It is being taught in Toronto schools. The curriculum is mandatory and there was no parental notice or option to withdraw children.

   Last year, parents voiced their concern over this same material. The Premier promised to withdraw the program, but it's still there.

   Would you teach your eight-year-old child that there are six genders?

Which of those genders would you say is best at multi-tasking?!

After physio and acupuncture, I felt tired, so I slept for half an hour, then headed out to do errands. In order to apply for LTD, I had to fill in five pages of forms, send them in, and send the specialist at the Edmonton Clinic a form to fill in, with a cover letter, which meant a trip to Staples, to get the letter printed, and then a walk to the post office and back, about eight blocks. While I was out, I dropped a class set of The Wars, by Timothy Findlay, off at my school, and some books for my Grade Fours, and said a quick hello to our head secretary.

   At the end of that little trip, which took about three hours, the walk home, about five blocks, was slow and painful. But the rest went well.

In the back of my mind, I've been contemplating a secret plan to get back to work sooner than the physiotherapist thinks I'm ready. Sometimes I think I can get in shape, get a doctor's note, and return to work by December 11, instead of going on LTD. And sometimes, like that last little walk I did, the final four blocks after standing in line at the post office, makes me think my plan is just a pipe dream.

After sitting for just a few minutes, making these notes, I felt fine, again.

Slept for an hour, woke up with a negative pain sensation in my right buttock, but it went away after I sat in front of the computer for a while, wearing my Dr. Ho belt. Even so, in the evening, when it was time to go to the late swim at the pool, I decided to take the physiotherapist's advice and rest, instead.

   On Netflix, I found a couple of documentaries that were worth watching. One was about the Gulf of Mexico, BP, and oil. The other was about the first female MP in New

Zealand and her view of economics. She said that the problem was the set of rules used by the United Nations, created by the Brits to pay for war, and followed by countries around the world, which were all about the GDP and did not take into account things like motherhood, housework, or a clean or beautiful environment.

Recovery Day 66: The End Of November

In the morning, the alarm woke me up, and before I got going I did an inventory of my body, as usual: Back, okay; legs, okay; feet, tight; toes, okay.

Lately, I haven't paid a lot of attention to my feet or toes. There is a tight band of muscles across the balls of my feet, but that odd feeling, like I was walking on bean bags, is long gone. My toes have had some feeling for a while now, but there was still some numbness and tingling. But this morning, the tingling was gone.

My feet were sore, as usual, but the only other pain was at the sciatic notch in my pelvis on my right side.

Not long ago, I didn't know what the sciatic notch was, or where it was located, or what it was called.

So, this morning, my aches and pains, from my lower back, down, were at their lowest levels in 66 days.

-- If only I had more energy!

There was no Early Bird swim this morning, so I didn't have to feel bad about missing it, but there was no way I would have made it. -- My energy was missing in action like Bruce Willis in Mercury Rising.

Our mercury was not rising. It was twenty below in Cold Lake, fifteen below in Edmonton, twenty-five below in Fort Mac, zero in Muskoka, and thirty in Australia.

Yesterday, where my sister lives, the temperature went up to 41. That's 105 fahrenheit and that's too hot for me and most people on the planet. My sister loves hot weather but on days like that even she relies on AC.

Of course, a lot of people using air conditioning, on hot days like that, uses a lot of energy, and in Australia that means burning more coal, and that has an effect on the atmosphere, so you could say that when you click on the air conditioning to make your room or house cooler, you are heating up the planet.

It's a vicious cycle, some say, that will kill us all.

Any other cheery morning thoughts? Ontario teachers were in the news, announcing strikes by elementary teachers all around the province, in December, with some secondary teachers to join them later.

What did Love Tree have to say?

Love Tree

If a person seems wicked, do not cast him away. Awaken him with your words, elevate him with your deeds, repay his injury with your kindness. Do not cast him away; cast away his wickedness. ~Lao Tzu

Let your smile change the world, and don't let the world change your smile.

-- That one was anonymous.

-- I'd have to work on those two things today.

-- If I found some energy.

Speaking of energy .... How did the Cold Lake Ice do at the Energy Centre last night? They played the improved St. Paul Canadiens, and they won by a score of 9 to 1.

Dallas Ansell led the way with a 5 point night and earned the games first star!

The win improved the Ice's record to 13 wins, 1 loss, and 1 OT loss for first place overall in the league.

-- On their Facebook page, they were referring to the hockey arena at the Event Centre as the Imperial Oil Arena or the Imperial Arena.

-- That had an ominous ring to it, I thought. -- It sounded like something out of The Hunger Games.

On my Novel Marathoners page, on Facebook, there was news about the Muskoka Novel Marathon, as it was listed in Open Book Ontario as a Literary Landmark as well as an Event.

Their map of Ontario literary landmarks made it look like the MNM was in Algonquin Park!

Open Book was an organization created to promote Canadian literature in Ontario and it was a project of The Organization of Book Publishers of Ontario, supported by the Ontario Media Development Corporation, which was an agency of the Ministry of Tourism, Culture, and Sport. It also got support from the Ontario Arts Council and the Canada Council for the Arts.

Funny .... I was thinking about this book and this region, the other day, thinking that this book about my healing adventure was also a book about Cold Lake and the Lakelands region of Northern Alberta, and it occurred to me that this had to be one of the best books about this place because, well, let's list all the other books about this place. -- Okay, I'm done.

This is not a bookish place, with a lot of writers, bookstores, writerly events, and so on.

Do I want to change that?

-- Maybe later.

Right now, all I want to do is get healthy and then get back to work.

Last year, I wrote a book of poetry about this place, called Celebrating Global Warming. This year, I am writing a book for kids, set here, called The CLAW, but in the past 65 days of recovery, or 90 days since the injury to my back, I have written about three pages.

For a guy who can write a novel in a weekend, that is NOTHING!

Instead of writing, I am focussed on healing. -- That takes all my time and energy.

My Scrabble score dropped below 1600 but went back up to 1611, with 231 bingos.

Canada Reads, on CBC, announced their new line-up.

The list did not grab me.

The theme was Turf Wars, with one book selected to represent each of five regions across the country: B.C. and Yukon, Prairies and North, Ontario, Quebec, and the Atlantic Provinces.

Open Book had an ad for the MFA in Writing program at the University of Guelph and it said "Contact Meaghan Strimas".

Was that MY little Meaghan Strimas?

It was!

She was in my Writers Craft class at The O.S.C.V.I.

Now she was running the MFA program at Goo U?

Awesome!

We were Facebook friends, so I sent her a little note saying, Congratulations!

That cheered me up!

Seriously, though .... In spite of the fact that I was relatively pain free and almost pain free in reality, I had no energy.

Online, I found "55 Ways To Get More Energy". -- But I was too tired to read it.

After sitting at my desk for an hour and a half, I decided to have breakfast and go back to bed.

Would I make it to the lunchtime lane swim in the pool at the base?

-- Who cares?

Physio wasn't until 2 in the afternoon. -- Maybe I would sleep until then.

At 12:30, I woke up, again, and felt about the same.

It was too late for the noontime swim, but I didn't mind.

In bed, I read a bit of The Monks and Me, the book about the French monastery, by Mary Paterson, a yoga teacher from Ottawa who ran the Lotus Yoga Studio in Toronto.

It was the kind of book you didn't want to read quickly. After ripping through the collected works of Paolo Coehlo and zipping through No Laughing Matter, I wanted to read Paterson's book slowly. After reading just one short chapter, I put it down and thought about it, for a while, and looked forward to reading another short chapter later.

When I checked out her website, I discovered she was downtown, not far from Mirvish Village, where my father and his father had lived, and where I often went on weekends during my Toronto decade. She had a YouTube video and a Facebook page. -- We had 20 mutual friends on Facebook.

Physiotherapy changed things, today.

First of all, I did my usual exercises, and the physiotherapist said, Good form, which is what we are after, and then she added a few new exercises, after testing my balance.

Standing on one foot was not good, but it was better than before, and better than two months ago.

She set up a block of wood between the parallel bars, on the floor, and got me to stand on it with one foot, keeping the other foot on the floor, and the exercise was to lift my lower foot by straightening the other leg.

That looked like it would be painful but it turned out to be easy and painless.

Lying on a mat, she gave me a length of rope and looped it over my foot so I could lift my leg, straight, and that wasn't hard, either.

Next, she showed me how to stand, with one foot in front of the other, inside a harness attached to ten pound weight, with both hands on a metal bar, and the exercise was to push the bar to lift the weight. After that came pulling the bar and the weight.

Aside from getting a zap of pain just getting my feet in the right position, that exercise was easy.

Even though all the exercises were easy, I was quite warm by end, and then it was time for acupuncture.

This time, she used a few needles in my butt and legs plus two in each of my hands. On the way in, I told her my pain was down but so was my energy, so she gave me acupuncture to boost my energy.

Walking home was a little painful.

When I got home, I got a call from Nola the neurology nurse at the Edmonton Clinic, with some questions from the specialist about how I was feeling, what I was doing in terms of recovery, and so on. So I told her the doctor at the Cold Lake hospital talked me out of the lumbar puncture, or spinal tap, at the last second, although I was ready to go, and he said I didn't need the EMG, either, as I was walking and recovering.

Since then, I told her, recovery was slow but consistent, although it was two steps forward and one step down, and I was walking, doing more and more in physiotherapy, exercising in the swimming pool, and felt I was going in the right direction even though the pace was frustrating.

Also, I told her I thought she was great, when I was at the Edmonton Clinic, and that if I was at 'zero' when I was there, I was now at 50 or 60 percent.

The physiotherapist asked me to contact my benefits people and get her the name of somebody to talk to in order to get going on the four hour assessment she wanted me to do and the daily two hours of physio she had planned for after the assessment.

She also said that it's understandable that I'm feeling tired, as long as I don't get lethargic, as physiotherapy is surprisingly hard work. It might not feel like it while you're here, but you are working on muscles that disappeared and it isn't easy to bring them back.

She added that I might notice my arms were tired, after today's exercises, so I should watch for that.

My arms were not tired. In fact, I did a little weight-lifting at home. -- Just three sets of ten arm curls with ten pound free weights.

Aside from a little soreness in my feet and a little pain at the sciatic notch on one side, it was the most pain-free day of the past 90.

My energy improved, a bit, but I did not feel like going swimming or even walking, but I did sit at the computer for a few hours, doing e-mail, updating this journal book, playing Scrabble, chatting on Facebook, researching this and that online, and I tried to use my physiotherapist's advice so all that time was exercise time as I sat in the right position, back slightly arched, practicing umbrella breath, and getting up every half hour to stretch.

I felt as though I could sit long enough to write a novel in a marathon. -- Too bad I didn't feel like walking or swimming or doing some other form of exercise.

It was almost the end of November and I felt bad that I had not gone in NaNoWriMo, National Novel Writing Month, and written another novel.

Recovery Day 67: Warmth, Tea, And Hibernation

On Friday, I slept in until 7:30 and missed the Early Bird swim. It was fifteen below, outside, but felt like minus twenty-five, and there was a little bit of snow falling gently on the black spruce outside my windows.

It wasn't winter, yet, officially, 23 days before Christmas, but Alberta had winter weather before Hallowe'en. The Prairies had even more cold and snow. It was slippery in southern Ontario as they had freezing drizzle that made driving dangerous. The area around Barrie, between Toronto and Muskoka, was the worst hit, as usual. Georgian Bay was starting to freeze over. It was winter in Quebec and the Maritimes, too. The weather websites had video of a school bus sliding off a highway in Newfoundland.

Vancouver and Victoria had sunny skies and highs of ten degrees above zero.

My sister said Adelaide's weather cooled down to 25 but a town nearby called Renmark set a record with 45.5 degrees.

In the news .... Ontario high school teachers were working to rule and elementary teachers were ready to go on strike on Monday but the Minister of Education said she would use legislative tools to keep them in the schools.

-- Good luck with that! I said to my computer screen.

Binnington, the goalie from Gravenhurst, was still the best in the OHL.

The OKC Barons were still the team to watch in the AHL, with two lines of future NHL stars, Canadian and European, plus a defenceman, and all of them developing chemistry.

In international news .... Canada recalled its ambassadors to the UN, Israel, and the Palestinian Authority after voting 'no' on Palestinian statehood in the United Nations. Canada and the U.S.A. voted 'no' and said that if the Palestinians used their new status as a state to attack Israel, we would cut off aid.

Canada had given the Palestinians 300 million in aid over the past five years, according to AP.

As for li'l ol' me .... Like yesterday, I was almost pain-free, aside from my sciatic notch, which gave me the odd reminder of my three month long adventure, but my energy was low, so .... I planned to keep a low profile until it was time for the Noon Hour Swim.

After spending a couple of hours at the computer, I set my alarm and hit the sack, hoping I would feel even better after a couple of hours so I could go jump in the pool.

Sleeping was heavenly as I was feeling less pain than before, for the past 90 days but it didn't exactly fill me with energy.

Instead of going outside and then going swimming, I made a list of things I would rather do instead of getting cold then wet then cold and straining my new muscles possibly to the point of being in pain again.

1. Make a pb& j sandwich
2. Drink tea
3. Read a book
4. Write a book
5. Stick a safety pin in my foot

Lately, I have been getting the message to drink tea from all over the place.

My cousin in LLB sent me an e-mail with a forward about cold drinks being bad for the heart and for weight-loss and so on. Mary Paterson's book about being in a monastery in France was full of tea and stories about slowing down, developing patience, and hot drinks in a cold climate.

Did I have any tea? That was another question.

In a cupboard, I found some Tetley's tea called Warmth. It was a Cinnamon Spice herbal tea somebody must have left here by mistake.

On the little can, it said, A comforting, aromatic infusion of cinnamon, orange blossom and sweet licorice with warm, spicy hints of cardamom and clove.

-- That sounded way too sophisticated for me!

At the Zen Forest, we had ginger tea quite frequently. To make ginger tea, you get a piece of ginger, slice off a tiny piece, peel it, chop it into tiny pieces, and pour hot water over it. When it cools a bit, drink it slowly, without engaging in conversation or doing anything else. Just drink the ginger tea. -- Taste it.

After several years of Zen training, I had no time for that ritual or hot drinks!

First, I had to find my kettle.

It's a light green plastic plug-in kettle somebody gave me years ago and I cannot quite believe it's still kicking around. How had it survived the past four or five moves I've made over the past decade?

The woman who gave me the kettle was something else. We met at a summer program for teachers in Stratford, where we were working on Additional Qualifications in Drama, which meant spending our days in a high school and our evenings at the Stratford Festival, for six weeks, so we did a lot of drama exercises together and saw all the plays at all the theatres in Stratford.

-- What a great summer!

She told me her husband was abusive and she planned to leave and take the kids and, after a little time on her own, move in with me.

When I took her and her kids up north, to see where I lived, she said it was perfect, except for one thing: Where's your kettle? I like tea!

So, she went out and got me this plastic kettle in an odd shade of green that didn't really go with anything in the world, and she said, This is for me, when I move in!

However, she never left her husband, so all I got was the kettle: No wife and kids or instant family. -- Just add hot water.

The kettle still worked, I discovered, and it took a long time, just short of forever, for the boiled water to cool off enough for me to taste the tea, but I exercised every ounce of patience I had and eventually had a nice surprise.

It was a smooth blend of cinnamon and rooibos. Could I detect a little anise and cloves, as well? My palette is not that sophisticated. But there was something there that tasted like licorice and Christmas. It was a little coppery and I thought it called for some dark honey to complement its taste and make it even more complex.

The second mug, steeped longer, tasted even better.

Do you think it's a disgrace to drink tea out of a mug? Only tea cups do it for you? Well, what can I say? My mug from the Stratford Festival had the word muse written on it, or scrawled on it, in an amusing way.

Something happened, part way through the second mug of Warmth.

How much caffeine was in that tea?

Of course I Googled it. And I found out Tetley's herbal teas are made of roots, leaves, seeds, flowers, spices, berries, and herbs, in endless combinations, and they contain no leaves from the Camellia Sinensis plant, so herbal teas are teas in name only and contain no caffeine.

So what was giving me a buzz? It felt as though I had just chugged one of those 5 Hour Energy drinks created by that Buddhist monk who made a fortune packaging a drink with no sugar and not much liquid to compete with the big cans of caffeine and taurine and ginseng, etc., that were such a huge hit in America.

Who gets really thirsty at the same time they need a shot of energy? he asked. And he laughed all the way to the bank, as we used to say.

According to Wikipedia, energy drinks generally contain methylxanthines (including caffeine), B vitamins, and herbs, carbonated water, guarana, yerba mate, açaí, and taurine, plus various forms of ginseng, maltodextrin, inositol, carnitine, creatine, glucuronolactone, and ginkgo biloba. Some contain high levels of sugar or are artificially sweetened. Energy drinks contain about three times the amount of caffeine as cola. Twelve ounces of Coca-Cola Classic contains 35 mg of caffeine, whereas a Monster Energy Drink contains 120 mg of caffeine.

That tea I tasted was something else, then. I felt like going in NaNoWriMo and doing the month-long novel marathon start to finish in three days.

What would I write about? All of a sudden, I had a great idea and a title to go with it: The Gospel According To Wikipedia And The Internet.

Sub-title: The Third Testament.

The idea was that the sequel to the Old Testament and New Testament or the Torah and the Christian Bible was being written in our era by spiritual writers in different places all over the world.

My novel would be about bringing them all together in cyberspace and what that would do for the world and for the writer who dared to put his name to the book that was the sequel to the Bible.

Why didn't Jesus write a book or two, leave us a message in written form, to be interpreted and misinterpreted over the centuries? -- I've often wondered about that.

He wrote in the sand, famously, but why didn't he put pen to paper, or ink on a scroll, and leave behind a diary or a journal or a message of some kind?

-- What was in that tea?!

After a little burst of energy and creativity, I felt like taking a long winter's nap.

What about exercise, physiotherapy, swimming my way back to work?

Well, they say a certain amount of hibernation is good for you.

Why do we act like it's summer all year when clearly it's cold outside and ....

-- I fell asleep in mid-sentence.

After a few more hours of painless sleep, I woke up feeling happy.

That was a change.

Although I'm an optimistic, up-beat, kinda guy, usually, I had not felt that way for the past few days.

But the feeling was back, and I was happy about that, too. -- I was happy I was happy.

You know?

It turned out to be the most pain-free day I'd had in three months.

Recovery Day 68: December Begins

November was over and it looked like December.

That familiar pain in the butt returned first thing in the morning.

Fortunately, it faded away while I sat at my computer and checked e-mail and Facebook, chatted online and played Scrabble, and tried a few different CBC Radio stations.

Sometimes I listen to CBC Montreal, Toronto, Vancouver, or Whitehorse, instead of Edmonton, and pretend I live there, or I'm visiting for a while. -- It's a very inexpensive way to travel and has a carbon footprint of zero.

After chatting online with Ontario for a while, I checked the news.

My friend with the double knee and hip replacement, plus chronic pain from arthritis, a survivor of sexual abuse, had just come back from a trip to Cuba. She posted a picture of herself riding a horse on a white sand beach beside blue water.

She was complaining about winter and the fact that her car stalled so she had to take it in to the shop. She drove into a ditch the day before that.

But she had some good news, too: Her agent wanted her to turn a novel she pitched into a series of crime novels with a title like Sex In The City.

She asked me if it was bad to use a title like that, so I told her that titles are not protected by copyright and she should 'go for it'.

The title her agent wanted: Sex Crimes In The City.

Alberta teachers were in the news: The Alberta Teachers Association asked the premier to accept a contract offer that would ease teachers' workloads in exchange for a wage freeze from now until 2014. The ATA represents over 40,000 Alberta teachers. Teachers were willing to accept a salary freeze if it meant their proposal to limit "assigned" hours to 1,200 a year — a move Education Minister Jeff Johnson previously refused.

After breakfast and herbal tea, it was back to bed for me. Although I thought I would rest for a bit and then go swimming, I slept until high noon.

It was fifteen below and snowing, again, or still, dry snow slowly falling on black spruce, and I didn't feel very energetic, so I thought I'd drag my butt to the grocery store and that might be it for the day.

-- Strange how something that gives you so much pleasure, like getting in the swimming pool, appears to be nothing but a pain in the butt on another day.

There was a report on Yahoo! that said polar ice melting and sea levels rising, because of climate change, could mean big trouble for Canada's coasts.

I noticed that it was now okay to report on climate change but nobody was using the term 'global warming'.

There was a cool story in the Edmonton Journal that asked this question: Does Edmonton have the highest quality of life in English speaking countries?

According to Numbeo.com the answer was -- yes.

The site described itself as, "The world's largest database of user contributed data about cities and countries worldwide, especially living conditions: cost of living, housing indicators, health care systems, traffic, crime and pollution". As of December 1st, 2012, the top cities in the world for quality of life were: Edmonton, Adelaide, Berlin, Austin, Zurich, Houston, Hamburg, Trondheim (Norway), and Boston and Chicago.

Calgary was #15, Toronto was 17 and Montreal was 18.

Vancouver was #23.

How about Cold Lake?

It wasn't listed.

While I was out and about, enjoying a winter's day in Cold Lake, I heard several people saying the weather was miserable.

I thought it was great.

Dry snow, not too cold, ice on the roads, no wind, and what would you expect?

On the other hand, I come from the land of the wet snow, in Ontario.

And for the past 90 days or so I've been under the weather, shall we say, so it thrills me be able to go outside and go anywhere!

Getting groceries in Cold Lake this afternoon was no problem.

My butt was a bit sore, so I put on my Dr. Ho belt, and drove to the store, walked all around, and drove home, no problem.

In fact, it made me feel like doing more.

I felt inspired!

Should I give it a rest and minimize the pain, or get some exercise, once again?

What I really wanted to do was write a novel and work on stuff for school. -- But I did not feel sharp enough for that kind of work.

In fact, it was hard for me to keep my eyes open.

Google gave me answers about pain and the brain and energy, confirming that pain has a serious effect on both.

Somehow, it reminded me of something Mary Paterson said in her book about her trip to France: "A forty-day sacred journey supports the philosophy of the ancient yogis. Within many cultural and spiritual traditions, the span of forty days is recognized as a key interval in which the unfolding and recognition of truth happens. Christ prayed and fasted for forty days in the desert to prepare for and understand his purpose; so did the prophet Muhammad in a cave. Moses was transformed by this time on Mount Sinai. In the forty-day Christian season of Lent, followers give up a pleasure or vice. And the Buddha enjoyed the peace of enlightenment under the Bodhi tree for a period just exceeding forty days."

-- My journey was forty days time two, and counting.
-- How had I been transformed?
Some say pain and suffering can be transformative. -- You can learn a lot.
What was I learning? What was I supposed to learn?
I found a website called whypain.org that had a list of a dozen benefits, each one with a Biblical reference, all about pain and suffering:

1- Pain and suffering can help us to learn important lessons in life.
Example: Mother eagle forcing her baby out of the nest to teach it to fly.
Scripture: Deuteronomy 32:10,11

2- Pain and suffering can bring about creativity, resourcefulness and courage.
Example: Parents who lost a child helping to pass laws or starting organizations to protect other children. Artists and composers sometimes do their greatest work during times of pain or loss.
Scripture: Psalms 18; 42; 63; 126

3- Pain and suffering can help us to comfort others who are going through similar pain.
Example: This is the benefit of support groups for various problems. People who have faced the same problems are able to help and encourage others. John and Phyllis Clayton have been able to help others because of their experience with diabetes and with a mentally retarded child. Jim McDoniel has been able to help others because of his experience with a handicapped child.
Scripture: 2 Corinthians 1:3-5

4- Pain and suffering can help to shape our character.
Example: People who have survived as prisoners of war or persecutions often have a strength of character which is admired by others. Gold is refined by the fire which heats it until the impurities come out.
Scripture: Isaiah 48:10; Zechariah 13:8,9: James 1:2-4

5- Pain and suffering can test us to show what we are made of.
Example: The patriarch Abraham was tested; Job, the ancient man of wealth was tested; the apostle Peter was tested; and the early Christians were tested. In all cases the testing showed the weaknesses and the strengths of their faith in God. Products which are sold in the marketplace are put through tests to find their weaknesses and to demonstrate their strengths.
Scripture: Genesis 22:1-14 describes the test of Abraham. The entire book of Job tells about Job's test. Matthew 26:69-75 tells about a test where Peter failed and he learned something about himself. Acts 4:1-21 tells about a test in which Peter was victorious and his enemies could see the strength of his faith. The testing of the early Christians is

described in various places such as Foxe's Book of Martyrs. In Matthew 7:24-27 Jesus beautifully illustrated the results of testing in a simple story.

6- Pain and suffering can lead to repentance and salvation.
Sometimes it takes pain and suffering to turn a person's life around and head it in the right direction.
Example: The Israelites who repented in times of persecution in the Old Testament. Saul who became Paul in the New Testament of the Bible.
Scripture: Judges 2:11-19 describes the cycle which the Israelites when through as they forsook their God only to be brought back to Him by suffering and then to forsake Him again when times were good. In contrast to that Acts 9:1-16 tells the suffering which lead Saul the persecutor to become Paul the apostle, faithful to his God until his dying breath.

7- Pain and suffering can sometimes help us to trust God.
Perhaps we are forced to turn to God because we have no other place to turn.
Example: There are numerous people who have made the decision to trust God because of their pain and suffering.
Scripture: Job 40:3-5; Job 42:2, 3; Lamentations 3:19-24; Daniel 3:16-18; Habakkuk 3:17-19

8- Bearing pain and suffering well can be an inspiration to others.
Example: Chet McDoniel was born with no hands, only one stub of an arm and no thighs, but he has been an inspiration to many as described in the book All He Needs for Heaven.
Scripture: The apostle Paul endured his "thorn in the flesh" but was able to take the message of Christ to many areas of the world and write most of the New Testament.

9- Pain and suffering can have a Divine purpose in preparing us for glory.
Example: The apostle Paul wrote that our suffering is "slight" and "momentary" compared with "eternal glory."
Scripture: 2 Corinthians 4:16-18

10- Pain and suffering can prevent us from becoming dangerously proud.
Example: Again Paul said that his "thorn in the flesh" was to keep him from becoming proud and arrogant.
Scripture: 2 Corinthians 12:7-10

11- Sometimes pain and suffering in the life of one person can result in the advancement of the gospel in the life of another person.
Scripture: Philippians 1:12-14

12- Pain and suffering can allow us to be like Jesus.

Example: We are allowed to share in Christ's suffering as we serve the one who suffered on our behalf.
Scripture: Philippians 3:8-11; Hebrews 2:9-11; 4:15; I Peter 4:12-16

Was I being tested? Was I supposed to react like Jesus, or Job? What was I made of?
-- So I Googled "what are you made of test" and found one on Mozilla.
It asked several questions and gave me my answer. It said, "The Storyteller: Tell us something we don't know. Seriously.
"Knock, knock. Who's there? YOU are! Always the one with a joke or anecdote, you're the type to fill in the details and make people excited about what they are reading or watching. You, my friend, are the consummate storyteller, narrating life as it comes."

The afternoon and evening was spent watching Owen Wilson movies, sitting in a chair, in a good position, using umbrella breath, and playing Scrabble, with a bit of e-mail. For Christmas, I sent my sister and her partner a hamper with wine and cheese and crackers. She sent me an 'e' saying she would like to Skype but she had a sore throat and needed to go to bed, so I told her I was too tired to talk and I was going to bed, too.
While I was at it, I sent out a dozen or so e-cards from the Jackie Lawson website. She had some new e-cards for Christmas.
To fall asleep, I listened to Deva Premal singing while the computer read me the last twenty pages or so of this book.
Friday had been a day without pain in my back or legs, just sore feet, and on Saturday the pain came back briefly, at that sciatic notch, but I didn't swim.
Would I be able to swim on Sunday?

Recovery Day 69: Countdown

In Hollywood movies classified as thrillers, there is almost always a big red clock with digital numbers counting down to an explosion that will kill somebody or many people or everybody. It's December 2012 and we are getting close to the date that many spiritual groups have claimed as time for the end of the world or a major shift in life on Earth or something like the apocalypse. But nobody is counting down the days or freaking out about it.

Twenty days.

People are counting down the days until Christmas and making plans for next year and the following years and the distant future.

This morning at 3:00 a.m. I woke up with somewhat intense pain in my right buttock and an odd thought combined with a premonition.

For the past ninety days, from the time I started losing my legs, I laughed at the idea that I would not be able to walk or go to work again, but this morning I stopped laughing and started thinking it might be true: How could I ever go to work with this much pain?

Even though I had a pain-free day and lately my days have been mostly pain-free, the pain comes back again and again and apparently it is having an effect on my brain.

Two months after seeing the specialist, and laughing, so he said, stop it, this is serious, I have finally stopped laughing.

In the morning, I woke up at 8:00 with a bit more energy, but a pain in the butt, so I had breakfast and sat in front of the computer for a couple of hours.

Shania Twain was in the news as she was launching a two year gig in Vegas so I found her channel on YouTube and listened to 45 videos as I chatted online, played Scrabble, returned e-mails, and read Christmas e-card thank-you notes.

In the afternoon, I pushed myself out to go to the pool and found I had time to get my car washed, too. -- You have to count all the small victories, the poet Richard Brautigan said, and getting the car washed was one of those. It was dirty and I had wanted to wash it for a long time but I had to prioritize everything in my little life for survival and a clean car wasn't up there with getting groceries and going to appointments for health issues.

But I have to say .... It does make you feel better to look after your car and make it look better.

At the pool, I did nothing but lengths, water walking back and forth, forward in the deep end, backwards in the shallow end, for a full hour. Once again, I was the first one in and the last one out.

One of the lifeguards called me by name and then explained how he knew I was: He used to go to our school and he checked things out with our current students. So .... That was nice.

Swimming went well enough. I had a little pain in my sciatic notch, but it wasn't bad enough to slow me down.

My legs felt heavy when I got out of the pool, as usual, and I had to walk slowly. And it hurt to sit down, to change clothes. But that was about it.

On the way out, I ran into one of my Grade 5 kids from last year, and her mom, and we had a little chat that lifted my spirits. The mom said her daughter really missed me at school. "When are you coming back?" she asked.

"Christmas, maybe," I said. "Oh," she said, "I'll be gone."

Her mom explained that they were taking her to Disneyland for the first and last time in her life.

Speaking of Disneyland .... It looks like I've been a little 'out of it' in regards to feeding birds in Alberta. Although I had lots of birds in my back yard in the summer, when I put out a lot of wild bird seed, and nothing horrible happened, it all changed in the wintertime.

First I put out a suet bell loaded with wild bird seed, and I was disappointed that no winter birds came to get it. But then I noticed that nobody in this town has a bird feeder. Driving around the little city, I could not see a single bird feeder of any variety. So I Googled it.

My brother lives in the country, outside of Penetanguishine, the French part of Ontario, near Awenda Provincial Park, and he has a big bird feeding station that is squirrel proof and attracts a lot of colourful birds: blue jays and cardinals, lots of chikadees, that will eat right out of your hand, and a few hummingbirds, too.

Once in a blue moon, a hawk will scare them all away, and then fly off.

What do Alberta birds eat? you ask. Is it the same wild bird seed that attracts bluejays? Well, yes, but it's a little more complicated than that.

Magpies are attracted to the smaller birds that are attracted to the wild bird seed. They will pick them off, one by one, kill them and eat them, in your back yard. And that's not really what you want to see at your backyard bird feeder.

You want to feed the birds, but you don't want to feed them to each other.

If you want to attract magpies, put out bits of raw meat, or raw eggs, or dead birds.

Why would you want to attract magpies? Albertans would want to know.

They despise the black and white flying carnivores. -- They say they will go for your eyes, if they get the chance.

After driving home, I felt like going swimming again.

While I was in the pool, I planned out a book of short stories, to be called Cold Lake Stories, with a dozen or so all in the first person but set at different times and featuring different characters to create the illusion that one person told all the stories in different eras as one entity in different bodies.

It started with an aboriginal person before Europeans arrived and ended in the future, a hundred years from now, when Cold Lake was a big city, still thriving, after a lot of the planet had flooded due to global warming and climate change.

It was Day 75 of the NHL lock-out and dozens of players who didn't go overseas or rejoin junior teams or go to the AHL were getting together in Phoenix for scrimmages agains the Coyotes. Sidney Crosby got a group of guys to go to Dallas and then Colorado for scrimmages and then they went down to the dessert, where the Phoenix team was selling tickets to fans to watch practices with exhibition games featuring some stars from other teams.

Recovery Day 70: Monday Ouch

At five in the morning, I woke up with a pain in the butt, on the right side. It must be my sciatic notch.

Moving, in bed, was painful, and so was walking.

As Lilly said in Blazing Saddles, I'm tired of waking up tired.

Is this because of swimming? -- If it is, I will never want to go swimming again.

-- Let's not get hysterical or jump to conclusions.

At eight o'clock, the pain was still there, and when I moved to get out of bed it hurt more. It was quite intense for a few moments and faded away as I walked and sat and had breakfast.

It came back as I walked to physiotherapy but went away when I was on the treadmill and then doing back extensions but came back while I was pulling weights and went away while I was pushing weights. The other exercises didn't hurt and felt easy. The physiotherapist said, I have an odd job because I have to say things like I like the way your butt is moving when you walk.

She explained that my core was more solid, my butt was less clenched, my back was arched, my stride was longer, so all-in-all I was walking better.

Just before acupuncture, we talked for a minute, and I expressed my frustration, recently, so the physiotherapist said, Don't worry, that's normal, this stage goes slow, so it is frustrating, but think of how long you've had this, so don't worry if it takes a while longer to fix it.

Although I had been feeling dozy for days, I woke up during acupuncture and I had a great idea for a book of stories set in Cold Lake called Lovers Forever. That would be the sub-title and the main title would be Cold Lake Stories. It would be a big best-seller locally, with that title, and when Hollywood turned it into a movie, it would be called Lovers Forever.

Was that title taken?

Unbelieveably, amazinigly, that title had never been used for a novel or a movie. It was the title of a song by Stevie Nicks that went like this:

Lovers forever ... face to face
My city, your mountains
Stay with me stay
I need you to love me
I need you today
Give to me your leather...
Take from me...my lace

When I got home, I found a novel on Amazon.com with that title, by an American romance writer named Shirlee Busbee, who had several New York Times best-sellers.

Cold Lake Stories would use my idea about a sequence of stories set in this place with different characters in separate stories as first person narrators but there was an added layer as all the stories would be about two people in love for centuries.

They would change ethnicity and genders and that would affect their relationship but they would always find each other, fall in love fast, and vow they would never forget each other.

I had a girlfriend like that, once! I had a long, on-again, off-again, romance, in this lifetime, and I got involved with a woman who said we would have that same sort of romance over many lifetimes. And she said, "Next time, I'll be the man, and you'll be the woman."

Shortly after she said that, I ended our relationship.

There was a quotable quote on my Facebook page, coincidentally, that fit perfectly with the new book idea: You cannot save people, but you can love them -- and that just might be enough.

It was signed lbyap, but I had no idea who that was.

It came from a Facebook page called Rising From The Illusion and a blurb on the page said that the illusion is that we are all separate. In other words, we aren't just connected, we are one.

Like Cloud Atlas, the new movie with Tom Hanks and Halle Berry, with Susan Sarandon, and Hugh Grant, it would be an exploration of how the actions of individual lives impact one another in the past, present and future, as one soul is shaped from a killer into a hero, and an act of kindness ripples across centuries to inspire a revolution.

Cloud Atlas was by the makers of The Matrix trilogy.

A character called Robert Frobisher says, I believe there is a another world waiting for us .... A better world. And I'll be waiting for you there.

Timothy Cavendish says, We cross and re-cross our old paths like figure-skaters.

Sonmi-451 says, Our lives are not our own. We are bound to others. Past and present. And by each crime; and every kindness we birth our future.

The movie was based on the novel, Cloud Atlas, by British author David Mitchell, which had six nested stories that take the reader from the remote South Pacific in the nineteenth century to a distant, post-apocalyptic future. It won the British Book Awards Literary Fiction Award and was short-listed for the 2004 Booker Prize, Nebula Award, Arthur C. Clarke Award, and others.

All of the main characters, except one, are reincarnations of the same soul in different bodies throughout the novel. They are identified by a birthmark that's a symbol of the universality of human nature. The title "Cloud Atlas" refers to the ever-changing manifestations of the Atlas, which is the fixed human nature. The book's theme is predacity, the way individuals prey on individuals, groups on groups, nations on nations, tribes on tribes.

The book's style was inspired by Italo Calvino's If On A Winter's Night A Traveller, a book I loved, which contains several incomplete interrupted narratives.

Mitchell's innovation was to add a 'mirror' in the centre of his book so that each story could be brought to a conclusion.

Italo Calvino was the author of Invisible Cities, another novel I loved, and he was considered a major contender for the Nobel Prize in Literature, but then he died.

Somehow I missed Cloud Atlas when it came out, maybe because I was busy working in Cold Lake, and I couldn't find it online, yet, so I watched The Time Traveller's Wife, again, instead.

That movie had more of an emotional impact this time, the second time I saw it, of course, as I am just more emotional these days.

In the afternoon, I spent hours contacting the benefits agency and going back and forth to Staples to get a permission form faxed to them, after talking to my brother for about an hour. He phoned to find out if we had a Mark's Work Warehouse in Cold Lake, so I guessed he was doing some Christmas shopping.

He said something about my back injury dragging on a long time, longer than somebody hurt in a car crash, he said, so that cheered me up a bit.

On my way home from Staples, which is beside Mark's, I stopped at the grocery store for some more supplies.

It was fifteen below in Cold Lake for the fifth day in a row.

My brother said he had so much snow he had to climb onto his roof and shovel it off, but then it turned warm, so everything melted, again. It was fifteen above, where he was.

My evening, as usual, was devoted to online media: Facebook with Scrabble and chat, and e-mail, with a movie playing, in the background. Edmonton phoned after I watched The Time Traveller's Wife.

Sitting down for hours made everything feel better. Sleep would be a reprieve. I looked forward to that, but I dreaded getting up in the morning because that intense pain in the sciatic notch would be waiting for me.

Recovery Day 71: Tuesday Morning Ouch

After several hours of pain-free sleep, for a change, I woke up happy, until I moved.

An intense pain in the sciatic notch greeted me like a life insurance salesman and would not let go for about twenty minutes. It was excruciating. It was like having your sweet dreams erased from your brain with a cheese grater.

Good morning to you, too, sciatic notch.

Why would it hurt so much, first thing in the morning?

Google THAT, I said to myself.

Medscape.com said Low back pain (LBP), is ubiquitous. An estimated 30-45% of persons aged 18-55 years have some form of back pain in their lifetime. LBP most commonly involves one of the following conditions: sciatic nerve entrapment, herniated nucleus pulposus, direct trauma, muscle spasm due to chronic or overuse injury, or piriformis syndrome.

Piriformis syndrome is characterized by pain and instability. The location of the pain is often imprecise, but it is often present in the hip, coccyx, buttock, groin, or distal part of the leg. The history and physical findings are key elements in differentiating the more common forms of LBP and piriformis syndrome. The literature and general knowledge on piriformis syndrome is limited, compared with that of sciatica or disc herniation. However, the common findings associated with piriformis syndrome are agreed upon.

And www.pain-clinic.org said Piriformis syndrome is a syndrome of low back and leg pain thought to be due to chronic contracture of the piriformis muscle that causes irritation of the sciatic nerve. The syndrome involves gluteal pain often accompanied by pain radiating down the affected leg in the distribution of the sciatic. It is commonly called "hip pocket neuropathy" or "wallet neuritis".

Two most common theories are 1) compression of the nerve between the inflamed muscle and the bony pelvis 2) compression of the nerve between the two inflamed fascicles of the piriformis muscle. Robinson observed that any inflammation or spasm of the piriformis muscle will compress the sciatic nerve whenever the leg is raised, producing the sciatica.

About.com had a lot of information on pilates,

If the sciatica is coming from a herniated disc, then we have to take all the disc precautions. Disc precautions include not going into unnecessary flexion, and sometimes extension. Avoid overusing the buttocks and the piriformis muscles. Avoid putting the nerve on stretch. Avoid too much flexion [forward bending] in the lumbar spine which could irritate the nerve if there is a disc lesion. Again, work from a neutral spine, get things to move and relax, and get the core strong.

Core strength goes beyond the surface muscles and asks us to utilize our deep internal muscles to maintain stability.

In the broadest definitions of the core, fitness experts include the whole central section of the body all the way from the pelvis and hips up through the midsection. A big list of core muscles might look like this:

Deep back muscles like the erector spinae and multifidus
Hip flexors and spine stabilizers like the psoas, iliacus, and rectus femoris
Hip adductors and abductors
The gluteus muscles (butt muscles)
The abdominal muscles from the surface rectus abdominis to the deep transversus abdominis

What you want to note about that is the above list is that we are talking about both surface and deep muscles as well as muscles of the front and back of the body. That's a lot more than just the abdominal muscles! In Pilates, it is called the powerhouse area.

The core muscles that are truly core are those that lie close to the center of the body. The psoas, a long muscle that runs down the front of the spine and attaches at the top of the femur; the multifidus and erector spinae, both deep spine muscles; and transversus abdominis, the deepest abdominal muscle are examples. Their actions have more to do with stabilizing than with the heavy work some of the more surface muscles do. When I think of core muscles, these are the ones I really think of. I might add in the pelvic floor and diaphragm as well.

So core exercises have to address a lot of muscles that work differently yet in concert with each other. It won't do to just focus on abdominal muscles. And it won't do to think in terms of isolations of muscle groups or brute strength. We need a variety of exercises that promote core strength and integration in different ways. Exercises that challenge our stability as we bend and move - making all the core muscles work together to stabilize the spine and maintain balance and freedom of the limbs - are typically top choices for core exercises.

Examples core exercises:
Variations of Plank Exercise
Abdominals Set
5 Back Extension Exercises
Side-lying Leg Kicks
Exercise Ball Exercises

Keep in mind that what makes stability exercises truly effective for the core is not just working the muscles. It's doing the exercises with excellent form so that the conditions are set up for strength where appropriate, balanced development, and overall integrative function. What good does it do really to train in an imbalanced way or without heading for the most optimal result? Good exercise instruction will always include tips on posture and alignment. You can educate yourself on the basics of good alignment and carry those principles into your workouts.

Tips for Good Alignment in Exercise:
Posture Check
Legs Parallel and Hip Distance Apart
The Box Image for Balanced Work
Shoulder Stability in Exercise

As usual, when I look for information online about anything related to neuropathy, ads for Neuropathy Support Formula pop up all over the place. -- It's called viral marketing.

After looking at those ads for three months, I finally placed an order, online, for a one month supply.

But I couldn't tell if the website worked, or not, as there was no confirmation my order had gone through.

Meanwhile, after sitting down for almost an hour, the pain in the butt had faded away. When I stood up, it hurt for a few seconds, but then it was gone, and I could walk again.

Hallelujah!

Now what? It's six o'clock in the morning, there's no swimming at the pool on Tuesdays, it's dark as night outside, it's fifteen below and snowing, as usual, for the past week, and I am feeling somewhat energized and awake.

Yesterday the physiotherapist said, I know it's frustrating for you because when the pain stops, you start to feel good, and you think you should be back at work, but the truth is, severe nerve damage and muscle loss requires a slow rebuilding of both.

She said, you are weak, you don't have much stamina, when your leg muscles get fatigued they hurt and they stop working, and it doesn't take a whole lot to fatigue those muscles.

-- That was embarassing!

She said, You guys want to power through it or just ignore it, but those approaches just don't work in this situation. You have to rebuild the muscles carefully, and it takes time.

While I was doing my exercises, I overheard the physiotherapist talking to somebody on the telephone, discussing a client, saying that he went back to work against her advice, as he had to make a living, he said, and he could handle the pain, but she believed he would only wind up injuring himself again, or further, working as a truck driver, hauling logs across Alberta.

It was easy for me to spend a couple of hours researching information related to my injury and recovery, or rehabilitation, sitting at my computer, and writing about my own experience, relevant to the info I found.

Each time I get up, every half hour or so, it's easier and easier to get up, stand up, and fight, fight, fight. -- I mean walk to the kitchen and back.

Yahoo! said above-normal temperatures are in the forecast for northern Alberta, Yukon and the Northwest Territories.

And I said Woo-hoo! -- My prediction, made last winter, was right: In Cold Lake, we create and celebrate global warming.

For about two months, people told me I was losing weight. For the past week or so, I feel I have been gaining weight. -- I know I have been building muscle as it shows in my legs. But what about my belly?

When I swim, it's flat. And that's that!

My goal for today is to swim at noon.

This afternoon, I have reflexology.

Maybe I'll swim again in the evening!

-- You see how I go? I get excited when I see signs of improvement and immediately get carried away and overdo it.

You know the old song: When will they ever learn? When will they -- ever -- learn.

After an hour and a half online, I spent some time doing something creative, for a change. When I put together my notes on Cold Lake Stories, I discovered I had a thousand words, already. -- I thought maybe it was going to amount to something.

Problem: The stairway at the pool was broken. Solution: Go to the hot tub and do your exercises there.

Problem: The supply teacher phoned, sounding stressed out, looking for books. Solution: Go to the library, borrow a book for her, go to the drug store, buy chocolates, and deliver the book and the chocolates to her.

The chocolates were a big hit. As well as our supply teacher, I talked to a few other teachers, and they all appeared to be stressed, close to exhaustion, and ill or injured.

'Tis the season!

Even so, they told me that somebody had made me a dinner and left it in the freezer with my name on it. Our head secretary went into the kitchen and got it for me, so I didn't have to walk that far, and we all tried to guess who made it or identify the handwriting that said, Chicken Carbonara for Mr. Avery.

While I was there, the high school student I asked to train my basketball team came out of the gym. He had a practice going with Grade 4, 5, and 6 boys, with the help of another student, and he told me they had a good defensive system and an offensive system set up, so I told him to work on a couple of set plays, including one to get the ball into play from the sidelines.

The teacher I asked to supervise them was helping out.

I felt as though I got the team going by remote control.

Reflexology, after that, was great, as always.

Mecell answered my questions about core muscles and what happened to me when I suffered severe nerve damage, so it made more sense to me. She said there was a lot of inflammation from my back down to my toes and the muscles do things to deal with pain and they remember pain and what they did before all the pain.

She said she liked what my physiotherapist was doing, getting me to be aware of my body, my muscles, my spine, even my nerves, and focus on building up the core, not by powering through a lot of exercises but by doing a few exercises correctly, or with the right form, to rebuild the core.

We also talked about Christmas and I may have mentioned that our concert had been canceled.

It turned out she was somewhat fanatical about Christmas and had a strong desire to make it happen for the kids at my school. She said she would get on the phone and find a Santa suit and somebody to wear it, she would get oranges and candy canes donated or paid for, and she would get it all into the school.

She told me she grew up in the north, by High River, and her father did a lot to build up their little town, including Christmas every year. A lot of people working in that town went back to wherever they came from, at Christmastime, but her father would find out who was left in town over Christmas and make sure each and every one of them got a Christmas dinner. And if there were kids, he made sure those kids got a visit from Santa Claus on Christmas Eve, with a wrapped present and an orange and some candy.

Honestly.

It brings a tear to the eyes.

She said that if she told her father there was a school that had canceled its Christmas concert, he would be telling her to get on it and make it happen. Get a Santa Claus into that school!

While extracating myself from the conspiracy, I may have implicated one of my colleagues, who is now teaching secondary school, but has a lot of love for the little kids at our school.

-- Just sayin'!

When I got home, I heated up the oven and put the frozen chicken pie in to heat up while I checked e-mail and Facebook.

Edmonton phoned while I was enjoying my supper and she asked me to tell me about my day because her day sucked, she said.

While I was telling her about a day that I thought was ho-hum, starting with pain, with stress from work, a pool that didn't work right, and cold temperatures on an overcast day, I realized that the pain went away, dropping in at work turned out well, and it had actually been a good day with a lot of things in it.

So, a day that started badly, with pain, turned out to be a fairly full day with hours of writing, pilotes in a hot tub, pain-relieving reflexology, lots of love from my school, plus frozen chicken carbonara, not to mention Scrabble, e-mail, and Facebook chat in the evening.

It made me wonder: What will tomorrow bring?!

Day 72: The Acupuncture/Write Method

A dreamy sleep, a pain in the butt getting up, a quick bite to eat, and then physiotherapy, followed by acupuncture and writing: that was my morning.

We added a new exercise and upped the weights on the push/pull machine.

Sitting on the big exercise ball, using five pound weights, I did 100 arm curls, using both arms.

The physiotherapist said, You are walking better, feeling more, and doing all the exercises better, so it's time we started you on a more intense daily program. -- If only the benefits people would hurry up and fund it.

She wanted me to do two hours with her every day.

I was happy about making progress like that.

When I told the physiotherapist that I slept well, and pain-free, but got a big pain in the butt when I got up, first thing in the morning, she nodded and said, I want you to do some of your exercises before you get out of bed, to get things moving, and then you won't start the day with that pain.

-- Genius!

I told her I'd been researching and reading all about the core muscles, as I discovered the term referred to a different set of muscles than I thought. She said the major muscles were the pelvic floor muscles (levator ani and coccygeus), transversus abdominis, multifidus, internal and external obliques, rectus abdominis, erector spinae or sacrospinalis, especially the longissimus thoracis, and the diaphragm. Minor core muscles included the latissimus dorsi, gluteus maximus, and trapezius.

Most people ignore the pelvis floor muscles and don't give good exercises for them, she added.

Her favourite were the ones that were the most fun to say, she said.

While I was lying on the massage table with acupuncture pins in my back, butt, legs, and ankles, plus one for energy, my face in the circle, pillows under my hips and ankles, for comfort, I felt the chi more than ever before.

The lights were shut off and Angela said, Have a nice nap, but I was more wide awake than I had been for months.

Instead of sleeping and dreaming, during acupuncture, I thought about my new novel, Cold Lake Stories, or Lovers Forever, got into dream mode, directed it a little, and let it roll, or unfold, or play like a movie.

This time, I saw the future, with the same couple of characters in the same place, about fifty years from now. Cold Lake was a booming city with an NHL team. The Cold Lake Energy, formerly The Ice, were defending Stanley Cup champions, trying to three-peat and develop a dynasty, in the Chinese Hockey League.

China had taken over the world and Canada was divided up into a dozen states, all of them designed, pretty much, to send resources to Beijing.

There were pipelines from Cold Lake and Fort MacMurray through the Rocky Mountains to the West Coast and under the Pacific to Japan and the Koreas as well as China.

When I walked home, which wasn't too much of a pain, I wrote about it, adding about a thousand words to my new project.

At 11:30, when swimming time started at the pool on the base, I felt sleepy, and I knew I couldn't actually get in the pool, so I made a plan to do something else in the afternoon.

The sun was shining for the first time in a week and a half but it was still fifteen below zero.

But the next day was supposed to be a lot colder.

And then it would get warmer.

-- More wild weather.

Instead of getting my computer to read this book to me, I got it to read my new novel in progress.

It was too bad I couldn't do NaNoWriMo, with writers around the world, but I was happy I could start writing a new novel in December.

Also, in my e-mail, I got a special offer from the University of Toronto for ten percent off a TEFL course, so I could save a hundred bucks on a certificate in Teaching English as a Foreign Language.

-- That would be useful when I went to China, or when China took over Canada!

For a change, I felt as though I had the energy to do the course online, write a novel, and get ready to go back to work, physically and intellectually.

I felt like Montreal Smoked Meat because, finally, I was on a roll!

Oops! I fell asleep, after physio and acupuncture, and did not get going again until two thirty in the afternoon!

When I got moving, I discovered I moved more effortlessly than ever before in the past three months, and that put me in a good mood. First I walked into the post office and out again, dropping off presents for Australia and Ontario, and then I went to Walmart for some exterior Christmas lights. Walking through Walmart was no problem, either. -- Just the odd reminder of the ghosts of pains past.

Outside in the parking lot, I took back a shopping cart for a new mother, who was having a smoke break, with her kid in the car, I was feeling so good.

All that walking went so well I felt inspired to throw myself a little celebration.

You won't believe it, but, for the first time in a decade of my life, I went to the McDonald's drive-thru. And I got the chicken nuggets with fries.

-- Do I know how to celebrate the small victories, or what?!

Putting up the Christmas lights, outside, was a little challenge, straining my new leg muscles, but I put on some Christmas tunes, on YouTube, and had some fun.

If you looked at it from the right angle, it looked as though the lights were actually on the huge spruce trees in the back yard, instead of just on a railing between the big windows at the back and the big black spruce.

-- I thought it looked great!

Day 73: The Return Of The Ouch

A little back pain in the evening, a pain in the butt in the morning, back pain on the treadmill at physio, and a good deal of pain like a tight band across the bottom half of my butt while trying a new exercise: That was the latest.

Two new exercises: Push-ups against the wall were easy and squatting to lift a weight and pick it up was okay until the fifteenth repetition.

That pain went away with some back stretches.

Acupuncture was different. The day before, I felt the buzz with each needle quite clearly, even in my ankles, but not today. All I felt were pinpricks.

Well, the physiotherapist said, Every day is different.

She said the pain in the lower back while walking on the treadmill was better than pain in the legs, so it was a good sign, indicating my core is stabilizing.

She was going to get me to do another lifting exercise after the squat and lift but when I indicated I was feeling a good deal of pain, she said to forget about it for now.

She also said it was time to cut back on acupuncture, do it once a week, to see how my body reacted.

Walking home hurt, but sitting down felt good.

People found the weather remarkable, as it was 20 below but the sign was shining in a clear blue sky, and I was happy to see the sun, watch the sunrise at 9:00, but I was pre-occupied with other things, like pain and doctors.

Was it time to see my doctor and get another note for work?

When would I be ready to return to work? I was certain I would be there before the Christmas break. Now, I wasn't certain I'd be able to return right after New Year's.

Yesterday afternoon, I felt so good. This morning, I felt so bad. After doing a simple exercise, squatting to pick up a wooden box with a five pound weight inside, fifteen times, I felt like I need to rest for an hour. Or a day.

It looked as though the NHL lockout would continue past Christmas and into the New Year and that was terrible for the NHL but great for the Canadian junior team and the World Junior tournament held every year in the days leading up to the New Year.

The Canadian junior team did well during the last two lock-outs. They would try to make it three-for-three this year.

In particular, the juniors hoped to have Ryan Nugent-Hopkins of the Edmonton Oilers, who was playing in the AHL for the Oklahoma City Barons during the lockout and would not be available for the Canadian junior team if the NHL season was "on". Having the NHL's #1 draft pick of the previous year, who should have been the Rookie Of The

Year, would be great for Team Canada. But, frankly, it would still be an awesome team without him.

In Cold Lake, it turned out that twenty below was the high of the day and the temperature was dropping.

How low would it go?

It probably wouldn't go as low as minus 30.

In Ontario, there are dramatic changes in temperature on a daily basis. In Alberta, it's a different story.

Edmonton got another dump of snow but not Cold Lake.

Today the sun is streaming in through my big window so it looks like summer. -- Until you see the snow that covers the ground and trees and buildings everywhere.

The snow sparkles like diamonds, millions or diamonds, sprinkled generously all over Cold Lake.

After physio and acupuncture, I slept for hours, and I was a hurting unit when I got up again in the afternoon. The pain level was not high but I felt that I should give it a rest. My feet were sore and my legs were stiff, so I walked like a zombie. But, really, I didn't feel horrible, or anything; I just felt like I should get horizontal and not move too much.

The book store called just before closing to say they had books for me. Edmonton called to say she had a rough day. Alabama posted a note on FB saying "Where's Martin?" because I missed our Scrabble date.

My Scrabble score had dropped below 1600 for a while but it finally climbed back to 1606, with 234 bingos, good for first place, amongst my Facebook friends.

I added around a thousand words to my new novel-in-progress, Cold Lake Stories, about the time white people first came to this part of the country. How strange it must have been for aboriginal people to hear about Europe and Europeans and then finally see and meet a few of them at Elk Point in a North West or Hudson's Bay company or on the North Saskatchewan River.

I fell asleep dreaming about the 1700s.

Recovery Day 74: Cold Lake Cold

It was 23 below in Cold Lake and 23 above where my sister lived, in the land down under.

At four in the morning I got up, as usual, and checked e-mail, Facebook, the news, and so on, and found an impressive new picture of the whole Earth by NASA called The Black Marble showing the planet at night, without clouds, with the lights that can be seen from space, indicating where our cities are located. It was clear to see which parts of the world had the bright lights.

As well as city lights, there were wild fires in Western Australia and gas flares in the Middle East.

It looked like Cold Lake had some lights on, too.

The map made Cold Lake look like one of the most northerly places in the world with lights. It was a NASA pic, so it must be true!

Edmonton called me before 8:00 in the morning, so I talked to her while she was on her way to work, and then I got going. The book store opened at 10:00 so I was there then, to pick up books for school, and then I drove them over, so my supply teacher could use them with the Grade 12 class. -- The Wars, by Timothy Findlay, on the approved list for Alberta.

First I helped a neighbour get his car started. His battery was dead, so he needed a boost. He has a walker as he has had hip and foot problems. He says he had to stay in bed because of his hip problem and now he has gangrene in his foot. He is now using my old walker.

My limbs felt better, the more I walked, so I thought I'd try the pool, but it was closed, for no known reason, so I did the Walmart walk, instead. That superstore is so big! You have to be in some kind of shape just to walk through the place.

Somebody at work phoned to ask where the purchase order or the receipt was for the books, so I told them it was tucked inside one of the books. Edmonton called at noon to say strange things were happening at work and that if her car blew up I was to call her boss and let everybody know what was going on with her.

How I wish I could reveal her identity and let you in on the mystery, but she swore me to secrecy, anyway, so you'll just have to live with that.

She sent me a funny clip from 22 minutes, on CBC, with the premier of Alberta, Alison Redford, toasting Canadians with a glass of Scotch, no rocks, saying she planned to spend Christmas the way she spent the rest of the year, having a drink with friends from the oil and gas companies, talking about ways to relax environmental regulations. She said, "Looks like another mild winter. -- You're welcome, Canada!"

It was just a joke, of course, like the title of my book of poetry, Celebrating Global Warming, and the claim that this is where we create and celebrate climate change.

On Thursday night the Cold Lake Ice beat the St. Paul Canadiens 14-0!

The Ice improved to 14-1-1 on the season.

Next up was Lloydminster Friday night in Lloyd.

The Bracebridge Phantoms made some trades, picked up a huge defenceman, won a tournament, went on a winning streak, and were now 15-3-2 for the season.

Their new stay-at-home defenceman was 6'4", 245 pounds, and was described as "rugged".

How I would love to see the Cold Lake Ice Jr. B team take on the Tier II Jr. A Bracebridge Phantoms!

My two favourite teams, right now.

Who would win?

You would have to think a Jr. A team could beat a Jr. B team. Right?

Bracebridge had hockey players from all over the place -- Canada, the U.S.A., Latvia, Serbia, the Czech Republic, the Netherlands, and Japan.

They had one guy from Bracebridge.

The Cold Lake team was all-Canadian and all local talent, with guys from Cold Lake and the closest town, Bonnyville, and just across the border in Saskatchewan.

-- My money would be on Cold Lake!

Recovery Day 75: Cold Lake Stories

On Saturday morning, I got up at 5:00 and added two thousand words to my Cold Lake Stories, so I was up to seven thousand words, with outlines and first drafts of half a dozen stories.

My Scrabble score, which had dropped below 1600, jumped up to 1629, with 236 Bingos, good for first place, again, amongst my Facebook friends.

I woke up with almost no pain and discovered walking wasn't much of a problem.

Now that my legs and butt, even my sciatic notch, had calmed right down, I was more aware of the pain in my feet. It had been there for months but was upstaged by bigger pains higher up.

The good news was that the numbness in my toes was almost all gone.

That strange tingling or pins and needles feeling had faded away a month ago leaving a tight band of muscle across the ball of each foot that had a lot of sensation. Sometimes it hurt so much that I didn't want to walk on it but other times it wasn't so bad and when I did walk on it, the pain went away.

After a lot of searching, I finally found the movie called Cloud Atlas, with its stellar cast trying to turn a difficult novel into something that would look good on the big screen. Reading about the movie filled me with positive anticipation. But seeing the movie was a major disappointment.

Rotten Tomatoes gave it 65%, so a lot of critics liked it, but I agreed with the 35% who said all the murderous plot strands did not add up to much.

Also, Tom Hanks and Halley Berry played characters that did not have any chemistry.

They would do much better in the stories I was writing, inspired by Cloud Atlas, called Cold Lake Stories, or Lovers Forever.

-- Could I get any more conceited? Arrogant? Or confident?!

It was hard for me to believe, but it was that time of year when various publications did their end-of-the-year summaries. I don't know about you, but I always like these things. And this year, especially for the past half year or so, I've been able to focus more on the news.

This year, 2012, was the year Obama got re-elected and the year of Superstorm Sandy.

It was the year Etta James, Whitney Houston and Donna Summer and Dick Clark died.

Peter Lougheed died. Neil Armstrong died.

Alison Redford surprised a lot of people by beating the Wildrose Party to get a majority government in Alberta.

There were mass shootings in Alberta, Toronto, and Colorado. The U.S.A. has around twenty mass shootings per year but doesn't discuss gun control. The Colorado horror happened at the opening of The Dark Knight Rises. Toronto's shooting took place at The Eaton's Centre. Edmonton's took place at the University of Alberta.

In other news .... Facebook had a billion users but did not have a good time in the stock exchange. RIM, formerly Northern Telecom, creators of the Blackberry, had a rough year.

At the movies, it was the year of The Dark Knight Rises, The Avengers, Skyfall, World War Z, and The Hobbit, not to mention Cloud Atlas.

Who got the big news headlines?

The Keystone pipeline, Quebec student protests, the Robocall scandal of the Canadian federal election, Kony, Pussy Riot in prison in Russia, bath salts, Jenna Talackova the transgendered Miss Universe Canada contestant, Quebec corruption, the God particle, China buying Nexen, iPhone 5, the iPad mini, the Quebec election shooting, the crazy Air Canada lady, Omar Khadr returned to Canada, Amanda Todd, Kate Middleton ....

The Olympics were held in London, England, the New York Giants won the Superbowl, the L.A. Kings won the Stanley Cup, the San Francisco Giants won the World Series, the Argos won the 100th Grey Cup, and The NHL season was prorogued.

On Friday night the Cold Lake Ice beat Lloyd 7-6 on the road.

After that game, they were off to Vegreville to play the Rangers.

After looking at the news, weather, and sports, I did some writing.

Cold Lake Stories was now over 9,000 words.

It was easy to write, like writing in butter, as they used to say; the story wrote itself. The framework came to me while swimming and more inspiration came after acupuncture. The rest came to me in my dreams as a I fell asleep listening to my computer ready me what I had written so far.

What was it Deepak Chopra had told me, when I went to one of his workshops on meditation? If you meditate, your life, like mine, will be a dream.

His advice always reminded me of that old song about the rowboat:

Row, row, row your boat, gently down the stream,
Ha ha, I fooled ya; I'm a submarine!

In the afternoon, I surprised myself by doing something I had been thinking about for months but knew I could not handle. When I couldn't walk, and when my back or legs were sore, I could barely make a bed, never mind move one. But, for a variety of reasons, I wanted to trade beds from one room to another, so I could sleep on a firmer mattress.

Moving the bed springs and the headboard were no problem and I moved the big mattress slowly and carefully so it presented no painful problem either.

It gave me a big feeling of satisfaction and accomplishment.

Lying down on the big bed with the afternoon sun streaming in for a few minutes did not inspire me to sleep, it made me feel like going out for some more action.

My shopping list only had a few things on it: milk, apples, and maybe some meat of some kind.

On the other hand, the bed, set up the way it was, looked pretty attractive ....

Did I mention the bookcase? As well as the bed, I moved a bookcase full of books and stuff.

After trying out the bed and not feeling any pan after moving that furniture, I got excited about finding out what I could do. And if I hurried, I could go outside before the sun set at 4:30.

After a couple of tours around the grocery store, I had mixed feelings.

My knee went weak for a moment and I had a flashback to the time I lost both knees and my ankles and so on, and wound up falling down on the ground

But it was just a momentary thing, followed by a pain in the butt, that shot down the back of my leg, all the way to my ankles.

But that passed, fast, too.

Aside from those brief moments, it was like that afternoon when I felt normal for a few hours, which turned me into Ecstatic Man

To celebrate, I drove to the lake for steak at the Eatery.

Cold Lake looked frozen over but's so big you can't actually see much of the surface. What I could see was frozen and covered with snow.

The walk from the parking lot across the street, with a few curbs and a couple of steps to the restaurant, was so smooth and easy, I felt ecstatic once again.

The restaurant looked beautiful, like Christmas, and the waitresses looked beautiful, like models, and the view of the lake from my table by the window was fantastic as a seat in a cafe by the river in Paris, and the food tasted fantastic, like a ten star restaurant.

A prayer was in order, I felt, sitting at the table in the restaurant, so I thanked God from whom all blessings flowed and I praised God abo e all heavenly hosts. I thanked the Father, Son, and Holy Ghost. And Mary. And Saint Kateri, too.

If only I could feel like that all the time. I would be the most annoyingly happy and enthusiastic person on the planet.

Maybe I have learned something from my health scare and the pain: As well as empathy for others with physical challenges, I had discovered a new theme for my writing, I realized. Cold Lake Stories was about love and death, war and peace, imperialism and evolution, from the time of the British Empire through the time of the American Empire to the start of the Chinese Empire.

Recovery Day 76: Run!

In my new bed, which was my old bed, I slept eight hours straight, without moving a muscle, had a touch of pain in my sciatic notch when I got up at six, but woke up feeling happy and stronger, as though I was a different person, maybe The Hulk, if The Hulk wasn't motivated by anger but got bigger after months of rest and exercise with good nutrition, reflexology, physiotherapy, and acupuncture.

Get the point?

My Scrabble score went up to 1629, with 236 bingos.

It was thirty below, but going up to minus fifteen.

The Cold Lake Ice lost a tough one in Veg by a score of 3 to 2, so they were now 12-2-1 for the season. -- You know what they say: You can't win them all.

After watching a lot of Will Ferrell on YouTube, I posted an Edmonton country singer's new music video on my Facebook page, called Hockey, Please Come Back, to help it go viral. It was released Saturday and already had 50,000 hits. The video showed a lot of guys in Oilers jerseys and the singer had a great voice. -- But it didn't look like hockey was coming back any time soon.

Everybody said sarcastic things about millionaires fighting billionaires over hockey profits and said the NHL was done like leftover pizza.

It was a sunny morning, so I climbed into bed to listen to my novel in progress, Cold Lake Stories, or Lovers Forever, and got up to wrap Christmas presents a couple of hours later. It was time to head for the swimming pool so I headed it, but with some reservations, as my energy was down.

When I got to the pool, I could see the stairs were still broken, so I decided to go for a walk, instead.

It was about 20 below zero, so I thought I'd walk indoors, instead of outside, so I went to the biggest store in town and walked the inside periphery. With a little meandering in the aisles, I stretched the walk out to about a thousand steps, and I had a little pain in the butt by the end of it.

In the store I saw a few of my students, and they got excited, so that was fun.

Even though I was hurting by the end of that walk, I didn't feel as though I was finished. After sitting down for a while, I felt like walking some more, so I drove up to the lake to try the broad walkway along the breakwall at the marina. And something strange happened while I was walking toward the middle of the big frozen lake.

I felt like running.

As well as writing about running, I had been dreaming about running, and thinking about running, and missing running. So I thought I would try a few steps.

After walking one hundred steps on the snowy breakwall, I jogged or ran very slowly for one hundred steps.

It felt fantastic.

So I walked another one hundred steps and ran another hundred and so on, so in total I walked 500 steps and ran 500 steps.

By the time I got back to the car, I felt as though I was done, as I felt a hint of pain in my rear end.

But I wasn't finished yet!

It was back to the big store for another periphery walk. And by the end of that one, I felt a little weak in the knees and sore in the rear end.

On the way out of the store, I leaned on a post, for a minute, and a group of three aboriginal woman asked me if I needed any help. -- I thought that was nice of them!

I said, No thanks, I just need a minute to rest and get a grip.

Are you sure? one of them asked.

-- I must have had a pained look on my face.

After that, I decided to head home, but I still wasn't finished.

I did a short run in the parking lot, just a dozen or so steps, and it felt good, again.

If there are two thousand steps in a mile, I must have walked a couple of miles.

Would I pay for it with pain in the morning?

Recovery Day 77: Good News/Bad News

Good news: The European Union got the 2012 Nobel Peace Prize for bringing decades of peace and democracy to Europe after the horrors and division of two world wars.

More good news: After running on Sunday afternoon, I felt no ill-effects on Monday morning. In fact, I felt pretty good. -- I woke up before the alarm clock went off, felt no pain in the usual places, discovered that a lot of things were easier to do, from getting out of bed to getting into and out of the shower to putting on socks.

Those things are not huge -- unless you have lost the ability to do them. Being able to do those things again made me feel like dancing!

It was the first day of strikes for elementary teachers in Ontario, the first day for secondary teachers to withdraw from non-classroom work, including Christmas concerts, and the day of a planned walk-out by high school students in support of teachers.

A buddy of mine sent me an e to say they were having an ice day, due to freezing rain, where we used to work.

He was on LTD for stress.

This was supposed to be my first of LTD, but I had no word from the benefits people in regards to that. However, I had faith in the system.

There were no school buses in Parry Sound, Muskoka, Haliburton, and Peterborough, or most of Central Ontario, my old stomping grounds.

The joy of discovering it's an ice day on a Monday morning, for teachers as well as students, is enormous. It would be amplified on a day like this one, when you anticipated labour strife and walk-outs. -- What a relief for everybody in that part of the country. I thought I could feel it, thousands of miles away, in Cold Lake.

Our buses were running. It was twenty below, as usual, but it was supposed to warm up a few degrees by the middle of the afternoon.

As for me, I was raring to go to physiotherapy first thing in the morning!

When I told the physiotherapist about running, on Sunday, and moving furniture, on Saturday, I also told her that I heard her voice in my head saying, Don't over-do it!

She said, That's not what I would have been saying!

At physio, I added something to each of my exercises. On the treadmill, I walked at 3 instead of 2.5. The new squat-and-lift-the-box exercise was easy, this time, with ten pounds added to the box instead of 5. We added the exercise we backed away form last time, lifting a wooden box with a ten pound weight from a waist-high shelf to a shelf that was shoulder-high.

No problem!

For the step exercise, I went from level 1 to level 2. Sitting on the ball, doing arm curls, with both arms, I used the 8 pounder instead of the 5.

She gave me another new exercise or two. Kneeling on all fours, on the exercise mat, I did the cat and the camel. For the cat, you lift one hand and then the other, like a cat kneading its bed. The camel arches the back one way and then the other.

Getting ready for acupuncture, taking off socks and sweatshirt, climbing onto the table, was a lot easier than usual.

"How are you feeling," she asked me.

"Great," I said.

"Okay then," she said. "Let's skip acupuncture today and see how your body responds."

She told me she had a dream about me in my school. She was walking through the school, in her dream, and out of the corner of her eye she spied me in my classroom, working away the way a teacher would normally work.

Right after physio, as soon as I walked a block or so, that pain in the butt returned, so I turned around and went home.

As soon as I sat down, for breakfast, I felt better.

More good news: Christine Sinclair, captain of Canada's women's soccer team, bronze medal winner at the London Olympics, and MVP with six goals, and the centre of a controversy over refereeing during a game, was named Canada's athlete of the year.

She was the first soccer player to be awarded the honour. It put it right up there with Bobby Orr, Wayne Gretzky, and Terry Fox.

Sinclair and Canada got ripped off, during the Olympics, by officiating that cost them a game and, probably, a gold medal.

Soccer is a great game that never gets enough attention in Canada.

I've always loved soccer. I played every summer, as a kid, and every fall, during high school, plus the springtime, when our team went to the All-Ontarios. As a high school teacher, I coached soccer for two dozen years. My first team went from worst to first, over five years, and had a perfect season, going undefeated all season, in the league play-offs, and district championships. My last team, in my old hometown, accomplished a smaller goal, defeating my old high school, in a higher division, their arch rivals, for the first time in living memory.

It's a great game.

I love 'the beautiful game' at the World Cup. And I loved K'Naan's soccer song, Wavin' Flag, in all its versions.

And Sinclair is an amazing soccer player. She scored six goals during the Olympics, which was an Olympic record.

The bad news was that more of the NHL season was cancelled. Close to half the hockey year was wiped out, including everything over Christmas and up to the day before the end of the year.

The season could start on New Year's Eve, but it didn't look good.

Oh well. Christmas and New Year's hockey belonged to the annual World Juniors Tournament.

This year, it looked like Ryan Nugent-Hopkins of the Edmonton Oilers would be leading Team Canada, with the goalie from Gravenhurst, Jason Billington, of the OHA's Jr. A Owen Sound Attack, in net.

The good news for me was that for the rest of the day I was pain-free. The bad news was that I didn't have much energy.

Recovery Day 78: Relapse

At three in the morning, it was twenty below, outside, and I was inside yawlping because I had a huge pain in the sciatic notch. Unlike the yell in the Walt Whitman poem or The Dead Poets Society, my barbaric yawlp was not about barbaric ecstasy, it was about sciatic agony.

    Okay, that may be overstatement. -- But it did hurt.

    What do you do if life hands you lemons? -- I took three ibuprofen, sat down and added three thousand words to my Cold Lake Stories, so it was now at thirteen thousand, and then I went back to bed until the alarm went off.

    But I was still tired, so I got the computer to read my new book to me, and went back to bed for a bit.

    At noon, I walked to the drug store, to pick up a parcel, after checking the mail, and that went alright, but walking home, afterward, was painful. The backs of my legs hurt. I thought I'd have to sit down in the snow but I pushed on and made it home without falling or stopping more than a few seconds.

    My feet hurt, before I left, but walking on them made the pain go away.

    Also, I felt foggy. I'm not a coffee drinker, but I felt like I needed a big, hot, coffee from Tim Horton's or Starbucks.

    Cold Lake needs a Starbucks!

After reflexology, I didn't need a coffee or a painkiller or anything except a place to walk. Mecell did her magic again!

    I walked in feeling sore and tired and walked out a new man ready to go shopping.

    How does she do those things she does?

    After reflexology, I went to the superstore, picked up some groceries, enjoyed walking around like a normal human being, and then headed home to make dinner and watch a movie.

    Mecell not only gave me reflexology, she gave me ....

    Hope.

My Scrabble score went up to 1639, with 237 bingos.

In the evening, I did a lot of work on a novel I wrote in the summer. There was a section or two of that back that I thought I might be able to use in Cold Lake Stories. There was a section on the time in the history of Cold Lake when the CLAW(R) was created -- the Cold Lake Air Weapons Range -- and there was a section on the time in the history of Cold Lake when oil became a big thing.

    It's funny, when you write a lot, and you leave a writing project for a while, and then go back to it. We call that 'putting it on the back burner' or 'in the refrigerator' and it usually works out well. However, sometimes an old project looks quite different when you look at it a few months later.

It reminds me of The Tenth Man, a novel by Graham Green. That's the British novelist, not the Canadian actor. Green wrote a novel and forgot about it.

One year, I used that novel for an English class I was teaching and I told my students the author of the book forgot that he had written it. They found that amazing. How could you go to all the trouble, take all the time, do all the work involved in writing a novel, and then forget you had done it? they asked, incredulous.

The truth is, after you've written dozens of books, it's easy to forget about a few of them!

There was a section of the novel I wrote in the summer that I remembered but could not find. I spent hours looking through the various drafts I had written but I could not locate the one I was looking for.

-- I hate it when that happens!

Oh well. I could always write it again.

Usually, when that happens, the rewrite turns out to be a lot better than the one that got away ....

Anyway, it felt good to worry about a literary project instead of some pain in the butt or back or legs and wondering if I would ever be able to go to work again.

Recovery Day 79: Almost 80!

After three or four good days in a row -- Friday, Saturday, Sunday, and Monday morning -- I had a relatively pain-free but low energy period from Monday afternoon through Tuesday. Until I went to reflexology at 4 in the afternoon.

In the evening, I was sharper. And then, when I woke up at three in the morning, I was wide awake and could not get back to sleep.

I tried to sleep while my computer read a novel to me, but my mind kept me awake. I was thinking about a section of the book that was missing.

After reading the novel that way, I got up and searched through my computer files, looking for the missing piece of writing.

By 5:45, I was still up, and not feeling sleepy, so I thought about going to the early morning swim at the pool.

The stairway at the pool was probably still out of commission, but I thought I could climb out of the pool without using the stairs or the ladder.

Well, my early morning ambition faded and I wound up having an unathletic day.

On the other hand, I did re-read and edit a novel that I wrote in the summer.

It was the night of the Sandy benefit concert at Madison Square Gardens, shown live in movie theatres and carried by two dozen tv networks, reaching two billion people, featuring Paul McCartney and Nirvana, Billy Joel and The Rolling Stones, Billy Joel, Bruce Springsteen and Bon Jovi, Eric Clapton and Alicia Keys, raising millions .... It was the day Ravi Shankar died .... And it was the day of a bizarre plot to kill Justin Bieber ....

It was 12/12/12.

Recovery Day 80: Team Canada

Up early, the usual morning pain in the butt is missing in action, thank goodness, and I feel fairly mobile .... A good start to the day!

Yesterday I was so unathletic, despite being relatively pain-free. Intellectually, I was busy.

For several hours I worked on book projects, and that felt good.

Normally, I have four book projects on the go. I write four books at the same time. Well .... Generally I have four book projects to work on but I focus on one until it's time to take a break and then I work on one of the other ones. That has been my modus operendi for decades.

For the past 100 days or so, I've worked on this book, this journal about my health adventure and healing, and nothing else. -- I didn't feel creative. I felt as though I had to focus on healing for survival, first, and then for recovery.

Recently, I've felt creative and capable of working on book projects, new and old, and I have a feeling that's a good thing.

But what about the physical part of my life?

My feet have been sore.

They have been sore all along, but now I notice them more.

Is it because the soreness has increased or is it because pain elsewhere has gone away.

I don't know. -- All I know is that they have been very sore.

Usually just walking makes the pain in my feet go away. Or it makes me stop being aware of it.

Recently my feet have hurt so much I did not want to walk. I just wanted to give them a rest.

It's confusing!

And what about LTD? Why hasn't the benefits group got their hands on the reports filled out by my doctors? Surely their paperwork was sent in weeks ago!

It's frustrating.

I'm confused and frustrated, my feet hurt, but I feel creative, and I'm happy to be relatively pain free. I am thrilled that the pain in my back and legs and butt has gone away.

Can one guy be thrilled, confused, frustrated, and creative, all at the same time.

Monday, I felt great in physiotherapy, but felt pain after, and rested most of the day.

Tuesday, I felt "out of it" most of the day, until reflexology.

Wednesday, I was mentally active, but did not feel like getting physical.

Where is the pattern?

My physiotherapist often says, Each day is different.

Is there a trend?

The optimist in me says the trend is toward the pain going away and my mobility improving.

And that's great!
But I'm worried about LTD paperwork.

My beard has grown in. After shaving off my Movemeber moustache and soul patch, my beard has grown in. That took thirteen days.

I wrote for half an hour and now I'm running a little late for physio! -- Better go!

Physiotherapy was good. No, it was very good. Okay .... It was great.
    Treadmill: 3.5 instead of 2.5. Box: 30 pounds instead of 20. Arm curls on exercise ball: 20 pounds X 10 X 3.
    Angela, the physiotherapist, said it was crazy good, but she told me to slow down.
    She told me that a few times.
    She laughed and said that by the end of our sessions together we would feel as though we were married.
    For a full hour, I kept on going, doing all my exercises and stretches, and at the end of it I felt like going for a long walk.
    We skipped acupuncture.
    As soon as I walked outside, my legs hurt a bit.
    Is it psychological?
    It's not logical.
    While I'm in the physiotherapist rooms, I feel good, but as soon as I step outside, I feel not-so-good.
    Maybe it's the cold, or the snow and ice, outside.
    Who knows?

While I was working out, I glanced at a magazine, in a bin, and I thought it had a picture of my physiotherapist on the cover.
    What?! I said to myself. My physiotherapist is on the cover of People magazine?!
    Upon closer inspection, it turned out to be Julia Roberts.
    Yes, my physiotherapist looks like Julia Roberts, but with a better haircut.

When she asked me how I was doing, I told her that I had a very unathletic day, yesterday, but I was mentally quite busy.
    She nodded her head sagely and said, Your body decided to put your energy into your brain, so you had mental energy instead of physical energy. -- And that's not bad. -- As long as you have energy, it's good.
    -- And that made me feel better.

Watching the Canadian Junior hockey team take shape in Calgary has been very interesting. The University of Alberta's hockey team beat them 4 to 1, which makes you wonder why we don't send the U of A team to the tournament, but Team Canada was resting their top players, so coaches could look at the other guys and make some cuts.

Jason Billington and Malcolm Subban split the goaltending duties. Subban let in three goals at the start of the game.

My Junior Team Canada has Binnington in net, the goalie from Gravenhurst, and Tom Wilson playing on the wing, with Ryan Nugent-Hopkins as the captain.

Wilson was drafted in the first round by the Washington Capitals from the Plymouth Whalers of the Ontario Hockey League. He won the Goal Medal with Team Ontario at the 2011 World U-17 Hockey Challenge, and was also selected to play in the 2012 CHL Top Prospects Game. He's a 6'4", 210 pound, 19 year-old from Toronto and he's the nephew of my friend Marty Avery, the woman who uses my name.

I feel as though he is my nephew or something.

Binnington is from Gravenhurst but news reports and other listings often say he is from Richmond Hill. He was born in Richmond Hill but grew up in Gravenhurst, my old hometown.

Nobody is born in Gravenhurst unless they have a home birth or get born in a taxicab or something as there is no hospital there. Most Gravenhurst people are born in the hospital in Bracebridge, like me.

Jordan Binnington, Tom Wilson, and Ryan Nugent-Hopkins will lead Team Canada at the World Junior tournament in Russia this time and they will win the gold. -- That's my prediction!

For lunch, I walked downtown, or uptown, or along the main street of town, so I could do some errands, and I stopped in at Beantrees for lunch. They recommended lasagne, and it was very good.

Somehow, sitting in the coffee shop, which was both spacious and cozy at the same time, it seemed to me, suddenly everything looked different. The anxiety that I felt building the last few days melted away. Instead of feeling worried about health and going back to work, I felt happy with physiotherapy and reflexology and my writing, and with the idea that I would be able to go back to work sooner or later.

Instead of trying my hardest to make it sooner, which had done no good so far, obviously, I decided to go with the flow and let the healing happen at the right speed and the right time.

It felt like a good decision.

Was it the aroma of the coffee beans, the atmosphere of the coffee shop, or the morning at physiotherapy that inspired me to change my attitude and focus on gratitude.

Or was it All of Thee Above?!

Maybe I was just feeling the spirit of Christmas.

A Facebook friend posted this quote as their status update: When you come to a difficult season, don't worry. God will give you the grace, the strength, the forgiveness to do what you need to do. Declare peace in your life.

In the late afternoon, I went to Sobey's, Zellers, and the pool. In Zellers, after walking all through the store, after walking through Sobey's, I got an odd feeling in my knees, a tingling sensation with a bit of burning, so I sat down for five or ten minutes, and then felt like new again.

It wasn't swim time, at that hour, but I drove over to the base just to see if the stairway was in the pool, yet, after being repaired. But it wasn't.

However, I did see several planes. It's not unusual to see a plane or two or even three in the sky above Cold Lake, but I saw seven, including one that was flying quite low over Zellers.

I stopped in the parking lot, where I was walking outside, and watched it roar into the big sunset.

At the same time, I looked around to see if anybody else was watching this spectacle.

Nobody else was looking up or at the awe-inspiring sunset.

It made me feel like buying somebody a big turkey for Christmas.

Recovery Day 80: Christmas Parties

The phone woke me up before my alarm clock went off so I talked to Edmonton and found out why I hadn't heard from her all week: She had serious health concerns, including an emergency hospital visit, but said she was back in the game and didn't miss a day of work.

And here I thought she was out dancing at Christmas parties all week long!

The phone woke me again a little later as the principal called to invite me to the staff party after school. They hosted an event for three other schools the day and night before, so everybody would be tired and they would want to blast off before long.

Of course, I promised to try to make it and said I would see if I could move a reflexology appointment.

My plan for the day was to go to the gym, instead of the pool, as soon as it opened. Oh well. -- An afternoon in the gym could be a good thing, too!

RNH was named captain of the Canadian Junior hockey team, big surprise. Subban was seen as the top goalie even though he had the worst training camp and Binnington had the best. And Wilson was left off the team but they decided to take a pair of seventeen-year-olds, supposed to be the top two NHL draft picks, and future superstars. -- Canada hoped they were right because Wilson was a big, bruising, first round draft pick for the Washington Capitals, and you're team better be incredibly good if you believe you don't need a guy like him.

Facebook said a new poll showed 4 out of 5 Americans now believed temperatures were rising due to climate change caused by global warming which was a result of pollution created by people on the planet.

Although Americans had the biggest carbon footprint, they were pointing the finger at China for becoming the biggest polluter in the world. China Daily reported plans to flatten 700 mountains in their huge country to get more coal to fuel their industries and pollute their air and affect the climate all over the world.

First I went to Hamel's, the famous meat boutique, suppliers of astronauts, and ordered a couple of deli trays, and an hour later I picked them up, to take them to the staff party.

Before I went, I got a call from the Glenrose, saying they wanted to remind me of an appointment I had never heard of before. After I said, Okay, I phoned them back and postponed it until they could talk to the specialist, since another doctor told me to forget about having an EMD, because I was recovering from whatever happened to me, and the nerve damage was going away, or getting better.

So, that was confusing, and upsetting, but I dealt with it and headed for the party.

It was good to see my colleagues again, and get a few hugs, but .... I felt somewhat estranged from them. I mean .... it felt like they had gone to war together, been in a battle, survived another semester, almost, while I was sleeping in every morning and having naps in the afternoons.

It was like going to a drinking party a little late, after everybody else is already fairly inebriated, and they are having fun together, acting silly, but they look rediculous to you. -- Ya know?

Outside the party room, I had a couple of good conversations with students. One was tutoring a little guy and the other was looking for me so he could get some help with a distance education course. And after I talked to them, I had a very good conversation with a colleague who showed up late for the party and appeared to be more on my wavelength.

After an hour of social interaction at school, I felt ready to blast off, and it made me wonder if I was really ready to go back to work and have several hours or more of that kind of action every day.

I went home, made a sandwich, checked e-mail, and I was ready to hit the sack, with a movie or two playing in the background.

On Facebook, I found out the Cold Lake Ice had won another one. Their status update said, Thanks to all fans who brought a unwrapped present for Santa's Unanimous at Friday's game!! The ICE took the support from their home fans and produced a 7-0 win vs the Saddle Lake Warriors. Next home game is Friday December 21'st.

-- Good writing!

The big news of the day was about a mass murder in the U.S.A. that moved President Obama to tears. A twenty-something guy with Asperger's syndrome opened fire in a school and killed 20 kids between the ages of 5 and 10 plus 6 adults at the school. And then he killed himself. But before he went to the school, he killed his own mother.

It was the second deadliest mass murder in America, after the Virginia Tech massacre of 2007.

Facebook carried a lot of expressions of empathy for the parents of those children.

Only a few people posted comments about changing gun laws in the U.S.A..

Julian Lennon said, My heart is broken, too.

Recovery Day 81: Saturday Sun

Yesterday the temperature went up ten degrees so it was only minus 20 and I took my fall coat off so I could walk around with just a long, warm, shirt. This morning the sun is shining in a clear blue sky. The smoke from a chimney across the way is going up, instead of sideways, which means there is a higher ceiling, so it must be a bit warmer outside than it has been for the past couple of weeks.

My sister sent me an e-mail with pics of her new sauna, and I just could not relate, but didn't know how to tell her, as I was surrounded by people with much more pressing needs than home renovations. So, I decided to tell her the story of my adventures as Santa Claus:

just got in from lac la biche, playing santa claus ... so much fun!
it was sunny, with a clear blue sky, and warmer, so i thought ... what can i do with the seven hours of daylight
that will get me outside!
        so i loaded up the car with presents -- two big boxes plus two big bags full of smaller pressies -- and that was a good workout for a guy who has trouble lifting stuff, these days ... went to Hamel's, this famous butcher shop that supplies Canadian astronauts -- it's also the place to take the moose or deer you shot, to have it dressed -- and i got a huge turkey. almost got a turducken! found a 26 pounder, young turkey, grain fed, from manitoba ... so it should be good!
        the drive over, and back, was spectacular! half the way, the trees were wrapped in white snowy ice after a winter fog there was nobody home, i thought, but shirley was in the shower, and didn't hear me, so i made like S.C. and left the bags and boxes and the frozen turkey in the back porch, hoping nobody would steal it all
        on the way back i saw the bald eagle and the murder of crows i saw last time -- very unusual. bald eagles go south, they don't stay in the frozen north for the winter. they are fish-eaters. i think this guy has a crush on a crow, but i don't know how to tell one from another!
people say the big black birdss are not crows, they are ravens, but they are either huge crows or small ravens, so i call them cravens.
        I got our cousins a computer chair, new hardcover novels by a Christian best seller guy for Shirley, in a bag, and for Lyn, chocolates plus a few hardcovers, including one by Joseph Heller called No Laughing Matter, about GBS, and that's what the specialist said to me -- this is no laughing matter (and i got ticked off!)
and i had books for Alex and James, plus a car flashlight for Alex and a Kumar christmas video for James, but then i found out Alex isn't here .... so i told Lyn to give James that stuff!
        and i got a foot spa for kezia, a good one, thinking the other women might use it, too

don't know what to get Liz. she might be a gift card person, because idk what she likes, besides books and cats, and i'm not buying a cat book -- not in this lifetime! lol

stopped at the grocery store on the way home and saw a group that's trying to fill up a sixteen wheeler with dry goods for the food bank.

so ... that was a pretty good workout, without going to the pool or gym ... and by the time i got home my butt sure was sore but after sitting down for a snack it was all good again!

they're doing a big food drive for local food banks, reserves, soldiers, etc., trying to fill up a sixteen wheeler, at Sobey's, so I'll go back tomorrow to give them something, i guess, if i'm walkin'!

My sister in Australia wrote back to say my story inspired her to sit down by her artificial tree and listen to her collection of Christmas music on dvd while she wrapped Christmas presents for friends in the land Down Under.

Edmonton phoned late at night as she spent the evening with her sister at a hotel near the airport and they had a Christmasy picnic while watching Hope Springs while her sister was in town on her way up north, where she worked as a nurse.

Despite the late night, I got up at six thirty on Sunday morning and added a piece to my Cold Lake Stories.

The new book was well over 15,000 words, my Scrabble score went up to 1629, and I planned to go swimming again, as I felt strong enough to get myself into and out of the pool without the broken stairway.

Yeah, that never happened! Instead of swimming, I went back to bed with the computer reading Cold Lake Stories to me. While I listened, I watched the ravens outside my big windows. They are unusual birds, these ravens, because they live in town, or our small city, hang out together, with magpies, and shop at Walmart.

The parking lot at Walmart always has a conspiracy of ravens. The TriCity Mall parking lot is loaded with Ravens, too.

It is the Northern Raven, which behaves a bit differently than the Common Raven.

That bald eagle and the conspiracy of ravens over by Lac La Biche is something that stays on my mind. I wonder what's going on. Do birds fall in love? They say birds of a feather flock together. Is there another reason for the eagle and the ravens to hang out together? Is it economics? Was there some reason the eagle couldn't fly south and that's why the ravens adopted him? Or is it confusion? Was the eagle raised by ravens like an ugly duckling?

-- I like to think it's love.

This morning I had a happy dream that I did not want to end, all about some house with a crazy design and me meeting the woman who lived there, and a marriage proposal. It was so vivid I thought I would never forget it, but by the end of the day, it had faded away.

What would I tell the physiotherapist in the morning? She always asks for a report on how I'm feeling and what I've done.

Friday, I did errands so I could go to a staff party and it was all good but after standing and yakking for an hour I was ready to head out. Saturday, I drove to LLB and back, loaded and unloaded the car, which included picking up a chair in a box and also a frozen 26 pound turkey. Sunday, I wrote and rested.

For long periods, I wondered about getting even more time off work, but sometimes I felt as though I could, possibly, be ready to return to work after the Christmas break, in about three weeks.

Day 82: Japandroids

Up at 3:00 after a dream about Eli Mandel, who died two decades ago, standing tall and giving a lecture on the work of Judith Fitzgerald, commenting on how well it would stand up over time, or not, and then walking away.

While I was up, I checked out the Rolling Stones list of top 100 albums of the year, and I realized I must be stuck in the past, or something, because the only performers I had heard of, on the list, were singer/songwriters from the Sixties who had new albums in the top ten: Dylan, Neil Young, and Springsteen. There was a Canadian band from Vancouver called Japandroids in the top ten and I had never heard of them so I found their greatest hits on YouTube and gave them a listen.

And then I went back to Dylan singing about a rival lover, with the line, I'll drag your corpse through the mud.

It all reminded me of a few things I wanted to add to the first one of my new Cold Lake Stories, so I did a little writing, adding another thousand words, and then went back to bed. Sixteen thousand words was about fifty pages, and that was the length of a book. But I wasn't half finished, yet.

Scrabble: 1639 with 238 bingos.

Physio from 9 to 10 was good, with a bad moment in the middle, but I walked out feeling even better than when I walked in. And walking home wasn't painful either, for a change.

In the middle of my workout, lifting the thirty pound box from a shelf at waist level to a shelf at shoulder level, I could feel some pain in my upper thighs, at the back, and my glutes, after I did fifteen reps. Angela got me to stop and do some back stretches and back extensions, instead, and that got me back in the game.

She joked about me wanting to stay for another hour, unlike most patients, who are happy to escape from physiotherapy as soon as their time is up.

When I told her I had a doctor's appointment on Wednesday, she said, Tell him I still want to do that four hour assessment and two weeks of follow-up but that I'm continuing to make progress.

Does that mean I'm not going back to work right after the Christmas Break?

Now what? It's ten in the morning and I don't know what to do.

The internet says that North Korea displayed the embalmed body of Kim Jong II on the first anniversary of his death. Gerard Depardieu was getting a lot of attention for moving to Belgium, about a mile away from the French border, for tax reasons, and writing a letter to a French newspaper, complaining about the Socialist government's new tax on the rich, around 75%, to solve a deficit crisis inherited from the former government of France.

Spanish unions planned a country-wide walk-out to protest austerity measures by the government.

President Obama made a strong statement about the mass shooting at the school in Connecticut. Many Americans were calling for changes to gun laws or mental health treatment in the U.S.A., or both.

In the Great White North, Environment Canada was predicting a green Christmas for most of Canada, except for the twenty-five percent of the population that lived in Alberta, Saskatchewan, and Manitoba. We were going to have a white Christmas and a white Easter, too.

Where I used to live, in the Greater Toronto Area, they haven't seen a white Christmas since 2008.

Environment Canada said, We are known as the Cold White North, but we're not as cold or white as we used to be.

There was no mention of global warming.

This is where we create and celebrate global warming!

According to the CBC, there was a horrible attack on school kids in China. A man stabbed an elderly woman and then 22 children outside an elementary school in China. He was subdued by security guards.

The attack happened in a village about halfway between Beijing and Shanghai.

The Xinhua News Agency quoted police who said the man was possibly "mentally ill".

Some of the children had their fingers or ears cut off.

There had been other attacks like that in China recently. Most of the attacks were carried out by mentally disturbed men involved in personal disputes or unable to adjust to the rapid pace of social change in China, underscoring grave weaknesses in the antiquated Chinese medical system's ability to diagnose and treat psychiatric illness, according to the report on CBC.com.

Chinese law prohibits the personal ownership of handguns.

Alabama, my friend down south, from my old hometown, had a different reaction to the news about the shooting in the American school. She said, Despite what was said by my Spicy Northern friends and my Teacher friends who are the Best in the world, I do believe a woman should be armed and dangerous. I would not go down quietly. I've got wasp spray, mace, pepper spray. I bought myself a slingshot today. Don't worry, the abundant squirrels are likely safe but if I kill one by accident I will eat it. Women: look into martial arts. Women, it is no longer cool to be sweet and vulnerable.

A little later, she added this: When do you hear of an armed woman found in a ditch? My dad told me he would rather get me out of jail than a ditch. I am not in any way condoning guns across the USA. Just letting you know..I have baseball bat, speargun, mace (permitted) and now sling shot.

So I sent her a note saying, You'd eat squirrel?   She said, If I killed it, I would boil it up and make stew, that's the law.

So I sent her some recipes and the list sounded like something Forrest Gump and Benjamin Buford "Bubba" Blue might cook up on a bad day because there was Bacon Wrapped Squirrel, BBQ Squirrel, Bushytail with Autumn Apples, Cajun Squirrel, Squirrel Camp Stew, Caribbean Jerk Squirrel, Chicken Fried Squirrel, Chicken Surprise, Citrus Squirrel, Crockpot Fried Squirrel, Daddy's Squirrel Stew, Easy Squirrel BBQ, Four Squirrel and Seven Spears Ragout, Fried and Baked Squirrel, Fried Squirrel with Mushroom Gravy, Hot & Spicy Squirrel Stew, Marinated Squirrel, Mesquite Squirrel, Redneck Squirrel Fry, Simple Roast Squirrel, Single Serve Squirrel, Sloppy Squirrel Sandwiches, Southern Squirrel Stew, Southern Style Squirrel, Spicy Squirrel, Squirrel Alfredo, Squirrel Bog, Squirrel Cacciatore, Squirrel Casserole, Squirrels in Cream Sauce, Squirrel Creole, Squirrel Croquettes, Squirrel Delight, Squirrel Dip, Squirrel & Dressing, Squirrel Dumplings, Squirrel Kabobs, Squirrel Jambalaya, Squirrel and Mushroom Gravy, Squirrel and Noodles, Squirrel Nuggets, Squirrel and Onion Gravy, Squirrel Pizza, Squirrel Salad, Squirrel Sloppy Joes, Squirrel Spit, and Squirrel Stew Too.

Although I felt better after physio, I gave it a rest after walking home. In the afternoon, I went out a couple of times, to do some errands, walking around the neighbourhood. But that old pain in the butt came back, so I did not feel like doing a whole lot more.

Recovery Day 83: My Guns

My Scrabble score jumped up to 1647 after several games with Alabama, who was still freaked out about the mass shootings in the U.S.A. She said her kids were worried about the end of the world at the end of the week, December 21, 2012. And guns, ammo, and food were flying off the shelves where she lived and across the American South, she said.

Google gave me the news that the FBI could not keep up with the number of background checks for new gun buyers, in the tens of millions, and that could represent even more gun sales, as each check could lead to multiple sales. When there is a mass shooting in the U.S.A., American gun sales go up. Also, President Obama has hinted he wants gun laws to change, so many Americans believe they have to get it while they can.

The U.S.A. had over ten thousand deaths resulting from gunshots last year, apparently, and Canada had a total of 52. Most countries reported double digits and nobody else came close to the Americans in this category.

Funny .... My peaceful brother owns a few guns and my violent brother has no guns other than his biceps.

Unlike my peaceful brother, I own no guns. But like my other brother, I like lifting weights.

Strange .... My old buddy Wally Lamb, who lives in the state where the mass shooting took place, and who wrote about the Columbine shootings, in The Hour I First Believed, has not posted anything on Facebook for days. Several people posted related comments on his Facebook page, like this one:

"Once this settles (the CT shooting) , I wonder if you'll have the energy to write another novel about a school shooting. You articulated grief and PTSD, trauma, dysfunction and overall psychology well. Maybe I'll try in about six years. It took you about 10 years to construct "The Hour I First Believed," correct? Wow.

"I'm not trying to be insensitive to this horrible set of events, but I just wonder. You really helped me to understand trauma and things of that nature more clearly after reading your novel. I wonder."

The teachers at the school were emerging as major heroes for their efforts to protect the students.

Edmonton called before physio, just to say good morning, so I was early for my physio appointment first thing in the morning. Instead of the usual hour, I did an hour and a quarter, with all the usual exercises plus a couple of new ones. And I learned a few things, too.

The new exercises were diagonals using red rubber bands, co-ordinated with breathing, and they were easy enough to do. The new goal is fluidity, the physiotherapist says. Normal motions are not just up and down, back and forth, straight lines, they are twisty and turny, too.

She said the other goal is consistent progress instead of taking two steps forward and one step back.

Also, she said that I will learn some new precursors to pain, or foreshadowing, so that the first thing I am aware of is not a pain in the butt. I will develop an awareness of my core muscles and they will tell me when they need attention.

Fascinating, I said.

And I booked appointments in the new year, up to mid-January.

Oh, dear.

At the end of physiotherapy, I felt like doing another hour.

On Facebook, I posted this question: What are your plans for Saturday?!

It was designed to go with a link I posted to a Yahoo news story that said, The world is ending on Friday and the time for panic is over. We must prepare. The world's biggest believers in the Mayan apocalypse have been preparing for years, some of them earning the title 'doomsday entrepreneurs' ....

In response, I got a few funny comments;

Carmela Ferriol: I will just duck and run what is all the fuss?

Richard Clarke: To be drinking responsibly on a beach. Well, I suppose if the world was ending I might not be so concerned with the responsibly part.Kelly Hearne: I have a cold and am damn pissed it is going to ruin my experience of the apocalypse.Jessica Clark: In theory, Saturday will not exist. So I suppose I will be doing absolutely nothing. But I hope to be finishing my Christmas shopping.

Bobbie Ann Mason and Julian Lennon shared a link to an article about how Australia tackled gun crime with tough restrictions on weapons. The government introduced a nationwide gun buy-back scheme which lead to 750,000 weapons being handed over. There was a gun registry and restrictions on different types of weapons.

The results? Since that time, there have been no mass shootings in Australia.

In the USA, where there are 25 homicides and 45 suicides using guns every day, they would have to buy back 45 million guns.

Elementary school teachers were getting a lot of respect, at last, in the U.S.A., after the mass shooting. In Canada, the story was a little different. Teachers were regarded more highly and paid better. However, in Ontario there were strikes and protests because the provincial government wanted to freeze salaries and make walk-outs illegal. The Elementary Teachers' Federation of Ontario said 35,000 teachers, including nearly 14,000 in Toronto, hit the picket lines in what some called "Super Tuesday" in the ongoing labour dispute with the province. It was a part of a series of one day walk-outs around the province.

The pool was fixed! It felt good to get in the water again. After an hour, I felt like doing another hour.

After reflexology, I felt good. Mecell made me laugh a lot. She told me about the magic of the number 8, or infinity, or the yin-yang symbol, which have the power to calm the hyperactive, take away headaches, send endorphins into painful places for relief, and give power or energy to weak parts of the body. We talked about students with ADD, ADHD, tourrettes, and other problems, too.

She gave me a cream to try for pain relief.

And she gave me her cell phone number to use over the Christmas Break, if I needed her for reflexology. The Wellness Centre would be closed but she said she could make use of the place if she needed to.

I felt like going to the grocery store to get a few things, but after that my legs were a little sore, so I just went home and played Scrabble.

My score went up to 1656 with 240 bingos.

By bedtime, I was walking like a zombie that felt pain. At midnight, it was the same thing. But at 4:30 in the morning, I felt better again.

I was worried about what the doctor would say.

What if he said, Get back to work again?!

Recovery Day 84: LLB

It's twenty below and overcast and very quiet in Cold Lake. Today I'm driving to LLB and back to see the doc to get another note to be off work, feeling worried about LTD benefits and what the doc will say. The physiotherapist said, Tell him I still want to do the assessment and follow-up.

The last time I heard from the benefits group was a week ago, so I sent them an e-mail.

I've been off work without coverage for a week.

Should I be pissed or should I just relax and have faith in the system?

The doc said the mail is slow so ....

He also said he wanted a note from my physiotherapist so it didn't look like I was slipping him twenties to get out of going to work.

On my way home, I dropped in at work to let them know what's going on with me.

And I dropped in on my cousins while I was in their town.

Seeing people, getting hugs, was great.

Driving was no problem. The scenery was spectacular. The black spruce forest was all white after a winter fog and a little snow with each spruce needle wrapped in ice and covered with the white stuff and the forest goes for miles in every direction, as far as you can see, to the distant horizon, a winter wonderland.

But I didn't see the bald eagle and the conspiracy of ravens I sometimes see near the turn-off for Lac La Biche.

The principal and I discussed a return date and the possibility of having it coincide with the end of the semester, at the end of January.

One of the teachers told me that the secondary school students are not happy these days and not getting along with each other, which was a marked changed from last year.

Today they had a ski day, so maybe that helped lift spirits.

Our ski hill is awesome.

I'd write more but I'm hungry for steak!

The steak was great but the temperature dropped below minus 22 so I headed home after that.

Driving and walking wasn't a problem all day. Aside from a little stiffness and soreness first thing in the morning, it was a pretty painfree day!

Recovery Day 85: Ancient Mayans And Chipewyans

American author Amy Tan, a Facebook friend, posted this on December 19, 2012:
CATASTROPHIC THINKING: Until yesterday, I wasn't aware that Dec 21st is Media-fueled Mayan Massive Obliteration Day. According to some versions, Dec 21st is just the beginning of the end. There will be 3 days of darkness before it's lights out forever. In anticipation of demand, stores around the world have wisely stocked truckloads of candles.

Why would anybody believe they and everyone else will die because another culture's timekeeping system was short-sighted. Had they ever paid any interest to Mayan culture before this? I scanned Facebook pages and it's surprising how many non-Mayans are getting, not only candles and food supplies, but assault weapons--I suppose, so that can protect themselves from an early death before the Universal Delete button is clicked.

What would people do differently if they truly thought the world would end ? Break their diet? Confess love? Confess hate? Tell off a boss? Leave graffiti? Spend a lot of money? Get drunk? Resume addictions? Get baptized? Call friends and family and talk until you've used up your free minutes and then keep talking because you'll never have to pay for the overage?

(Hey, if the ancient Mayans nailed it, please don't laugh at me as we are consumed in fire).

Amy Tan's posting got one response: Nasa discovered a hole in the earth's magnetic field 4 years ago. If the the projected solar storm of this year comes, indeed most of the earth's power grids will go down. This is scientific and it's well supported. Giant Breach in Earth's Magnetic Field Discovered - NASA Scienc at science.nasa.gov. NASA's five THEMIS spacecraft have discovered a breach in Earth's magnetic field ....

What would I be doing on the final day of life on Earth as we knew it? Would I be meditating? Going to a Christmas party at my school? Writing all day long?

Playing Scrabble?

My score went up to 1664.

First thing in the morning, I posted a two minute video I made of the trip from Cold Lake to Lac La Biche. It called for a comment, so I wrote, Heavenly highway 55: the frozen forest of black spruce turned white in northern alberta where prairie grassland meets the boreal forest.

That drive, that forest, that landscape, after a winter fog and a little snow, was spectacular.

It was spectacular in summer and autumn, but it was something else again in winter with a little snow frozen to it like frosting.

As I drove through this landscape for over an hour, I kept thinking, I hope this is what heaven looks like.

Eventually I realized that I was already in heaven, of course, so a piece of heaven does, in fact, look just like this.

Highway 55 is also known as the Northern Woods And Water Route -- or it used to be -- as it was the road the furthest north. It is close to the 55th parallel of latitude.

What's north of Highway 55?

-- A lot of frozen forest up to the tree line and then eventually you would reach the Arctic Ocean. -- Also frozen!

This was Dene country, from here to the Arctic, in the days before Thompson, Mackenzie, the North West Company, the Hudson's Bay Company, the fur trade ....

That's a big country!

And you know what they say: In a big country / dreams stay with you / like a lover's voice / fires the mountainside ....

(Big Country, a Scottish band, big in the 1980s)

Facebook had a feature called 2012: See Your 2012 Year In Review. If you clicked on it, I think it showed a sampling of the photos you posted over the year. I didn't try it.

I thought I might try it at the end of the year, or never, as my memories of this year are quite clear, and I like to look at the picture's in my mind's eye.

Global doomsday hot spots around the world were getting a lot of attention. The Pic de Bugarach mountain in the town of Bugarach, France, in the French Pyrenees mountains, was one of them. Stalin's tomb was another.

For $1,500, you could stay in Stalin's underground bunker in central Moscow, over 200 feet below ground, designed to withstand a nuclear attack — with a 50 percent refund if nothing happens.

In England, people have already converged on Stonehenge for an "End of the World" party that coincides with the Winter Solstice.

Mount Rtanj, a pyramid-shaped peak in Serbia, was another hot spot, along with Cisternino in Italay.

In China, the group called Almighty God, or Eastern Lightning, which preaches that Jesus has reappeared as a woman in central China, was warning people about the coming apocalpyse.

There are sixty million Christians in China and 300,000 of them believe Jesus has returned as a plain-looking, 30-year-old Chinese woman who lives in hiding, has never been photographed, has written a third testament to the Bible, and has composed enough hymns to fill 10 CDs.

A sect called Lightning From The East teaches that Christians who join her will ascend to heaven in the coming apocalypse.

On my return trip to Lac La Biche, I did not make a video of the winter wonderland. It was overcast and dull on the way over and dark on the way back. My appointment was for 3:00 but I didn't get in to see the doc until after 4:00 so I was on the road from 4:30 to 6:00. It was dark and there were long sections without lights. Although I kept my speed up, everybody passed me, going faster at night than during daylight.

The doc signed a note recommending the test the physiotherapist wanted and a note saying I should be off work until February 1st, to get all the physiotherapy follow-up work done.

A month earlier, I would have been enormously upset about not going back to work. But I have learned that this kind of recovery takes as long as it takes.

When I got home, there was a letter from the specialist's office saying he didn't fill in forms and I should get my family doctor to do that.

THAT was upsetting, but a friend of mine who has experience with LTD and health benefits told me the doctor's form should be enough. And, she added, if it wasn't enough, I could appeal, and appeals were usually successful.

-- The things you learn!

Sitting in the doctor's office, I started coughing. -- Just a little cough every ten minutes or so.

The cough didn't go away. I coughed all the way home.

On the way home, I listened to CBC Radio One from Edmonton, all about their turkey drive, which exceeded their goal of 230,000 bucks by a long shot. At the end, they were trying to get it up to 300,000, to set a new record.

They reported on the Idle No More movement, which, they said, in a matter of days, has become the largest, most unified, and potentially most transformative Indigenous movement at least since the Oka resistance in 1990. They interviewed two women from the Hobema Reserve, south of Edmonton, who said it started with Chief Theresa Spence of the Attawapiskat First Nation going on a hunger strike and grew quickly to include protests in Edmonton, including one at the West Edmonton Mall.

They said that since 2008, the Harper government has cut aboriginal health funding, gutted environmental review processes, ignored the more than 600 missing and murdered Indigenous women across Canada, withheld residential school documents from the Truth and Reconciliation Commission, abandoned land claim negotiations, and tried to defend its underfunding of First Nations schools and child welfare agencies.

Chief Alan Adam of the Athabasca Chipewyan First Nation says Ottawa's omnibus budget legislation weakens environmental protection in Canada.
He says oilsands projects have already sullied rivers and lakes in the area and the budget bill — quote — "gives the green light to destroy the rest."

Adam's comments came as he joined a highway blockade north of Fort McMurray that was part of the aboriginal Idle No More movement.

A mass rally was planned for Ottawa on December 21st, the day of winter solstice.

The NHL was in the news as well as the end of the world. The NHL lockout would reach day 100 on Monday, and they announced more cancelations plus the fact that mid-January was the deadline for starting a shortened season or cancelling hockey for the year.

The radio announcers joked about the end of the world.

When I got home, I saw my cousin had posted a note on Facebook saying that if it was the end of the world, we should blame the NHL.

Canada's World Junior team lost their first game. Finland outshot them and beat them 3 to 2.

Subban was in net. The game-winning goal was scored at the end of the game.

Binnington would be in goal for the next game, against Sweden, on Saturday.

At 4:00 in the morning, I woke up, and I stayed up to watch 4:15 roll in, meditating and monitoring the internet, to see if anything radical happened. Instead of the apocalypse, with massive destruction around the world, it turned out to be just a shift in consciousness for some people.

Bhagawan, at Oneness University, in India, claimed he had awakened 70,000 people, which was his goal by this date, and that meant they had achieved some sort of critical mass, with tens of thousands waking up each month from now on.

As for me .... After sleeping another couple of hours, I woke up at 7:00 feeling quite different. Instead of feeling sleepy and tired, I was wide awake, even though I had a cough, after going to the doctor's office.

What would I do with this day?

Swimming was out, visiting kids at school was out, a lot of things were out, as I wanted to get rid of my cough and I did not want to spread it around. I had to get my doctor's note into the school, so I planned to do it while they had the kids out at a Christmas event.

After that, I would phone the physiotherapist in Lloydminister to see about setting up the assessment my physiotherapist and doctor wanted me to do.

It was still very dark at 8:00 in the morning. As Friday's dawn began sweeping around the globe, there was no sign of an apocalypse.

Yucatan Governor, Rolando Zapata, said "We believe that the beginning of a new baktun means the beginning of a new era, and we're receiving it with great optimism."

Even before the baktun's end, hundreds of spiritualists from Asian, North American, South American and European shamanistic traditions mingled amiably with the Mexican hosts at a convention centre in Merida on Thursday. Dozens of booths offered people the chance to have their auras photographed with "Chi" light, get a shamanic cleansing or buy sandals, herbs and whole-grain baked goods.

This is the beginning of a change in priorities and perceptions. We are all one, some said. No limits, no boundaries, no nationalities, just fusion."

The Oneness University.org website sounded ecstatic: "Throughout cultures of old and new, great seers and mystics have spoken of an upcoming age for humanity where we as a species flower into higher states of consciousness. December 21st 2012 marks the birth of this new age of Oneness known as the Golden Age.

"Sri Amma Bhagavan's mission and vision has been to assist humanity in this transition. The mission was to create a critical mass shift in human consciousness by awakening over 70,000 people. This mission is accomplished!

"The great awakening for humanity is now fully underway as we witness tens of thousands of people awaken each month in ever increasing numbers. To celebrate and inaugurate the birth of this new age for humanity, there would be a very special webcast event to meditate and receive Blessings for Awakening directly from Sri Amma Bhagavan. As a preparation for this special meditation for Awakening with Sri Amma Bhagavan, you could take part in the live/online OM and receive Deekshas from the Awakened ones all through this month. invite all your family and friends for all the programs and deekshas the entire month and be a part of this historic event!"
Namaste!

Here's the physiotherapist's report for the doc:
Re: Comprehensive Rehabilitation for Mr. Martin Avery

At this time Mr. Avery has made significant recovery from his severe back dysfunction. He now manages his pain effectively without medication, using postural correction and extension exercises to centralize his pain. His gait i improved significantly, no longer rquires an aid, and rarely limps. Mr. Avery's walking tolerance is approximately half an hour, and standing greater than half an hour results in pain. We continue to focus on syumptom management along with progressive functional tasts such as pushing, pulling, lifting, balance, and general conditioning and core work.

It is recommended that Mr. Avery undergo a formal Functional Abilities Evaluation to objectively determine his current status. The purpose of this evaluation is to compare his present status with his job demands. Following this assessment, recommendations can be made regarding return to work, what modifications or changes can or should be made, as well as provide an outline for a comprehensive functional rehabilitation program. The intent of a comprehensive program is to effectively prepare an individual for safe return to work following prolonged impairment and absence as well as minimize potential for regression or re-occurrence. Mr. Avery has applied for long Term Disability. It is desired by his benefits provider that the recommendation for Comprehensive Rehabilitation be requested by a physician.

In the morning I spent an hour at my school to see a Christmas presentation by the students. It was moved from a location on CFB Cold Lake to the school. To avoid a mob scene, I timed my visit to coincide with their absence, so I could drop off my doctor's note and blast off again without coughing on anybody.

Instead, the students ran over as soon as their Christmas presentation ended and I got around 100 hugs and then had short conversations with two dozen students in Grades 6 to 12.

It felt great to get all that love.

When I got home, I discovered that visit was exhausting, so I hit the sack for an hour. And when I got up, I got the news: At least five schools in Alberta were closed today due to what officials are calling potentially threatening situations. Classes were cancelled at St. Dominic Catholic High School and West Central High School in Rocky Mountain House. As well, the Red Deer Catholic School Division said on Friday that a student told another student he was going to take a gun to school. In Medicine Hat, Crescent Heights High School was closed as a precaution after an incident that the school is calling a low risk to students and staff. Classes at two schools in Devon near Edmonton were also cancelled following a threat made online.

The Alberta incidents came to light just one week after a mass shooting in Newtown, Conn. Twenty elementary school children and six staff members at Sandy Hook Elementary School were killed by a 20-year-old gunman, who then killed himself.
In the afternoon, I went for a short walk to the drug store to get cold and cough medicine. On the way back, my legs hurt so much I had to stop every twenty steps or so to stretch my back and try to get rid of the pain in my thighs at the back of my legs. It was fairly severe and it scared me a little bit.

How can I go back to work like this?

Recovery Day 85: Binnington

On Friday night, I signed up with TSN to watch the World Junior Hockey tournament, saw condensed versions of the Canada versus Finland game and the U.S.A. versus Sweden game, and got up on Saturday in time to see Canada play Sweden.

Binnington stopped 60 shots and two out of three shootout attempts.

The IIHF counts shots at the net, rather than shots that reached the goalie. They counted 60 shots but Canadian hockey analysts counted only 30. -- Even so, he faced a lot of shots and he looked good all game.

Would that be enough to make him Canada's #1 goalie?

Suban let in three goals and lost when Canada played Sweden and Binnington let in one goal and won when Canada played the defending championship team from Sweden.

Go Binnington!

Ryan Nugent-Hopkins scored the game-winner in the shoot-out, after overtime.

The USA lost to Finland by a score of 5 to 1.

TSN followed up with a special, after the game, counting down the top 40 Canadian juniors at the world tournament over the past 40 years, with highlights of their performances. Gretzky, Crosby, Getlaff, Jamie Storr, Shane Corson, Lindross, Doughty .... So many great players and great moments!

The Cold Lake Ice won 16-3 over the St. Paul Canadiens. Ice came from behind last night down 6-4 in the third to win the game with 10 seconds left as they beat the Vermilion Tigers 7-6. Dallas Ansell had 5 points and Matt Laboucane scored the winner on his 4th goal of the game. The Ice we'll be in for another tough battle tonight as Wainwright makes their way to the Energy Centre.

The ICE finished off the first half of the season tonight with a 5-1 win over the Wainwright Bisons! Matt Laboucane had another hat trick as the Ice head into the break first overall in League standings with a record of 19-2-0-1. The Ice wish all their fans and sponsors a very Merry Christmas and a Happy New Year.

My Scrabble score jumped up to 1671 and I got a challenge from the person in the #2 place on my list.

In the afternoon, I had to go out to get some groceries. And in the grocery store, I ran into one of our teachers, who told me she quit her job, so the last day of classes before Christmas was her last day of classes ever.

She told me she informed the staff only on the last day of classes.

So, I was fighting to get back in and she was in a hurry to get out!

After the grocery store, a clothing store, and a drug store, my legs hurt a bit, at the back, so I got to my car before I fell down and drove around for a little while. While I was in Cold Lake North, to see the lake, I noticed that gas was down to 1.00, so I filled up my tank.

My legs were okay by then.

When I got home, I watched the Canada/Sweden game, start to finish, cheering for our juniors, even though I knew they were going to win. -- It was still fun.

The job of escorting Santa the North American Aerospace Defence Command (NORAD.) But it's more than monitoring the jolly man's flight on radar: Santa also gets a fighter jet escort during his time in North American airspace. Videos released on NORAD's website reveal two of the four Canadian fighter jet pilots given one of the most special, secret missions around: escorting Santa's sleigh during his Canadian deliveries "like a small parade." After Santa's flight through Eastern Canada is complete, the Quebec-based pilots will hand off to CF-18s from 4 Wing in Cold Lake, Alta., somewhere around the Ontario-Manitoba border. The western pilots will escort Santa to the border with Alaska before handing off to their American counterparts.

In the evening, I didn't feel like doing anything except exchanging e-mails, chatting online, and playing Scrabble. There was a hockey game at the arena, in town, with the Cold Lake Ice playing the Wainwright Bandits, but I thought it made more sense to stay home and get over my cough and cold or whatever it was.

So, I won several Scrabble games, lost one, and had a couple of bingos.

What would my score be, after that?

It takes a day for online Scrabble to calculate your overall score and the standings.

What I really wanted, instead of a higher Scrabble score, was to wake up without a cough and with a pair of legs that worked without any problems.

Was that too much to ask for?

Recovery Day 86: Shakin' All Over

Oh, what a night! Although I stopped coughing, more or less, I shivered and I shook when I went to bed around eight o"clock. And then I got hot, so I threw the blankets off, tore off half my clothes, and put my feet on the big glass door, which was as cold as ice. -- It didn't feel nice.

    I was wide awake, at three in the morning, so I got up and wrote this.

    Although I didn't want to spend the night writing, I did have a lot of ideas for my Cold Lake Stories, including a name for a hockey team that was new to the NHL: The Fort Mack Tarzans. Or Tarzens. With Tarzan's body printed on the jerseys and socks and hockey pants that looked like his leopard skin loin cloth.

    "We're not wearing that," the team said with one voice.

    "Then you're not playing on this team," the coach said.

    "Let me have another look at that," the team captain said.

Scrabble: 1653, with 243.

It was 23 below in Cold Lake, 33 below in Fort Mack, only 19 below in Edmonton, minus 2 in Toronto, minus 6 in Montreal, and plus 6 in Vancouver. In Adelaide, it was 22 above and going up to 42.

    In Britain, National Weather Service issued more than 500 flood alerts and warnings for England and Wales.

    The windchill factor in Cold Lake and Edmonton made it feel like minus 30 and that meant frostbite could happen in minutes.

Sunday morning CBC was great, from about six until noon, while I was listening.

    Hallelujah!

At noon I traded novels with a writer friend back home. She was a crime writer, not a literary writer, but she had written a non-fiction novel about sexual abuse.

    I sent her my book about the killer Santa Claus, a Catholic novel, partly fact and partly fiction, like Santa Claus, a walking contradiction.

On Facebook, YogaDawg posted a picture by Ai Weiwei showing Santa Claus and the Laughing Buddha side by side.

    A couple of years ago, I wrote a novel called Santa Claus and the Laughing Buddha.

    The two figures have a great deal in common, but not everybody knows about it.

    Everybody in the West knows the Laughing Buddha as that fat guy sitting by the cash register in your favourite Chinese restaurant. -- But there's a lot more to the story of the Laughing Buddha!

Recovery Day 87: Cough, Cough

It was the day before Christmas and a cold one it was, minus thirty in the morning but no reason to fuss. They say you should plug in your car when the temperature goes down to minus fifteen but my car has started, no problem, when it was minus 20, so I haven't plugged it in, yet, -- Maybe I should have plugged it in last night.

On the other hand, I don't have to drive anywhere today, so ....

My cousins keep sending me messages about going over to see them for Christmas dinner, but I can't see it happening while I have this cough. They say they can't stand the thought of me spending Christmas alone. However, driving from here to there, hanging out for hours, talking a lot, while coughing and coughing, makes Christmas alone sound like a good idea.

Anyway, as Oscar Wilde said, I think it's very healthy to spend time alone. You need to know how to be alone and not defined by another person.

Since I've had this cough, which started a few days ago, in the doctor's office, I've spent more and more time in bed, which means I haven't noticed the pain in my back or legs, so much. There has been no back pain at all and no leg pain or pain in the butt, either. When I get out of bed and walk to the kitchen, I'm a little stiff, but that's about it.

Sitting has been no problem, either. I've spent some time on the computer, checking e-mail, Facebook, and online news. My Scrabble score is now 1665.

The novel I read yesterday, for a friend, turned out to be a good book. It's a non-fiction novel about her history of abuse, starting in childhood, with her brother, and ending with recovery or healing recently as an adult, age 47. And I make an appearance in the book as a Zen guy who teaches her how to meditate, takes her to the Zen Forest, and treats her like a human being, helps her with her writing, and is unlike the other men in her life. The guy she married turned into an alcoholic. Her father was unable to deal with the fact that his son was abusing his daughter.

She is a murder mystery writer and writing non-fiction was a new thing for her. The book was short, the writing was tight, and it had a lot of momentum. It was like the last part of a roller-coaster ride. I gave her some positive feedback and an idea I didn't think she would like. I suggested she could turn it into fiction and add a murder at the end. If the person she was describing was not herself but a fictional character, it would be very satisfying for the reader if there was a bigger climax than a positive statement about healing. There would be a bigger bang if she killed someone.

If it was a crime novel, the female protagonist could kill her husband or go on a murderous rampage and kill a lot of men. She could be the female equivalent of the guy responsible for the massacre of women in Montreal.

I haven't forgotten his name, I just refuse to write it.

Wouldn't that make a great book! Imagine the story of a woman who gets revenge for the Montreal massacre by getting some guns and going on a rampage, killing only men.

It's amazing this story has not already been written and turned into a major Hollywood movie.

In real life, as in fiction, the number of female mass murderers is much much smaller than the number of males. A few Australian cases come to mind. There was that prof at the University of Alabama. A few others killed their husbands, ex-husbands, and kids. A mother drowning her kids in the bathtub is big news. A woman picking up a gun and targeting men may be a story whose time has come.

I almost had a run-in with the woman known as 'the mother of all female suicide bombers', a Palestinian terrorist hiding out in Lebanon, as she hijacked a bus I missed, years and years ago. But she wasn't out to shoot men, she was out to kill Israelis and Jews.

In the afternoon, I felt well enough to venture out for a quick trip to the grocery store to get some supplies. It was twenty-five below and my car wasn't plugged in but it started, no problem, and cruised along, over the ice and snow, to the packed parking lot at the TriTown Mall. The big grocery store was so busy it actually ran out of a few things, like Coke, and had some bare spots on the shelves.

Everybody was friendly, though, and nobody acted all stressed out, the way they do in Toronto on the day before Christmas. The only person who appeared unhappy was the young woman working as a cashier. She was shivering and frowning, so I suggested she get a coat, since she was near the front door, and a lot of cold air was getting in, but she said, I don't need a coat, I need a new job!

The young guys whose job was to retrieve shopping carts from the parking lot huddled together, just inside the store, to remark about how incredibly cold it was out there.

It wasn't actually that terrible if you were dressed for it and only had to walk from the handicapped parking spot to the front door. All I wore was my fall jacket and I felt fine.

Christmas Eve was spent with some of my greatest loves: CBC Radio, Facebook, e-mail, and online Scrabble.

As It Happens, on CBC, featured some golden oldies with Barbara Frum, which was great. CBC replayed an old reading by Alan Maitland with the story of an airplane making an emergency landing in England with the help of an old plane called a Mosquito and the pilot finding out that there was no Mosquito or any of the other help he imagined. -- That story always gets to me.

The movie I didn't like before, called Cloud Atlas, looked a lot better when I saw it for the second time. As I fell asleep watching it, I was thinking about going to dinner with my cousins on Christmas Day, but doubting I would be able to make it because my cold was still quite bad.

My cold was bothering me so much, I almost forgot about my back, butt, and leg pain. But I could not forget about my feet, because they were quite sore.

As I hit the sack, I realized I had walked all around the grocery store without even thinking about my injured areas. My mind was busy with other thoughts about Christmases past and present and future.

In the morning, around 4:00 a.m., I saw my Scrabble score jumped up to 1672, with 244 bingos, good for first place on my list of Scrabble players.

88: Christmas!

The sun came out for Christmas and I managed a very brief walk in the winter wonderland
but I decided against going to LLB for Christmas dinner with my cousins and their
families because I was still coughing and the violent coughs gave me a headache and I
didn't think I would be the best company!

My cousin told me she had the big turkey cut in two so she could cook half of it for
Christmas and the other half some other time, maybe New Year's, because it was so huge.

It's the biggest turkey I've ever seen! she said to me.

Who? I said. Me?

When I went out for a walk, it was twenty below or a little colder and the sun was
shining brightly in a clear blue sky. There wasn't much moving outside, other than ravens,
or crows.

In the afternoon, after A Christmas Carol, with Jim Carey, I had a sudden shot of energy
and used it to clean up the place. First I unfroze the window door and brought in a few
things I had stored out there, then I picked up all the Christmas wrapping paper, and then I
worked on the kitchen counter, and I made dinner. Those little jobs took a few hours and
it felt good to get them done because I had been wanting to do a lot of them for a long
time.

Get over a cold while recovering from severe nerve damage on Christmas Day while
listening to CBC Radio shows and checking up on Canada's team in the World Junior
tournament: That was my day.

How I wished I could play!

One of our boys hurt his back at hockey practice in Russia.

Somebody provided a comparative cost analysis for the World Junior Tournament in
Russia versus Canada. Last year, to go to the games, you would have to win a lottery and
spend over two thousand bucks. This year in Russia, you could see them all for under
eighty bucks. And you could stay in a relatively good hotel for the whole week for just
over three hundred bucks. -- You could spend three hundred dollars a night, in Canada, if
you wanted to go first class.

The first game of the tournament was in the morning at some ridiculous hour, but I thought
I might get up and watch it anyway. Canada versus Germany. It should be an easy win for
the Canucks. But they were starting Subban in net, instead of Binnington, and they were
down a couple of forwards to injury and suspension. -- Even so, it was expected to be a
cakewalk.

In Ufa, Russia, a city of one million, two hours from Moscow, where it was 23 below, just
like here, Ryan Nugent-Hopkins had a goal and four assists to lead Canada to a 9-3 win
over Germany to open the 2013 world junior hockey championship.

Mark Scheifele scored twice and Jonathan Huberdeau had a goal and two assists for Canada. There were eight different goal scorers for Canada.

Subban, in goal, looked nervous in the service of his country, but came out alright.

The Canadian juniors were in the pool of death with gold-medal contenders Russia and United States.

The defending champions, Sweden, were in Pool A with Czech Republic, Finland, Latvia, and Switzerland.

-- Hard to predict who would win Pool A!

Even after losing three defencemen, Sweden would likely have an easy time of it and then be an easy target for Canada in the semi-finals.

Getting up early to watch hockey night in Russia was not a problem. -- I'm usually up at that time, anyway. But I don't usually stay up.

The next game: Russia versus Slovakia at 11:00 a.m.

In other news .... There was a big fire in Edmonton on Christmas Day with twelve million in damages. Apparently the water kept freezing in hoses used by firemen, which made fighting the blaze a bit of a nightmare. -- They use special steam units to melt the ice in the hoses.

A big storm hit Alabama and the coast from Florida to New Orleans, with hail and snow and possibly a tornado. My Alabama Scrabble partner said it was warm, around 70, with rain, where she was, and she sent her kids and their kids home when they got the report that the weather was turning bad.

Slovakia surprised Russia at the World Junior Tournament, and the rest of the world, as they did not get blown out, like Germany playing Canada, and tied the game, at the very end, but lost it in overtime.

That means 1 point for Slovakia, 2 points for Russia, 0 points for Germany, and 3 points for Canada.

Recovery Day 89: Cranky

Edmonton called to say she had a cold but she had to go to work and we chatted in the evening, online, at which time she said, Sorry for being cranky but this cold is making me miserable.

Likewise, I said.

Instead of just a sore throat, I was sneezy and grumpy and felt like a bunch of dwarfs had used my head as a drum. I coughed so hard it gave me a headache.

Although I tried to do a few of my physiotherapy exercises, in good moments, mostly I slept the day away.

My cousins sent me nice notes about Christmas presents. They said the turkey I got them was so big, they had to get somebody to saw it in two, so they could cook it, and even though they only cooked half, they still had leftovers and made soup. They also said it was the tastiest turkey ever.

It was a young, grain-fed, Manitoba turkey from Hamel's.

They liked all the books, thought they were great picks, and they thought the chair was a good idea, too, as it was something they needed but were unlikely to get. My cousin's daughter put it together, without instructions, in about ten minutes, which was pretty impressive. Her mom said that her ideal job would be putting together IKEA furniture.

I sent her a cartoon of a job applicant at IKEA with a manager pointing at a bunch of furniture pieces on the floor and saying, Have a chair.

In the evening, I played a little Scrabble. At night, I put on movies but fell asleep and had my computer read to me but fell asleep. I heard most of the new novel after going through it a few times. I watched the day's hockey games a few times. TSN had the full games and also compacted versions of the games online. I thought that was a great idea. You could watch Canada beat Germany 9 to 3 again and again in about twenty minutes. That would be a terrific feature for a great game in the tournament or for the greatest hockey games in history.

Most of the day, I felt too tired to do anything but too awake to sleep.

The good news was that my miserable cold symptoms made me forget about my nerve damage issues for a big part of the day.

Recovery Day 90

My Scrabble score went up to 1650 and my nearest rival dropped to 1596. A few of my games with Alabama were very close, with the lead changing back and forth between us numerous times, right up until the last couple of moves. But I knew she liked to save an 'x' for the end, if she could, so I tried to save an 'a' to go with her 'x' and that's how I beat her on one game.

My cold symptoms were hanging in there and my sleep had been interrupted several times, so I was happy the U.S.A. was going to play Germany at 7:00 in the morning. The Germans said it was a special honour to play against Canada any time and something else this year because of the NHL stars. They would play a different game against the U.S.A.

The team from Germany did well last year at the U18 tournament, beating Russia in one game, and they had a few guys who were playing in the OHL, including one or two who were on their way to the NHL.

The American team captain was expected to be drafted second or third in the first round and he had done some trash talking, saying the U.S.A. was the team to beat.

So .... We hoped everybody beat them. Like Finland.

In the pre-tournament exhibition series in Finland, just before the World Junior Tournament, Finland beat the U.S.A. by a score of 5 to 1, which was huge.

The Americans looked big and they had cool uniforms but they still played like Americans.

Canadians played like they wanted to be stars in the NHL. The Americans played like they hoped to have their college scholarships renewed.

Recovery Day 91

Canada had a slow start in the game against Slovakia, fell behind 2 to 0, but battled back to win by a score of 6 to 3. A couple of Canucks got suspended for rough play.

Russia beat America in the next game, by a score of 2 to 1, and that meant Canada was in first place with a perfect record as Russia lost a point to Slovakia and the U.S.A. lost three points to Russia.

Go, Canada, Go!

My cold was still horrible, but the drugs helped. My objective for the day was to take more drugs and sleep more and that was about it.

My Scrabble score went up to 1658.

It was thirty below outside, in Cold Lake, just as it was in Ufa, Russia.

Recovery Day 92: Go Canada!

Canada beat the U.S.A. 2 to 1 at the World Junior Tournament in Russia in a tough game. The Americans hacked and whacked at the Canadians until their penalty box was full. It was not a lot of fun to watch, except for the fact that the good guys won. RNH got the first goal and Ryan Strome got the other.

Canada led 2 - 0 for most of the game on the goals by Ryan and Ryan.

Subban was named the player of the game for Canada.

The Americans actually outshot Canada, but Team Canada outscored Team U.S.A.

So, Canada was perfect and in first place while the Americans, who said they were the team to beat, lost 2 to 1 to both Russia and Canada, so they could be facing relegation.

The ads that go with the program were getting to me. The one for Buckley's cough medicine, at the end, was appreciated, because I still have a cold, or bronchitis, or something. The Canadian Whiskey ads, with the guy who says, "I'm the CEO .... You're welcome" and 'the only thing that tastes better than whiskey is smuggled whiskey' are getting to be very annoying. So is the weird one about taking away hockey in Canada with the girl in goal saying, Take away my pads and I'll just wear thicker socks', and the little boy threatening to move to Russia, are just weird.

There were reports of wicked winter weather in Europe, like Ufa, with close to a thousand people dying because of cold conditions and snow on the ground in the south of France, in Algeria, and even the Sahara Desert.

Cold Lake warmed up, a bit, to minus 15, and the forecast was for warmer weather to start the New Year.

I had a physiotherapy appointment on the last day of the year but didn't know if I would be able to make it because of my cough and cold.

Yesterday I studied Cloud Atlas. There was an interview with the author of the book on CBC Radio and there were interviews with the cast as well as the directors online. I watched them all. How many times have I seen the movie? Just three. And I think my book will be much better. The structure and the story lines get a lot of attention, in Cloud Atlas, but they don't really add up to much. Do they? And the themes of reincarnation, slavery, and people as predators, are weak.

The author, an English teacher from England who worked in Hiroshima for several years, said he doesn't believe in reincarnation but he would like to.

-- Lame.

Scrabble: 1658.

There was a video report out of Utah about American teachers being trained how to use guns.

A female elementary teacher said schools are easy targets for shooters if the teachers aren't armed.

This seems completely wrong-headed, to me.

Teachers with guns?

America, what is happening?

Wally Lamb returned to Facebook with this message: Wanted to say thanks to those of you who checked in with me in the wake of the Newtown tragedy. As some of you know, the research I did while writing my Columbine book, The Hour I First Believed, took its toll. "Vicarious traumatization," my therapist called it. Nothing compared to what the victims' loved ones endure, but difficult nonetheless. So the recent school killings at Sandy Hook silenced me and knocked me for a loop from which I'm now emerging. I appreciate your concern and continue to pray for Newtown's victims, their families, and all of those deeply affected by that sad and terrible event. Take good care.

One thing the World Junior Championship shows, which doesn't get nearly enough attention, is the greatness of the Ontario Hockey League. Most of Canada's team and the top players of the other teams play in the OHL. That includes the captains of the American and Russian teams and the goalie for the U.S.A.

The Canadian Hockey League today announced that a total of 68 CHL players are listed on official International Ice Hockey Federation rosters and will be competing in the 2012 IIHF World Junior Hockey Championship beginning today in Calgary and Edmonton, Alberta.

The 68 players includes a representative from all 10 participating countries, including Canada who leads the way with 21 CHL players followed by the Czech Republic with ten players, Slovakia with eight players, Switzerland and the United States each with seven players, Russia with six players, Sweden with four players, Finland and Latvia each with two players, and one player from Denmark.

The Ontario Hockey League led the way with 32 players competing in the tournament followed by the Western Hockey League with 20 players, and the Quebec Major Junior Hockey League with 16 players. Of the CHL's 59 clubs, 39 are represented in the event. The OHL's Niagara IceDogs have the most players of any OHL team in the tournament, with four playing for Team Canada.

Hockey people are in the know about the OHL, obviously, and it's a well-known part of life in Ontario, but .... The OHL should be shouting from the rooftops, advertising on tv, sending press releases to all kinds of media, proclaiming its greatness so all OHL arenas are full to the rafters for every game and there is more coverage, including televised games, for the OHL.

The same is true for the other junior hockey leagues in Canada, too, but the OHL is clearly the leader and it should get the attention it deserves.

Here in Alberta, not too far from Edmonton, in Oilers country, we are looking forward to the Canada/Russia game at the World Juniors like everybody else in Canada, and around the hockey world, but we have an added interest as the captains of the two teams are both Edmonton Oilers.

Ryan Nugent-Hopkins was the captain of Team Canada and Nail Yakupov was the mercurial of Team Russia.

They were both first picks in the NHL draft, in successive years. RNH should have been the Rookie Of The Year in the NHL last year and 'The Nail' might have a shot at it this year.

Nuge comes from B.C. and played junior in Alberta and Yak comes from Russia but played junior in the OHL.

Recovery Day 93: Fireworks

On Monday morning, I was up before the alarm, had a shower and so on, not that I had anything to do or anywhere I had to go, the day of New Year's Eve. My sister in Australia sent me a long e and I responded to it, as it was already 2013 in the land down under. So, I was all ready for the big game, Canada versus Russia, at the World Junior Tournament, which started at 7:00 a.m. Cold Lake Time.

 Canada juggled its line-up for the game against Russia.

 They were undefeated in the tournament but they were making changes. -- Why?

 The U.S.A. beat Slovakia earlier in the day. They were playing for third place or relegation.

 So, Germany and Slovakia were out.

 Sweden and the Czech Republic were in first and second.

 Canada won the Spengler Cup with a team loaded with NHL stars.

 It has been 25 years since Canada won gold in Russia.

 On the other hand, the Juniors have won all but a handful of games this decade or century or millennium.

 Team Canada moved Jonathan Drouin, age 17, up to the first line, to play with RNH, because he proved to have an extremely high hockey IQ, they said, and he was fast, too, plus he had a great scoring touch.

 That would be a gifted line, but would they be tough enough?

 Drouin would be the number one draft pick for the NHL, for sure, now. Seth Jones, the American hyped as his competition, did not have a great tournament.

 Before the start of the big game, in the pre-game show, there was a lot of reminiscing about the history of hockey and games between Canada and the USSR when the Soviet Union was a totalitarian military dictatorship and the games provided opportunities for propaganda about whose system was best, Soviet or West.

 Canadian hockey people were interviewed against a background of Red Square with the Soviet army marching past enormous portraits of Lenin and the Canucks complained about the fact there were soldiers with machine guns and dogs all over the place and the people all dressed in drab colours with a lot of black and brown and grey.

 The U.S.S.R. was a thing of the past, thank God, but the new Russia was still a mystery. It wasn't a closed society, as it was in the past, but it was still a different culture, and it was seen as a rival or competitor rather than a like-minded friend or ally.

 The Russian hockey team did not play in the old style of Soviet teams, which was army-like, robotic, the opposite of creative, with five man units attacking, attacking, and attacking. The new Russian style was to let individuals do their own thing and go for long passes, breakaways, sneaking behind our defence, to score lots of goals.

Personally, I felt like hell, as I didn't get a good sleep, since I kept coughing, violently. At 9:15, I had an appointment for physiotherapy, but I had sent a couple of e-mails saying I had a bad cough and didn't think I should show up to infect everybody.

My predictions for the game? Canada would win.

I wasn't crazy about the TSN hockey commentators, any more than the Nike or whiskey ads that went with the series. They sounded as though they liked the U.S.A. more than Canada during their big game. They would probably praise the Russians too much, too.

Two of the officials, for the game, were Americans, which could be a good thing.
    Subban was in net, again, going against Andrei Makarov. And even though Subban was great in the game against the U.S.A., I'd still feel more confident if Binnington was in net.
    Makarov played for the Saskatoon Blades, during the regular season, and Saskatoon was just a few hours away from me, south and east.

Canada got off to a good start and then Russia took a bad penalty, hitting from behind, that gave Canada five minutes on the power play, which resulted in two goals.
    Russia got one back in the first period but Canada went up by two, again, in the second, and added one more for good luck in the third. The Canadians out-hustled, out-worked, out-shot, and out-scored a high-flying Russian team.
    But I had to miss the end of the game to go to physiotherapy.
    No worries: TSN would have it online.
    I wanted to watch the whole game again.
    -- A couple of times.
    It was a good one.

Boon Jenner got to wear the cape for Canada but RNH was awesome, collecting three assists, and I must admit that Malcolm Subban was great. He handled two dozen difficult shots and made it look easy. Canada had 46 shots on net.
    Seventeen-year-old sensation **Jonathan Drouin**, bumped from the second line to the first, scored Canada's lone even-strength goal in the second period on a wraparound. **Dougie Hamilton, Mark Scheifele and Jonathan Drouin scored for Canada with captain Ryan Nugent-Hopkins assisting on all three goals. Jonathan Huberdeau added an empty-net goal.** RNH had a tournament-leading 11 points in the first round of the tournament.
    The final score of the game was 4 to 1 for Canada.

Next, Canada would play the winner of the U.S.-Czech quarter-final. Sweden finished first in their division and would play the winner of the Russia-Switzerland quarter-final.

So, it would probably be Canada versus the U.S.A., again, and then Canada versus Russia, again, for the gold, with Canada winning.

While the U.S.A. and Russia looked like they were running out of gas, This Team Canada was turning into a juggernaut, getting better with every game.

At physio, I did ten minutes on the treadmill and then the physiotherapist checked me out, to see if there was any improvement or backsliding since the last appointment, before Christmas.

She tested my leg, foot, and toe strength, which was all good, and then tested my balance, which was not so good. She got me to do some exercises to improve balance.

She said we were still aiming at a return to work at the end of January.

Back presses should be your best friend, she said, and it would be good to get back into the pool, when you get over your cold.

She also suggested I go to the doc about the cold and get checked out for strep throat as she heard it was going around. Although I nodded my head, I had no intention of going to the doc or to emerg; all I wanted to do was go back to bed. -- With the hockey game on.

Wednesday at 1 w Dr. Birkill.

Pope Benedict XVI marked the end of a difficult year Monday by saying that despite all the death and injustice in the world, goodness prevails. He celebrated New Year's Eve with a vespers service in St. Peter's Basilica to give thanks for 2012 and look ahead to 2013. He appeared tired during the service and used a cane afterward — an indication that the busy Christmas season may be taking a toll on the 85-year-old.

In his homily, Pope Benedict said it's tough to remember that goodness prevails when bad news — death, violence, and injustice — "makes more noise than good." He said taking time to meditate in prolonged reflection and prayer can help "find healing from the inevitable wounds of daily life."

Tibetan spiritual leader His Holiness the Dalai Lama has said the recent incident of a brutal gang rape of a young girl in New Delhi is "very, very sad" and expressed his concern over the degeneration of moral values in the society. The Dalai Lama was speaking to a major Indian news channel NDTV in the Indian capital. "Firstly, such incidents are really very, very sad," the 77-year-old Tibetan spiritual leader said. "India is a huge country with a long history and a very, very civilized cultural heritage of non-violence - ahimsa."

The Dalai Lama noted that although material development with modern education is helpful and necessary, he urged concerned people to pay more serious attention in developing India's own ancient traditional values.

Reading the latest about the Pope, the Dalai Lama, and the news across Canada, I had an odd thought: Say China took over Canada .... What would they do when the First Nations protesters blockaded the railway or refused to have an oil pipeline built on their territory so China could get oil from the tar sands?

From teeming Times Square to an Asian capital hosting its first public New Year's Eve countdown in decades, the world looked to the start of 2013 with hope for renewal after a year of economic turmoil, searing violence and natural disasters. Fireworks, concerts and celebrations unfolded around the globe to ring in the new year and, for some, to wring out the old. lavish fireworks displays lit up skylines in Sydney, Hong Kong, Shanghai, and in the United Arab Emirates city of Dubai where multi-coloured fireworks danced up and down the world's tallest building, the Burj Khalifa. In Russia, spectators filled Moscow's iconic Red Square as fireworks exploded near the Kremlin.

About 90,000 people gathered in a large field in Yangon, Myanmar, for their first chance to do what much of the world does every Dec. 31 – watch a countdown. The reformist government that took office in 2011 in the country, long under military rule, threw its first public New Year's celebration in decades.

London was dry and clear, for a change, and the familiar chimes of the clock inside the Big Ben tower counted down the final seconds of 2012 and a dazzling display of fireworks lit the skies above Parliament Square. People cheered as the landmarks were bathed in the light of the display, which included streamers shot out of the London Eye wheel and blazing rockets launched from the banks of the River Thames.

Across Canada, there were outdoor festivities with music at Saint John's Market Square, Montreal's Place Jacques-Cartier, Toronto's downtown Nathan Phillips Square outside city hall, Queen Victoria Park in Niagara Falls, and Vancouver's Robson Square.

In Cold Lake, there was a thing on the base -- CFB Cold Lake -- but I was too sick to go.

It was the quietest New Year's ever, for me, spent playing Scrabble while listening to As It Happens on CBC Radio.

The controversial and ineffective Kyoto Protocol's first stage came to an end today, leaving the world with 58 per cent more greenhouse gases than in 1990, as opposed to the five per cent reduction its signatories sought.

From the beginning, the treaty that was adopted in 1997 in Kyoto, Japan, was problematic. Opponents denied the science of climate change and claimed the treaty was a socialist plot. Environmentalists decried the lack of ambition in Kyoto and warned of dire consequences for future generations.

I played Scrabble, chatted online, checked e-mail and Facebook, got a bunch of New Year's messages, and watched a movie called New Year's Eve.

It was the quietest New Year's Eve ever, for me, and I didn't feel great, because of the cold and cough, but the good news was that the physiotherapist believed I was on track to return to work at the end of January. (Woo-hoo!)

Recovery Day 94: Happy New Year

New Year's Day 2013 started with surprising weather: the temperature shot up to zero in Cold Lake and Edmonton. It was fifteen below in Muskoka and twenty above in Adelaide, Australia.

At six thirty in the morning, CBC Edmonton had an interview with an astrologer who predicted a year of wild weather, natural disasters, the discover of new bugs, widespread diseases, and so on. Itr was the year of the snake, they said. -- The water snake.

The U.N. said 2013 was the year of water cooperation and also the International *Year of the* Quinoa".

The Toronto Blue Jays made major deals and people were saying this was their year. They traded future prospects and paid a lot of money to add enough players to turn them into a serious contender for the World Series, once again. The hard thing for the Blue Jays was to win their own division by beating the Yankees and the Tigers. On paper, it looked as though Toronto could take on New York and Detroit.

The NHL season was still dead, because the players were locked out.

In Edmonton, it was the year of the Fab Five, as the number one draft pick joined the Fab Four, and the Oilers were all set to turn into winners for years and years.

As for me, well, I was happy to have survived 2012. My back, legs, and feet were recovering from severe nerve damage. My cold and cough symptoms were slowly fading away.

The fiscal cliff came and went, in the U.S.A., and it wasn't the end of the world.

The anniversary of the War of 1812 continued as that war heated up in 1813. Laura Secord and the British Empire won the war with the U.S.A. and Canada was, in a sense, born.

Prime Minister Stephen Harper ended the year by listing his government's achievements in 2012, but the Opposition is pointed to several low-lights. In a statement, Harper said Canada entered 2013 with some of the strongest economic growth among the Group of Seven richest nations. And he said the world has taken note, with Forbes magazine ranking Canada No. 1 in its annual review of the best countries for business.

The Opposition New Democrats said Harper's decision to allow the $15-billion takeover of oil giant Nexen by a Chinese state-owned company was the worst Tory low-light from 2012.

The prime minister says his government helped Canadians by aggressively pursuing trade and investment agreements and streamlining the review process for major economic projects.

Critics have called that code for gutting environmental protections.

Playboy honcho Hugh Hefner tied the knot on New Year's Eve, as expected. Hef, 86, and Crystal Harris, 26, got married at the Playboy mansion.

My Scrabble score went up to 1666. -- That did not sound like a good number!

In the morning, I got my computer to read my new novel to me, again, and I had a strong desire to add a couple of chapters, but I didn't think my brain would cooperate. In the middle of the morning, I had a coughing fit that I thought would wake up my neighbours across the street, it was so violent. It left me feeling dazed and confused.

CBC News told me that I was one of close to four thousand people in Canada suffering from flu symptoms, which was a huge number, compared to previous years. Also, influenza was showing up a lot earlier in the year than usual.

Team Canada, Junior, had the day off and CBC said they were just hangin' out at their hotel. There were some group events planned for them but they were having fun just chillin' together.

Canada scored the most goals in the preliminary round (21) and tied with Sweden for the fewest against (8). Canadians were four of the tournament's top five scorers after pool play. Nugent-Hopkins led the tournament in points with three goals and eight assists in four games. Jonathan Huberdeau has seven points and Mark Scheifele and Ryan Strome led Canada in goals with four apiece.

At noon, the temperature was zero, so I headed out, and while I was walking the sun started shining. After weeks of minus twenty weather, it felt almost tropical.

When I drove up north to look at the lake, I saw several ice fishing huts and a few trucks on the ice.

The price of gas dropped down to 98 cents, which was something I hadn't seen for years.

The only store that was open was Walmart, so I went in to walk around and pick up a few things. I ran into somebody at work and we had a nice little chat about school and when I'd be back. "Groundhog Day sounds like a good date to aim for," I said.

When I got home, I hit the sack and slept like a groundhog for a couple of hours, but then I got up and felt like writing.

I added a chapter to Cold Lake Stories with a seven page piece set in World War Two. The main character falls in love and gets married just before going overseas for the war, has a good time working on radar and then liberating The Netherlands, including a concentration camp, returns a war hero, meets his son, reunites with his wife, says goodbye, and dies.

When the computer read it to me, I decided I liked it.

In the evening, I exchanged e-cards with my sister, on her birthday, turning 60, I chatted online with Edmonton, Ontario, and Alabama, played a little Scrabble, watched an odd movie called Hannah, set in Europe, about a little girl trained to be an assassin, who kills her father, and several other men, plus her mother's double. The girl reminded me of someone I dated when I was 19, physically and otherwise. But that girl got attacked by a grizzly bear, years later, and did not win that battle.

Recovery Day 95: LLB LTD

The United States and Canada would battle for a spot in the 2013 World Junior Hockey Championship final after USA demolished Czech Republic 7-0 to earn a place in the semifinal and a rematch against their rivals, Thursday at 1:30 a.m. MT.

My Scrabble score went up to 1670.

My cough was still bothering me. Although I coughed less often, it was still rather violent, and left me feeling dizzy, shaky, and hallucinating, for a few seconds.

After getting up at 4:30 and writing for an hour, I decided to go back to bed, with the computer reading Cold Lake Stories to me as I fell asleep again.

The book feels finished but it's just over 50 pages and 23,000 words. -- That's about 66 pages: a very short novel!

Canada did not win the big game over the U.S.A. In fact, they got rocked. The final score was 5 to 1 and Canada's lone goal was scored about a year after the whistle. The American's came out flying and the Canadians came out flat. The Yanks looked like they were still playing their last game, a 7 to 0 victory over the Czech Republic, and the Canucks looked like they were still enjoying their day off. -- Without any of the joy or enjoyment.

The Ruskies lost their game, too, by a score of 3 to 2, to Sweden.

So, it would the U.S.A. versus Sweden for gold and silver, a game nobody wanted to see, and Canada versus Russia for the bronze.

What happened to Canada?

After four goals, they pulled Subban and put in Binnington. But it wasn't Subban's fault. The team in front of him did nothing for him. Binnington did look better, but it was too late, by the time he finally saw some action.

At physiotherapy, I got a good report from the physiotherapist. She said I was walking a lot better, my balance was improved, my back was more mobile, and my butt was looking good. She said it was no longer tucked under and it looked muscular.

At the end of my hour-long routine, I told her I felt like doing more, so she put me back on the treadmill, and I did another ten minutes.

As I walked home, Edmonton phoned, on her way to work. And she was quite flirty, I thought,

My cold was driving me to distraction. In the night I had a few violent coughing fits that left me feeling dizzy. I coughed a couple of times at physiotherapy. So, even though I felt like going out for steak and eggs to celebrate a good physio session and report, I decided it would be better to rest in bed, drink plenty of fluids, and blah blah blah.

Recovery Day 96:

Even though I still had a cough, I woke up with the alarm feeling as though I had some energy. I was tired of waking up tired. I was sick of being sick. But this morning, I felt like getting up and going for a walk. Or something.

Outside, it was quiet and only ten below zero. So, it would be a good time to go for a walk.

Friday morning. Alberta teachers and students were still on their Christmas Break. Ontario teachers were in the news because their Minister of Education imposed a contract on them so they were heading back to work ready for more political action. -- I felt for them. That is so hard. The job itself is demanding enough. Trying to educate the public while fighting the government on top of teaching is way too much.

Yesterday, after getting a good report in physiotherapy, and doing well at physio, I felt optimistic for the rest of the day.

The Canadian junior hockey team did not play at 1:30 a.m. Mountain Time this morning. They would play on Saturday morning. Would they beat the Russians and win bronze? Did anybody care? Their coach was quoted all over the place saying it is no longer Canada's right to win gold.

Sweden dominated the Russians and they were the defending champions who would be the hosts of the tournament next year. It looked as though they would three-pete.

One cranky Canadian said it was time to treat our national junior team as kids and hockey players rather than stars. -- Did they really need a five star chef to travel with them plus a team of relaxation specialists?

They were way too relaxed in the big game they lost to the Americans. And the Americans were the opposite of relaxed. They played as though they were desperate and anxious to win.

They played Canadian.

Before I headed out for a walk, I added two or three thousand words to my new novel, Cold Lake Stories, and by then it was 11:00 in the morning.

Writing is like that, sometimes. Time flies and you forget to eat breakfast but you don't notice until it's time for lunch.

The Ice won in Saddle Lake by a score of 5-3.

Recovery Day 98: Missing

My entries for the past two days are missing! Nothing happened on the weekend. My cough was still bothering me so I tried the old cure: rest in bed, drink plenty of fluids, and take plenty of cough syrup.

This morning, Monday, was back-to-school day for teachers and kids across Canada and the U.S.A. and the Western World, but I got to sleep in.

But I didn't sleep in for long as I had physiotherapy at 9:00 a.m.

The NHL Lock-Out is over. Canadian hockey fans are expected to get right in there as though nothing happened and they are starved for hockey action. They are not so sure about American hockey fans.

In physio this morning, I did the usual exercises without any trouble, walked a bit faster on the treadmill, 3.1 instead of 2.5, and got a new routine or two to do on the exercise ball: hold a small ball, about five pounds, in both hands, and move it from side to side, with arms extended, and with the ball in one hand or the other reach down and up on both sides as though reaching for something in a drawer at a desk.

No problem.

When I told the physiotherapist I felt as though my range should be greater, she tested me, and said it was good.

She told me she planned to go ahead with the four hour assessment on Wednesday, since I suggested paying for it in advance, so the benefits people can pay for it later, and she gave me a heads up about what we would be doing. I'll take you to your limits, she said, so don't plan to do anything afterward or for the next two or three days, because you could be sore all over. We'll have you lifting weights, sitting, standing, stretching, turning, and we'll increase the weights or the resistance until you can't do it.

She said I want you to keep going even if it hurts, to see if you can manage to do things with pain.

And she asked me to brainstorm the five areas that could be a challenge at work. Hmmmmm.

Sitting is no problem. Walking more than twenty minutes could be a problem. Standing up and sitting down is no problem. Standing for more than half an hour could be a problem. Lifting the box of weights, thirty pounds, from waist level to shoulder level, is a bit of a challenge. More weight than that would be a real challenge.

How about running? How about jumping?

So that's my list of five: running, jumping, walking for half an hour, standing for an hour, and lifting weights.

And what about balance?

Today I tried standing on one foot and then the other. I could do it for about thirty seconds, which is up from zero, but .....

After physio, I walked home, had a shower, put the laundry in, walked to the drug store to mail a letter, walked home, took a load of laundry in, moved a load of laundry, went for a drive to the lake for inspiration, went home, returned several e-mails, checked Facebook, walked to the drug store to mail another letter, walked home, and walked to the coffee place down the street, but didn't get anything, and walked home again. My goal is to keep going for four hours or more, to get ready for the physio assessment.

My Facebook status is about the price of gas, which is 96.9 in Cold Lake and 119.9 in Muskoka. A Facebook friend in Quebec said it was 139.9 where she was.

In the evening, I walked five blocks to a restaurant, and back again, then went to the grocery store, walked around it five times, and stood outside talking on the phone for over half an hour.

In the evening I did a bit more writing.

After two and a half weeks with a cough and cold, it felt good to have a fairly full and active day.

Day 99:

When I woke up this morning, I was thinking about something my cousin said, in an e-mail, the other day: You have endured so much over the past five months.

When I read that, all I could think of was the fact that so many people have to endure so much more for so much longer.

And I kept laughing, even when I couldn't walk, because I believed it was only temporary. -- I had a lot of faith.

This morning, the physiotherapist was at work an hour early, she told me, because she was thinking about my case. Yesterday she talked to the benefits people and she said she came away with the feeling we might not be able to do the assessment the next day, as planned, as they had to give their approval, and that wasn't going to happen right away.

However, while I was on the treadmill, she returned a call to the benefits people and had a discussion that turned it all around.

We were back "on" for the next day.

What did she say.

She made the point that we were trying hard to get me back to work and off LTD.

So, she changed up my routine a bit.

She got me to do some basketball moves, starting with the basic stance used in many sports, then doing a side-to-side shuffle.

We played catch with a small medicine ball.

It was fun.

Sure, she said, but just wait until you see how you feel afterward.

(I feel fine!)

She phoned me when I got home and asked me to write up my version of what happened to me when, with calendar dates, in regards to numbness, function, sleep, and so on. I asked her if she wanted a novel or a short story. She said, One page, bullet points.

Here goes ....

August 29, 2012
Back to work, the week before classes, putting locks on lockers, I felt a sudden and severe pain in the back that doubled me over and dropped me to the floor. So I went to a chiropractor. After an adjustment, the back pain went away, but I had trouble walking.
September 3 to 7
At work, walked awkwardly, leg muscles contracting and spazzing. Went to emergency and a doctor gave me anti-inflammatories. By the end of the week, I was falling down and could only walk with 2 canes.

September 10

Went to emergency again. Was given a doctor's note to be off work. Could no longer walk with canes, so I got a walker.

September 14 and 15

Could not walk at all. Crawled. Feet and legs numb, no feeling, no strength.

The first doctor I saw prescribed a pain killer, an anti-inflammatory, and ice packs for my back. He also suggested taking a hot bath and drinking whiskey.

He said I would be doing the tango in three to five days.

He said the chiropractor probably hit my sciatic nerve and after some bed rest I'd be back in the game.

He said my leg muscles were alternately contracting and spazzing, and that's what made me walk like I was swimming.

The second doctor prescribed a strong pain killer, anti-inflammatories, a hot pack for my back, plus anti-depressants, and wrote me a note so I could take seven days off work, and recommended physiotherapy at a place downtown.

She said I should stay flat on my back, except for going to physiotherapy, and I would be able to go back to work after some serious bed rest.

She sent me down the hall for x-rays and reported that I had one disc that was thinning slightly, but it was above the small of my back, and was not likely the cause of my pain or inability to walk.

The third doctor said, There's something going on down there, but I don't know what it is. He got a lab technician to take some blood, sent it to a neurologist in Edmonton, and said the neurologist would contact me sometime in the future, and then I would have to go to Edmonton for an appointment.

Doc #4 said, The neurologist is the expert, but judging by the way you walk, plus your x-ray, I'd say it looks like DDD, degenerative disc disease.

September 22: Day 1 of Recovery, I call it, as I stopped getting worse and started feeling better.

September 29

After a lot of rest, I can walk a few steps. After walking or just standing for five minutes, I feel wobbly, tired, and feel as though I am going to fall over.

My feet have that needles and pins feeling.

Yesterday, I couldn't move my big toe, on either foot, by itself, but today I can

October 3

Back pain starts

October 5: Neurology Specialist, Emergency, Edmonton Clinic

Feet, calves, completely numb, no muscle strength or reflexes. Knees very week. Zero balance. Could stand up for a few minutes. Walking with walker.

Doctor ruled out the big things: cancer, ALS, MS, AIDS, syphilis, ordered more tests, EMG and NCS in Edmonton, lumbar puncture in Cold Lake, physiotherapy to work on balance, and said activate std then ltd

October 6

Walked 1000 steps with walker

October 8: Some back pain, stiff legs.

October 12: Back pain, legs stiff, sore.

October 13: Lumbar Puncture, Cold Lake Hospital

Prepped, ready for spinal tap, but the doctor talked me out of it, said I didn't need it or the EMG, as I was walking and recovering.

Walked 2000 steps with walker, 20 steps on my own.

Walked more every day: 100 steps, 500 steps, 1000, 1,500, 2,000

October 16: 4000 steps

October 17: Physiotherapy in Bonnyville

Treadmill 10 minutes was my limit, balance 0, a few leg raises and I fell on the floor. Traction felt good.

October 18: Physiotherapy, same, fell on the floor.

October 20: Worked out for 2 hours, felt shaky, slept the rest of the day.

October 22: Physiotherapy #3: treadmill 10 mins., 0 balance, could not lift my legs when on all fours.

October 23: My legs look and feel muscular, again, for a change.

October 25: Physiotherapy was very painful. Could not do back bridges, lift legs with 10 pound weights, could not do resistance exercises. Exhausted, stiff, sore. My toes feel numb, my feet have that pins and needles feeling, which is not good.

October 26: Physio in Bonnyville, doctor in LLB, dropped in to school to see volleyball game, stood for half an hour.

October 27: Walking LESS. 1,500 steps.

October 29: My feet felt cold -- first sensation in my feet in 60 days.

October 30: New physiotherapist. Changed the way I breathe, sleeping position, walk.

October 31: Sore glutes or butt starts here and continues for about two months.

November 1: Returned the walker to Primary Care Network. Physiotherapy exercises hurt, so we backed off. Nerve stretching ex was also too painful. Numbness in feet faded away, pins and needles in balls of feet. -- Slept well, for a change. Sore glutes.

November 2: Walking caused pain in my legs, at the back of my thighs.

November 4: Walked up and down eight stairs, was in a lot of pain for the rest of the day.

November 6: Proprioceptivity is off. I don't know where my body is, in space, and I feel out of touch with my back, legs, feet.

November 8: After acupuncture, I felt okay, but sleepy. Was sore and sleepy for the next two days. Sitting on a hard chair was painful.

November 9: Glute and back pain stopped but came back while showering. Feet feel sprained. I spent two days in bed, resting, with back pain, butt pain, leg pain.

November 11: After physic and acupuncture, back bends went better than before. Got coaching on how to walk, heel to toe, with a longer stride, and it went well. After walking through grocery store, leg pain returned.

Walked up and down 24 stairs at the arena and around the top and in and out. Went to a hockey game that night, for a little while.

Three or four days with no pain.

Started normalizing my sleep patterns, instead of sleeping a lot during the day and being awake at night.

November 13: Set-back. Relapse. Back, butt, leg pain returns. Walking stiffly, again.

November 17: Water walking in pool went well.

November 20: Swimming the second time was painful. Walking afterwards was difficult.

November 21: Swimming the third time went well. Walking afterward was okay, for a while, and then quite painful. Very stiff and sore, especially leg muscles. Physiotherapy went well but after acupuncture I was exhausted. Lots of pain. Recovery is up and down, not in a straight line. Feel I'm at about 50% of my usual strength, flexibility, stamina, etc. But balance is still very bad. Was able to do some writing, sitting for an hour, thinking creatively, for a change.

November 22: Physiotherapy was painful. Back extensions -- very painful.

November 23: Physiotherapy went well.

November 24: Lots of pain. Used a chair as a walker.

November 25: Lots of pain in the butt. Could barely get out of bed or walk a few steps. Exhausted.

November 26: Pain extends down my legs.

November 27: In the pool for half an hour only. Walked a little bit but my back hurt, knees felt week, back was sore. Lots of pain in my right leg.

November 28: Pain lessened. Walked on toes for the first time. Balanced on one foot for half a minute.

November 29: Physio went well. Treadmill at 3 instead of 2.5. Did all the exercises bigger and better. The pool was good, too.

November 30: Physio went well, was told I could swim every day, not just the days I didn't have physio.

Pain in feet and in sciatic notch but not back, butt, legs.

December: Pain comes and goes. Sciatic notch usually hurts. Swimming and physio goes well for a week.

December 10: Pain returns. Intense. Very tired. Acupuncture woke me up.

December 11: Pain-free sleep but intense pain in the morning. Sciatic notch.

December 12: Feeling better. Added exercises with weights in physiotherapy. Start 2 hours of exercises per day.

December 13: A little back pain in the evening, a pain in the butt in the morning, back pain on the treadmill at physio, and a good deal of pain like a tight band across the bottom half of my butt while trying a new exercise: That was the latest. Two new exercises: Push-ups against the wall were easy and squatting to lift a weight and pick it up was okay until the fifteenth repetition. That pain went away with some back stretches.

December 15: After physic and acupuncture, slept for hours. Sore feet, still legs.

December 16: Numbness in toes fades away. Pain comes and goes.

December 17: I moved furniture.

December 18: Ran a few steps.

December 19: More weight lifting. Stop acupuncture. Pain from walking fades with sitting.

December 20: Relapse. Intense pain in butt.

December 12: A week without pain and with some mental energy.

December 13: Physio went very well but walking afterward was painful.

December 14: Standing and talking for one hour at Christmas event exhausted me.

December 15: Lifted turkey, chair, presents, drove to LLB and back, no problem.

December 17: At physio, lifting 30 pounds from waist to shoulder level, too painful. Otherwise, good.

December 18: Physio went well. In the evening, stiff and sore. Morning, better.

December 21: Caught a cold, started coughing. Lasted for 2.5 weeks. Lots of time in bed. No back or leg pain, just foot pain.

December 28: Physio was okay. A bit stronger. Balance still bad.

January 7: After lots of time in bed, physic went okay. Had an active day: walked 20 blocks, wrote for hours.

January 8: Added squat exercises and ball catching at physic.

That was way too long!

Summary:

August 29:

Intense back pain. Went to chiropractor. Back pain gone. Walking awkwardly.

September:

Muscle spasms and contractions for a week at work, walking awkwardly.

Off work, walking with 2 canes for a week.

Walking with a walker for a week.

Not walking at all for 3 days.

Walking with a walker for another week.

Walking with canes for another week.

Feet and legs, numb, no muscle strength, no reflexes, and balance was zero.

Knee strength started to come back.

So, I went downhill for three and a half weeks and then started to recover.

October

Walking 20 steps, then 100, 50, 1,000, 2,000, 4,000

Started physiotherapy in Bonnyville and had a relapse. Fell a few times.

No balance, exhausted after 10 minutes on treadmill, leg raises very difficult, foot muscles very weak.

Walking less.
Sensation returned to feet October 29.

November.
Physiotherapy in Cold Lake, returned walker to Primary Care Network.
Gentler exercises. Slow improvement.
Pain comes and goes. New pain in legs and butt.
Pain returns with walking, standing, sitting. Back extensions are impossible. Balance is bad.
November 25: Lots of pain in butt, extends down legs.
November 28: Pain lessens, balance starts to return, exercises increase, water walking in pool begins.
November 30: Pain in sciatic notch comes and goes. In the pool more.

December
Pain comes and goes.
Intense pain in the morning. Back pain in the evening. Tired. Sleeping long hours.
December 15: A good week, little pain, walking more, ran a few steps, lifting 30 pounds.
December 20: Relapse. Intense pain in butt.
A good week. A bit stronger. Balance still bad.
Caught a cold, spent a lot of time in bed. Back pain, leg pain, fade away. Sore feet.

January
Cold fades, walked 20 blocks, wrote for hours, no pain in back, butt, or legs. Sore feet.
Added exercises in physiotherapy.

hey there janisistar. long time no e. how's it going?
after my physio cleared me to ski on sunday, i decided to take a trip to edmonton, get some skis, stay in a hotel, see zhou-zhou, and have a little adventure. and it was a blast. except seeing zz. that was depressing! oh well.
i talked to bill a few times while i was at mountain equipment coop, getting the cross country ski package, on sale, and he sounded good. linda has new skis but they have no snow and it was 10 above, so they were going skating in the arena, instead.
i'm looking forward to trying that!
the drive down was spectacular with great scenery and a fabulous sunset for 2.5 hours i was on the road. i found a great little green hotel on whyte avenue, the cool strip by the university, with free valet service, and a teacher discount, so i spent 2 nights, instead of the 1 i planned.
took zhou zhou to lunch and then had diarrhea! found a fantastic pizza place. bought a cool hat. hung out at chapters, got some books, including cloud atlas and the new alice munro.

had bacon at breakfast, two mornings! bacon! that's a first in about three and a half decades!

i got tired while i was there, so spent lots of time in bed, trying to find something on tv. good ol' cbc. and juste pour rire.

the drive home was uneventful, which was what i was up for.

it's 20 below, too cold for skiing, today, sunday, but monday is supposed to be above zero, and the ski place won't be so crowded.

i want to try downhill on the small hill with the t-bar, maybe half an hour.

the only part of my physical test i didn't nail was balance, and skiing is a good way to build up the muscles needed for that.

i looked at the cross country trails here and they look easy so that should be okay. - and it will be great to get into the forest again.

i love edmonton, had a fantastic hotel, got great skis at a cool place on sale, found the best hat and a great pizza place, so i should be happy! but i'm pooped. and ZZ was draining, emotionally, so ... think i'll have a nap in my own fabulous bed!

it was good to get out of town and it feels good to come home, so .... aren't i well-adjusted! i slipped on the ice in the city but caught myself, so my legs hurt for a bit, but i sat down and ate pizza and it was all good again. but by saturday night, i was a little sore, so I took it easy until it was time to head home the next morning.

Solid effort from the ICE as they pounded the Rangers 8-1! Dallas Ansell scored a Hat Trick and Christian Nypower and Niko Bourget answered with a pair of goals and Dylan Sharun scored his first of the season in his 2nd game with the ICE. Next up home game Wednesday vs St. Paul.

today i went to the ski hill, even though it was closed, just to case the joint, get my nerve up, and all that. it's closed mon and tue but wednesday is supposed to be warmer, so that's my goal. downhill skiing on wednesday! hey, if that goes well, i'll start dreaming about skiing in jasper and banff!

woke up with a bit of butt pain so i went back to bed and slept until the noon hour swim, which went well. but ... i gots de blues!

so, i drove over to saskatchewan and i saw a moose!

then i took myself out for a steak. first i walked the breakwater .... a bit of pain, but it went away when i sat down to eat steak!

might go to LLB tomorrow to see the cousins, and take a feast from the Wok Box. they don't have a restaurant like that, over there.

frankly, it's better than anything in edmonton, never mind llb. this plan depends on weather, as it's supposed to go up to zero,

which is good, but if it's ice or slippery, 4get it!

ttys
jtm
m

Scrabble: 1681, with 253 bingos.

Recovery Day X: I've Lost Count

Scrabble: 1684.
This morning I didn't feel like swimming so I went back to bed but I dragged myself to the pool for an hour at noon. Out of sorts, the opposite of ecstatic, a little sore, I didn't feel like doing much. But swimming helped.

For a full hour, I was active in the pool, walking and swimming and pretending I was skiing.

Cold Lake set a record with a high of six degrees. The previous high was 5 and the previous low, a long time ago, was minus 45.

Is it the lingering cough that is bothering me, the odd little reminder of sciatic pain in that pelvic notch, or is it something else?

My energy is down.

Maybe it's the overcast skies or the change in temperature.

Whatever it it, I hope it goes away soon.

Tomorrow I have physiotherapy again, after five days off, after the big assessment.

Maybe it's withdrawal I'm feeling.

I miss reflexology, too.

My novel is coming along, slow but sure.

Somebody at work called to invite me to dinner yesterday and she complained about being fat, living to eat instead of eating to live, and having a huge waistline, and then she said, You and me both.

One little insult at a low moment can really bring you down.

I wanted to drive to LLB with a feast from the Wok Box today but CBC weather and news had warnings or pleas, saying please stay off the highways if you can as they are slippery.

Cold Lake snow is melting but people are worried about the meltwater freezing and making driving conditions deadly again.

All I want to do is crawl back into bed.

Recovery Day 101: Return Of The Physio

Up a little early, I wrote for an hour before going to physio. And it went well.

At physio, for the first time in five days, I told Angela I had a couple of days with very little energy, after driving to Edmonton and back, and felt disappointed, as I felt the assessment day was a turning point.

She said it really was a turning point and it was very good that I didn't have a relapse afterwards but she was a little surprised about the low energy.

I told her I woke up with the intention of going to the early bird swim but didn't make it, so she said to make it. Fight the lethargy and get in there. If you were working, she said, you would have to get up and get going in the morning, so get into the habit.

Frankly, for the past couple of days, I had to drag myself into the pool at noon, never mind early in the morning.

She gave me some new exercises to try. One was walking on a balance beam just a couple of inches off the ground. That was a challenge, but it got easier as I went along. Walking the plank side to side was easier. Doing it backwards was a real challenge.

The lower to the ground I got, the better it went. Also, just holding onto the physiotherapist's hand, even a finger, made it a lot easier.

Proprieceptivity, she said. Your toes, feet, and legs aren't as sensitive as your fingertips. You get a lot more feedback from your fingertips than from the lower half of your body.

Another new exercise involved lifting a box of weights and walking the length of the room, and back. When I did that with 35 pounds in the box, my legs hurt. So I did it with 20 pounds in the box, and that went better, but it still hurt my legs a bit. So I stretched like a cat on all fours and tried it again. And that time, it was no problem.

She also got me to do some basketball exercises with a small, weighty ball, going through the motions of shooting a basket and of throwing a chest pass.

That was easy.

I confessed that I had not been skiing, yet, and she said, You can try a little cross-country but don't even think about downhill skiing.

-- I WAS thinking about downhill skiing!

She said she was going to modify my report, following my assessment, with today's feedback, and recommend two hours of physio with a partial return to work.

-- I don't like that idea.

I'd rather go back to work full-time when I go back.

Why?

I don't know. -- It's just a feeling I have.

Sometime after physio but before swimming, my spirits started to lift. And swimming lifted my spirits more and more. After an hour in the pool, I drove over to the cross-country skiing place on the Base, to check it out, and it looked flatter than the trails at Cold

Lake Provincial Park, as they were on a golf course. So, that was the place for me to ski, to begin with.

Would I go on this particular day?

The conditions were good: fresh powder snow on a solid base with a set track that's been there since Hallowe'en.

The sky is cloudy but the temperature is only minus 7.

Lethargy does not describe me. The physiotherapist said I had to fight the lethargy. But lethargy means a state of sluggishness, inactivity, and apathy, or a state of unconsciousness resembling deep sleep.

 is synonymous with **lassitude, torpor, torpidity, stupor, languor ....** Those nouns refer to a deficiency in mental and physical alertness and activity. *Lethargy* is a state of sluggishness, drowsy dullness or apathy: *The war roused the nation from its lethargy.*

Lassitude implies weariness or diminished energy such as might result from physical or mental strain: *"His anger had evaporated; he felt nothing but utter lassitude"* (John Galsworthy).

Torpor and *torpidity* suggest the suspension of activity characteristic of an animal in hibernation: *"My calmness was the torpor of despair. Nothing could dispel the torpidity of the indifferent audience.*

Stupor is often produced by the effects of alcohol or narcotics; it suggests a benumbed or dazed state of mind: *"The huge height of the buildings . . . the hubbub and endless stir . . . struck me into a kind of stupor of surprise"* (Robert Louis Stevenson).

Languor is the indolence typical of one who is satiated by a life of luxury or pleasure: *After the banquet, I was overcome by languor.*

A better word to describe me, I would say, is exhausted.

Exhaustion means extreme tiredness; fatigue; the condition of being used up.

After the four hour assessment with the physiotherapist, followed by a busy day, on Thursday, with swimming for an hour and half followed by driving to Edmonton on Friday, with lots of driving and walking in Edmonton on Saturday, then driving from Edmonton to Cold Lake on Sunday, I wasn't exhausted, but I was pooped.

On Monday and Tuesday, I did a one hour workout in the pool, each day. On Monday, I drove over to Saskatchewan, checked out Kinosoo Ridge skiing, and got some groceries. On both days, I got a good deal of writing done.

So, don't call me lethargic. Call me tired, fatigued, used up.

And another thing: I was fighting a cold!

Come on! Gimme a break!

Highway 55, the road to Lac La Biche, was blockaded, today, by people from the Cold Lake Reserve, part of a national day of protests, in support of the Idle No More movement, started in Alberta, now a national phenomenon, with some whites standing alongside First

Nations, to protest an omnibus bill belonging to the government of Stephen Harper, which could be characterized, I would say, as anti-environmentalist.

White folks aren't complaining about the roads and railways blockaded by First Nations people, this time, because Harper's omnibus bill is disliked by just about everyone, except, possibly, Big Oil.

Anyway, I heard on the radio, CBC One, in the car, that the police in Lac La Biche were asking everybody to stay off the slippery roads in the area.

The Globe And Mail said: Another blockade was planned along Highway 55 near Cold Lake, a small city in eastern Alberta that's near the southern edge of the oil sands, and is home to a major Imperial Oil facility. Organizers there are pushing for Bills C-45 and C-38 to be rescinded. "We want them to get the message about what they're trying to do," said organizer Mervin Grandbois, 55.

It also was billed as a highway shutdown, but traffic is being let slowly through and the demonstration has remained peaceful, RCMP said.

The Vancouver-based Fraser Institute has issued another health-care report card that's bound to create controversy. The free enterprise-oriented think tank released its 2013 Provincial HealthCare Index. It concludes Quebec and Ontario residents get the best value for money from their public healthcare system. But if you live in Newfoundland and Labrador or in Prince Edward Island or Saskatchewan, you're not getting great bang for your medical buck, according to the Fraser Institute. The institute's ranking stacks up like this: Quebec 10, Ontario 7.43, New Brunswick 5.87, Nova Scotia 4.73, B.C. 4.12, Manitoba 3.66, Alberta 3.35, Saskatchewan 1.17, P.E.I. 0.48 and NL 0. Yes, zero.

The study used data from 2010, the most recent year available, to measure the provision of healthcare in comparison to expenditures, the report's executive summary says. It looked at 46 indicators pulled into four groups: Availability of resources; use of resources; access to resources and clinical performance of medical goods and services.

Alberta did not score at all well! However, in my personal experience, I would say it is actually a bit better than Ontario.

In the evening, to prove I'm not lethargic, I went swimming, again, for another hour. It went well. -- What can I say? I swam lengths, did a lot of water walking, did the sideways shuffle basketball drill, and did a sideways traveling step, put one foot in front and then behind the other. The music was good, the crowd was small, and a good time was had by all.

In the pool, I hatched a couple of ideas. One was to write one more story for the Cold Lake book, about 5,000 words, to get the word count up from novella to novel. The other was to cut out a lot of stuff from this book, my back book. And when I got home, I chopped it down from 222,000 words to 183,000. -- Close to 40,000 words. And I thought I could cut out more some other day. That was over 500 pages. A book about a bad back should be about half that long, I felt.

Recovery Day 105 Or So

Although I made it to the pool in time for the morning swim, the pool wasn't ready for me. There is no Early Bird Swim until the 21st. And that's alright with me!

It was good to get up and get out early in the morning, like a normal human being. After driving over to CFB Cold Lake, I drove by Ecole Voyageur, but there was nobody there, yet.

Well, actually, it looked as though one of the elementary teachers was already at work, setting up her classroom for the day.

Physiotherapy at 9:00 turned out to be a bit different. I was early and did my ten minute warm-up on the treadmill before 9 and then Angela informed me she had some unusual things planned for me: burpees and a sled.

To do a burpee, you drop to your knees, on a mat, and then do a push-up, or press-up, get to your feet, again, and then do it over and over. My first burpee was very awkward, as I haven't done anything like that for more than five months, but they got better after I did a few. After a dozen, or so, it was time for the sled.

It looked like a dog sled with no dogs and chrome parts instead of wood with flat runners instead of skis. On the sled was a box with weights in it and the whole thing weighed about 70 pounds. My job was to push it the length of the room, on a carpet, so there was lots of resistance, then turn, and push it back the other way, and repeat. After I did it a few times, the physiotherapist hopped on for a ride, so there was a little bit more weight to push but, fortunately for me, she doesn't weigh much, so it felt just about the same.

After ten trips back and forth, pushing the sled, I got out the balance beam, my nemesis, and walked forwards, then sideways, and then, finally, backwards. Because my balance was so bad, that was harder than pushing the sled or doing burpees.

I did a lot of back extensions and cat stretches on a mat on the floor, between exercises, to keep my pain under control. And that made me sweat a good deal.

The only other exercise I did was the one that hurt me the day before: carrying a box of weights back and forth, the length of the room. But it proved to be a lot easier today than yesterday.

At the end of 40 minutes, Angela said, Okay, you're done. That's a radical change and a big step up, so let's no push it any more, today.

Yesterday I did 70 minutes. Today I was happy to stop at 40 because I could feel the strain. It wasn't pain, exactly, but I could feel a lot of sensation in my lower back, my butt, and my legs. It felt as though I had just finished the first football practice of the year.

Okay, truth to tell, I never played football. It was like the first soccer practice of the year, the first hockey practice, the first lacrosse practice, or the first day of cross-country running. It hurt a bit but it felt good.

It was snowing outside and the temperature was dropping below the forecasted low, but it was not as cold as Ontario, which always makes me smile. When I got home, I checked the roads and the weather forecast for Lac La Biche, saw predictions for freezing rain, and decided not to go.

Maybe I would hit the pool a couple of times, for another 2.5 hours, after my 40 minutes of physio.

I was making progress, again.

Due to poor road conditions the game vs St.paul and your Cold Lake Ice tonight has been cancelled. S no game at the Events Centre tonight.

Day 106: Snow Day

There was no morning swim and no physiotherapy so I slept in a bit and then went skating.

At the rink at CFB Cold Lake, I skated for about fifteen minutes. And I had the rink all to myself.

It was a snow day, so I thought there might be a lot of kids there, but it was just me.

It felt like an accomplishment but it wasn't a lot of fun, like swimming, or a very good workout. But I was happy that I could do it.

After skating, I took my skates and went to my school, and there were a lot more students in the building than I thought there would be, on such a snowy day. They all came running out to see me, so that was a lot of fun.

I stood for two hours, yakking to teachers and kids.

In the evening, I joined a dozen of my colleagues for dinner at OJs restaurant. For a long time, I was the only guy there. It was a good French immersion experience. And I walked 4 blocks there and back.

So, skating for 15 minutes, standing for 2 hours, talking for 2.5 hours, and walking 8 blocks .... That was my workout for the day.

Day 106 Or Whatever

On Saturday, I felt like being lazy. Lethargic, as the physiotherapist said. But I didn't feel like fighting it. I felt like writing and I felt like watching a little hockey.

It was the first day for the new hockey season, after the lock-out, with a lot of games scheduled, including the Leafs versus Canadiens, in Montreal, on Hockey Night In Canada, complete with the celebration of an old ritual in the new Habs hockey rink, with former Montreal captains passing a live flame from one to another, from generation to generation, to the current captain, who then led his troops to defeat on national television.

The Leafs looked a bit better than the red, white, and blue, and won the game by a narrow margic, 2 to 1, after leading on the scoreboard for most of the game.

I played Scrabble, with the game on in the background, and my score jumped up to 1693.

I saw the start of the Vancouver Canucks game, too, but it looked like the Ducks were going to beat Canucks, so I lost interest.

It looked as though the Canucks and the Ducks could both clobber both Montreal and Toronto.

I liked the results of some of the games in the U.S.A. The Penguins beat the Flyers, the hawks beat the Kings, the Panthers beat the Canes, the Stars beat the Coyotes, and the Wild beat the Avalanche, but I didn't like the Senators beating the Jets, the Blues beating the Red Wings, or the Ducks beating the Canucks. I didn't care about the Devils beating the Islanders, so much, or the Lightning thumping the Capitals, or the Blue Jackets beating the Predators.

All I really wanted to see was the Oilers beating everyone with their young guns playing run and gun and good luck to the goalie and the defence.

The Leafs and Canadiens played a sloppy game and looked as though they hadn't been on the ice for the first five months of the season.

I skated badly, when I tried out my new legs at the rink on the base, but I thought I could help the Leafs or Canadiens, even though I was old enough to be the father of the younger players!

My novel was progressing nicely. I wrote as though inspired and took the word count over the 50k mark, so it was no longer a novella, or short novel, or Brautigan, like some of my other books; it was a full-length novel.

And THAT gave me a good shot of energy on Sunday morning.

Day 107: Thirty Below

It was thirty below but felt like 40 but I went swimming and it felt like the tropics. After an hour in the pool, I went to the grocery store and spent an hour and a half walking around, loading and unloading the car in the cold, and it all lifted my spirits quite a lot. Not that I was sad, or anything, but the exercise, the cold, and the sun all served to inspire me to do more.

Scrabble: 1699, with 255 bingos. I played Scrabble with the Oilers game on the radio. And it was a good one.

In their first game of the season, the Oilers played the Canucks, Canada's best team last year, and engineered a thrilling come from behind victory, falling behind 2 to 0 but battling back for a tie by the end of the game and then outscoring the Canucks 2 to 0 in the shoot-out, so it's called a 3 to 2 victory.

The young guns were in the game but not on the scoresheet as the Oilers older guns led the way.

Day 108: Twenty-five Below

Got up at 4 to plug in my car so I could go swimming at six. It's the 21st, the day the morning swim starts, so it would be good to go. With the early bird, noon, and evening swims, I could spend 3 or 4 hours in the pool. Skating is something else. Swimming is the best workout and is best for the back, so why not go with that?!

There were a lot of new things at physiotherapy. First, I did the old things. After ten minutes on the treadmill, I carried the box, pushed the sled, did arm curls, walked the plank, did the balance board, and stepped onto the #2 step. But then Angela told me she used to be an aerobics instructor and she got out a step used in aerobics classes. I did several different exercises with the step. First I just touched it with one toe and then the other, then I stepped on it and back, and then I used it width-wise, stepping on and off, and then lifting one leg backwards as though skating.

In the end, she gave me a sit-up exercise, keeping a straight back, lifting not too far or high, and finally a leg lift, one side and then the other, without using hips. And she said, You are going to hate me tomorrow.

None of the exercises felt challenging, for a change, as long as I did back extensions, standing up, as I went along.

The physiotherapist is recommending going back to work three days a week, half-time, for a week, and then five days a week, half-time, and then full bull-goose looney.

It's time to call the doc's office and set up an appointment so he can sign the papers saying 'back to work'.

Twenty physio visits plus four with the first guy equals 24 X 60 = not much considering it saved my life.

My legs were fatigued, as my physio says, after the morning session. In the afternoon, I walked to the post office and around downtown, about ten blocks, and could feel it at the backs of my thighs like a little fire. But I kept on going until I got home.

My Scrabble score: 1706

Recovery Day 109

Got up in time for swimming, tested legs, went back to bed.
Physiotherapy was "ouch".
Well, the physiotherapist said, we found your new limit. The aerobic step exercises,
yesterday, felt good at the time, but they made you sore later on.

Swimming went well. In fact, I felt better after physio and swimming. That is, I felt better
each time. I felt better walking out than walking in.

In the afternoon, I had an odd experience: I put my winter boots on, for the first time, and
walked to the post office and back. It's only about three blocks.
    Walking over was no problem but walking back got to be so painful I thought I'd
never make it. My thighs burned, at the back, and my knees started to get weak. I had to
stop every 100 feet or so to do back extension.
    I saw myself walking in a glass window and thought I was watching a 200 year old
man.
    As soon as I kicked my boots off, at home, I started to feel better, and when I sat
down to write this, I felt good to go, again.

In the evening, driving was no problem, and walking a short way was no problem, but
after going to Mark's Work Wearhouse and the then walking through No Frills, my
hamstrings were hurting. Until I sat down.

Day 110

That's a lot of days.

Swimming made me feel better. In the morning, I was a little sore, but after swimming I felt like doing more.

For well over an hour, I walked around Walmart.

No problem!

Will I swim again tonight?

Or is it too cold?!

ICE win!!! 4-3 over Killam. Dallas Ansell gets the winner on his 2nd of the night.

Day 111

Although I felt fairly good in the morning and during the treadmill warm-up at physio, I lifted 30 pounds and went for a walk with it, and that hurt. And after that, everything hurt. It was a pain in the butt.

The physiotherapist gave me traction and a pirifirmis stretch, which helped.
Ouch.

so, atfter that 4 hour eval, the physio recommended going back to work, and yesterday the doc signed the form saying GET BACK TO WORK
but it's for half time 3 days a week, for one week, and half time every day for the next week, and then full-time

i dropped in on my cousins to celebrate and took dinner for 4, just a hundred bucks, from boston pizza -- broken lasagne, pizza, garlic bread,
salads, drinks ... my brother bill is famous for dropping in at dinner time and saying, what, oh, you're just sitting down to eat? what a coincidence

so we had a lot of fun. they are crazy funny

then i told them i was driving home in the dark but i went to a bad hotel in lac la biche and had a good time watching a great hockey game,
the oilers won in the last seconds, with a goal by my new hero, ya ya yakupov, and then i had a bad sleep in a bed that was not so great,
and woke up a little sore :(

Day 112

so i drove to cold lake, 2 hours, and gave the doc form to the principal and she said, oh, okay, i'll have to see if the supply teacher will work
half days ... and then she said, can you come in next week so we can have a discussion ...
and then she said, i'll have to evaluate you as soon as
you are back fulltime

and she booked ESL for my prep periods

so, as usual, grumpy girl rained on my parade !!!

grrrrrrr !

Day 113

Low energy, no swimming, just grocery shopping, driving around, beautiful day, only ten below, blue sky, bright sun, but .... What can I say? No spark, no energy.
    Looking forward to the Battle of Alberta on Hockey Night In Canada, the Edmonton Oilers at the Saddledome to take on the Calgary Flames.
    Calgary has owned Edmonton for a long time but the Oilers won a couple last year, in the Battle of Alberta, so .... Things were changing.
    Calgary was winless and would not want to go 0 and 4. The Oilers were riding a wave of excitement after beating the defending Stanley Cup winners with a dramatic win in overtime after Yakupov tied it with four seconds left after Nugent-Hopkins got a goal that was waved off.

Calgary beat Edmonton 4 to 3.
    Sad!

Day 114
ICE win their 25th of the season with a 5-4 victory over the Vegreville Rangers. Next game this coming wednesday at home vs Saddle Lake

Swam for an hour. Sore after. Frustrated.

Day 115: Energized!
Woke up at 5, wrote a play for the junior drama group at school, went to the pool, swam for an hour, then went to school, stood and talked to kids for an hour, then did physiotherapy for 45 minutes, all good.

But then I helped a blind Chippewa guy walk down the street, four blocks, and I had to stop twice, because my butt hurt so much.

The physiotherapist watched me walk in to physio this morning, without me noticing, and she said she didn't like the way I was walking. She said, Get better boots and buy some of those ice grip things for the bottom, so you don't have to worry about sliding and falling on the ice.

The boys invited me to their basketball game at 5 and the girls at 4. And I'd love to go.

The principal liked my suggestion of getting the junior drama group going, but she said, I don't think you'll be going to the drama festival, so we can keep it an all-French experience.

That's a-okay with me! It's a tonne of work, 24 hours a day, all weekend, and there's no time off school, before or after, to get there or to recover. And there's a lot of running around while you are there.

In the afternoon, I went back to the school for about three hours and had a meeting with my supply teacher, watched a basketball game, and sat on the bench with my basketball team for their game. And we won!

So, that was a pretty big day: Swimming, an hour at work, physio, walk, three hours at work, for a total of about seven hours of activity.

Day 113

Physiotherapy was challenging but good and was followed by acupuncture, for the first time in a long time.

I came home and wrote a 3,000 word story to finish off my novel, Cold Lake Stories, and it all came to me during acupuncture.

While I was writing, the principal phoned to say I can't go back to work and she has hired the supply teacher for the days we discussed for my return, and it didn't even upset me.

A report from the benefits people is needed.

So I drove over to Staples and faxed them the doc report and sent an e-mail. They responded with an e-mail promising an update tomorrow.

So what's the big deal?

While I was out, I went to Mark's Work Wearhouse and found found a fantastic pair of winter boots and bought the anti-skid or anti-slide gripons marketed as snow tires for your feet.

And then I drove up to the Tim Horton's window but I couldn't get my window open, to order, et cetera, because it was frozen shut!

The window woman at Tim's laughed and said, You and everybody else.
So I laughed, too.

Acupuncture included a needle at the top of my head. The physiotherapist said, I had one of these while I was giving birth, so I deduced it was for pain.
Day 146

Physio went fairly well but I had a lot of sensation in my hamstrings. That's code for pain. After acupuncture, I felt good. There was less pain and I felt sharper, more awake, and had more energy.

Leaving physiotherapy, I saw a blind guy walk into a snowbank and the wall of the liquor store next door, so I offered to help him. He was happy to get some help. We walked four blocks, to his bank, and crossed the street, but I had to stop twice, to stretch my back, as my hamstrings hurt. I thought we made quite the pair: a guy who could barely walk helping a blind guy go down the street.

His name was Steve and he was several years younger than me, around fifty, but could have been a hundred. He said he got picked up by the cops, the night before, for being outside late at night, and spent the night in the drunk tank, but he wasn't drunk. He had a beer in his coat pocket, he said, and he drank it right in front of the cop shop as soon as he got out. And then he threw the can at the cop shop.

I asked him about the differences between Chipewyan and Cree and he said, There's a lot of Chipewyan around here.

At Mark's Work Wearhouse, I bought a big pair of winter boots with big treads for traction. In addition, I got the strap on snow tires for the feet.

In the afternoon, the principal called and canceled our plans for my return to work on Monday. She said the benefits group has to talk to my employer first. The benefits group sent me an e-mail saying the same thing.

Day 115

At physio, I avoided the heavy lifting, to avoid pain.
Acupuncture again.

My physiotherapist is leaving for Mexico and the benefits group has not approved her request for secondary physio, two hours a day, so we are taking a hiatus.

Day 116

Friday, February 1, the day I was supposed to go back to work, the date the doctor approved, the deadline the physiotherapist had me working toward, was a day of disappointment. Instead of going to work, I went to the pool, swam for an hour, then drove around the city aimlessly, feeling my feelings, for a while.

After driving to the lake, I decided to make the best of the situation, so I drove home and went to work on books.

Day 117

On Saturday, I felt very tired, so I slept in, but I made it to the pool for an hour.
ICE lose wheat kings win 3-2

Day 118

On Sunday, I was tired, again, so I slept in and missed swimming. -- That was a first.

At noon, I walked around Walmart until my hamstrings got a little sore, had a snack at the Wok Box, and headed home. After sitting for a few minutes, my legs felt better again. But I didn't feel like going on a walkathon, or anything.

When my legs or butt is sore, it is very tiring.

In the afternoon I wrote 15 poems and in the evening I wrote another 15, all about hockey, for a book I want to do, to be called Ya Ya Yakupov!

Although I felt tired, I couldn't sleep. Usually, I have no trouble sleeping.

I watched One Thousand Words and The Answer Man, again. And The Silver Lining Playbook.

The thrill was gone.

My computer read my new novel to me, again, and that made me happy as I fell asleep.

Day 119

Monday, a day I was supposed to be at work, the alarm went off so I could make it to the pool in time for the early bird swim, but I turned it off, rolled over, and went back to sleep for a couple of hours.

At 8, I got up to have breakfast. At 9, I felt like doing something.

I've been thinking of booking an appointment for reflexology and I've been thinking about joining a gym.

The physio suggested going to the gym at the base, since I swim there, and doing all the things I do at physio.

The Oilers lost in OT versus the Canucks, a reversal of their last game, with The Nail getting another assist on the first goal.

-- Where is Nugent-Hopkins?

120:

Tuesday morning at the pool, I watched the sun rise, while swimming, then drove up to the lake, looking for steak and eggs, but decided to go to the best restaurant in town, my place, for an egg scrambler with cheese.

Unusual roads this morning .... Snowplows left a white triangle about two feet high in the middle of the roads with a break at some intersections so it was more like a maze and I made a mistake first thing in the morning turning left to get to the swimming pool as I could not cross the street to drive on the right side so I drove on the left or wrong side for a block. Fortunately, there was nobody else on the road in my neighbourhood at that hour in the morning. It reminded me of the movie Amarcord, by Fellini, with those funny shots of sidewalks with snowbanks over your heard.

Lac La Biche roads are reported bad, again.

Tomorrow I travel to Edmonton for the teachers conference on Thursday and Friday and I'm staying Friday night so I could drive home in daylight on Saturday.

This morning, my pain levels were down. My butt and upper thighs, at the back, get sore, but I worked on them in the pool, arching my back, walking like a T-Rex, and walked normally when I got home. -- That's a big relief.

After swimming again at noon, I went to reflexology, one more time, with the gifted Mecelle, and it was out of this world fantastic, as usual.

She looked at my feet and told me all about my health and fitness.

You're building muscle, developing balance, and your proprieceptivity is better, she said.

You had a bad cold or bronchitis for a long time.

There is still pain in your back, between your shoulderblades.

Your hips are even, for the first time in a long time.

I pointed out that the hair on one leg had vanished in one area. She said that is a sign that your sensitivity to pain or your tolerance has changed. When you were in a lot of pain and could not walk, you felt no pain. Now that you are getting back to normal, your sensitivity to pain is normalizing, too.

Her next client was a no-show, so she kept going. My treatment lasted well over an hour. It was closer to two. When I walked out, I felt as though I was walking in heaven, without pain, for the first time in months.

Mecelle answered my nagging questions and she made me laugh. And those things were great, too.

I asked her about the pain in my feet, told her they felt as though they were going to explode, for most of the month of January, but had calmed down recently.

She said it was a sign of back pain and plenty of it.

I told her that I noticed something in the pool. There are thick black lines on the bottom of the pool, indicating lanes and I used those lines to practice walking in a straight line, to develop balance, forwards and backwards, and today for the first time I felt with my feet that they were not lines of paint but lines of tiles. My feet were sensitive enough to give me the feedback about a line of tiles in the pool.

Developing sensitivity is a very good thing, she said.

I told her that walking had been difficult.

Have you changed your boots? she said.

A few times, I said.

Each time you change your shoes or boots, it changes the way you walk and it affects your muscles from your feet to your legs and back and beyond. For you, in your condition, it's dramatic. When you are back to normal, you can change shoes whenever you feel like it and you won't notice. Right now, it will cause a lot of change and pain in your body.

Fascinating, I said, several times, while I was in her big comfy chair.

I worked on my new book of poetry. Thirty-eight poems I like about hockey, hockey, hockey.

121: Sleep-In Before Bon Voyage

In the morning, I got out of bed when the Early Bird swim was over.

It was a strange night. I felt as though I was going to throw up, had trouble sleeping, and I had a lot of dreams.

Drove to Edmonton, got to hotel, no problem, had dinner, and hit the sack. First I did some writing and then I watched the Oilers game. Also, I worried about making me way to and through the conference as my feet were sore and numb and my legs were stiff and I hadn't done anything like a conference for a long time.

122: Edmonton

In the morning, instead of waiting for a cab for twenty minutes, I walked for ten, to get to the Shaw conference centre.

Walking was a challenge as my hamstrings hurt and I had to sit down on a bus bench close to the Shaw.

On the way to the Shaw, I saw a gorgeous sunrise over the river valley and a dozen magpies. -- What a sight! Black and white magpies with a technicolour sunrise. Beautiful!

After two morning sessions, walking through the Shaw, sitting for hours, walking and stretching between sessions, I walked back to the hotel to get something to eat and to hit the sack for a bit as I felt very tired.

In the afternoon, I missed a session at the conference because I was sleeping but I made it back in time for the final session and then walked through the big area with booths selling things for teachers. And then I walked back to the hotel.

In all, I walked a dozen city blocks plus about the same in the conference centre.

Perhaps the pain was caused by wearing different shoes and new shoes. Instead of the big snow boots I was getting used to at home, I wore a new pair of running shoes. The hamstring pain was quite sharp but went away after I sat down for a bit.

When I got to the hotel, I found my colleagues at a big meeting and sat with them. So that was several hours of sitting for conference sessions and the meeting of teachers. Afterward, I was stiff and sore, so I didn't go to the dinner and dance.

Instead of going out for the evening, I hit the sack, after doing a little writing, and tried to find something on TV.

I haven't watched television for decades and every once in a while I check it out and get reminded why I don't watch TV. It's so slow and boring, it feels like a big insult to my intelligence, and the advertising is disorienting.

On the other hand, I watched a show called The Manhunter, about this guy on horseback who tries to track guys running through the bush. The episode I saw had two American hikers trying to make their way through the wilderness of Northern B.C. while the manhunter tried to track them down. He almost caught them and apparently he usually wins but in this episode his prey got away.

The BC wilderness looked great.

123: Edmonton Again

The second day of the conference went a bit better. Again I walked three or four city blocks to get there, and back, and walked through the Shaw centre, and dealt with hamstring pain by sitting down for a few minutes. In the display area, I found some good resources for my younger classes and bought a U of Alberta hoodie for just twenty-five bucks.

At lunchtime, I talked to a teacher I met in one of the sessions. She was part Chipewyan but looked Scottish and was living up north in Great Slave Lake. -- That's a town that is quite far south of the lake with the same name.

She said her father went through the residential school system for aboriginal people and had a horrible time but managed to get his life together as an adult. He was institutionalized for dimentia and had the sundowner syndrome, she said, which I had never heard of. Around sundown, he got very cranky, to say the least, in spite of medication.

She was a teacher who planned to retire and took her boxes of lesson plans and resource materials home with her just in time to lose everything in the big fire that destroyed a lot of Great Slave Lake.

So, she was teaching again, without her resources, and happy in the classroom.

We had a good chat and thanked each other several times.

In the evening, after walking back to the hotel, I took myself out for steak, then called it a night, after some writing.

123: Home Again

The drive home took four hours as I got a little lost in Edmonton, trying to find a way through the downtown area. That added a half hour to my journey. And then I got stuck in a convoy of big trucks heading to Fort Mack.

The convoy was going slow because it was snowing and the highway just north of Edmonton was snow-covered, so I was happy we were all traveling at a safe speed. Usually there are white pick-up trucks passing you on that road, no matter how fast you are going. Most people fly along at around 120 but lots of trucks pass you, going so fast that you feel slow.

After I got home and unloaded the car, I had to go out to get some groceries.

Walking around Cold Lake was a lot less painful than Edmonton. -- Why was that?

Maybe I was getting used to the new shoes.

In the evening I talked to my brother for a long time on the telephone. He thought it was odd that I wasn't back at work on the date the doctor specified but understood the business between the benefits group and the employer could take a while.

I told him I was disappointed at first but also relieved as I wondered if I could handle going back to work since my back was still giving me referred pain and I was glad I had some more time to get stronger.

Instead of going swimming, as planned, I listened to the second half of the Oilers game on the radio. I caught the first half on the car radio and the second half in bed.

124 Sunday

ICE beat the Tigers 12-2 in the final regular season game. This is the 2nd time in the teams history finishing first overall in league standings with a record of 29-4-1 with 59pts. Game 1 of playoffs vs St.Paul is Friday the 15th right here at Imperial Oil

Place. Thanks to all the sponsors and fans for another successful season of Cold Lake Ice Hockey!

Swimming was great. It went faster and I went faster than ever before in the past half year.

After swimming, I walked around Walmart until my hamstrings hurt, sat down for five minutes, and was good to go.

Drove up to the lake for some inspirational beauty, stopped for groceries on the way home.

And my back now has some sensation, my hamstrings are a little sore, and I think it's time to do some hamstring stretches while the Oilers game is on the radio.

125: On Monday I Joined The Gym

i finally joined the gym, after my swim, and then i went to see 2 bball games, on a whim, and was invited to
sit on the bench again. after the game, i made a move to pick up a desk that had to be taken out of the gym and 4 kids
boxed me out, basketball style, while a kid from quebec grabbed the desk .... lol ... they wouldn't let me pick it up!
somebody has coached them gooooooooood.

126: Pancake Tuesday

today i went to the gym and worked out a lot harder than in physio, and longer, about an hour and a half.
it is on the second floor, has huge windows, a great view, and all new equipment, unlike the one at cfb cold lake.
the view over snowy fields to a series of ridges with evergreens and snow reminded me of .... you know!
so i felt right at home.
the fitness instructor said, oh, hi, mr. avery. she has a kid at my school. i had no idea who the heck she is.
there's also a padded walking track at the energy centre, so i did that, too.
i did step exercises and everything i do in physio PLUS the elliptical machine and the skating simulator.
i only did them for 5 minutes each as i have learned new things hurt me -- later.
and after all that fun, i drove over to the pool and swam a lot more and walked a lot less.
i did 90 lengths in 90 minutes!
the lifeguard from walkerton, ontario, i always talk to said, That's pretty good!

127: Ash Wednesday

Although I spent 3 hours in the gym and pool yesterday, I spent just half that time today, with half an hour in the pool, doing 30 lengths in 30 minutes, and about an hour in the gym, using the cross-trainer, the skater, a series of machines for legs and back and so on, plus arm curls and leg squats.

First I brushed thick, wet snow off my car, and my neighbour's car.

Afterward, I got a few things at the grocery store.

On my way in, I helped my neighbour get his walker out of his car and lock his car doors and get other doors open.

He always zings me with an insult, which deflates me for a little while, so today I called him on it and he said, Sorry.

I told him I coach basketball and he said, You must do it by lecture rather than by example.

Nice.

This morning I got an inspiration for my hockey poetry book and added five sections of questions for 500 marks as a study guide to complement the poems so my English colleague at work and other English teachers can make use of it and have some lesson planning done for them.

While I was working on it, I missed the Early Bird swim and morning at the gym, but I didn't mind. Since I did a big workout yesterday, I thought I'd give it a rest today. They say that's the way to build muscle. However, I felt like doing some exercises, so I swam for half an hour and then went to the gym.

128 Valentine's Day

After an hour in the gym, and over an hour in the pool, I feel better than I did in the morning.

In the gym, I did 15 minutes on the treadmill, 10 on the cross-trainer, 200 arm curls, ten toe lifts with 10 pound weights, and then tried the hamstring stretcher. That was a first. It showed I need a lot more flexibility there, as I could only get it to 35%.

In the pool, I did a mile in 50 minutes and 90 lengths in 80 minutes, which was better than 90 in 90, which is what I did on Tuesday.

I think I could do it faster.

Lunch was great: an egg scrambler with tomatoes, onions, cheese, oregano, garlic powder, milk, and four eggs. Sooo good!

With an Arizon tea, with ginseng, to go with it.

129: Family Day

A funny thing happened in the swimming pool today. After I swam one mile in one hour and did some water walking, I quit, feeling uninspired, but while I was walking out, a woman with red hair stopped me to say, Hey, you are strong in the water!

When I asked her, she said she was a swim coach and the coordinator for swimming on the base.

She said, And I know you are a teacher, too.

That was it: A quick chat. And I was out of there, feeling like taking on the rest of the day.

130 Saturday
131 Sunday
134 Monday

Tired from swimming 5 miles in 4 days, and going to the gym every day, I took the weekend off.

I wrote a lot and went out to do errands but that was about it.

135. Tuesday

After an hour and a half work-out at the gym called the Wellness Centre in the Energy Centre, doing everything bigger, I swam for an hour and a bit and got a new PB.

Last week I swam 90 lengths in 90 minutes at the start of the week, 90 lengths in 80 minutes at the end of the week, and today I did 90 lengths in 70 minutes.

At the gym, I did 15 minutes on the treadmill, 10 minutes on the skater, 15 on the cross-trainer, went up from 30 pounds to 60 pounds on all the machines, did 200 arm curls with 10 pound weights, up from 150, and did 100 toe raises holding 24.6 pounds, instead of 10.

Those are big improvements.

And I felt better walking out than walking in.

Also, I dropped 20 pounds since the last time I weighed myself, last week.

136 Wednesday

An hour in the gym, and an hour and a half swim.

137 Thursday

PB in the pool, 100 lengths in 80 minutes, then a felafel, and an hour in the gym.

138 Friday

Tired. Stayed home. Why? Was it swimming 4.5 miles in 3 days?
Biceps: 19 inches. Forearms, 15.
Calves: 19 inches. -- Used to be bigger.

139 Saturday

After an awesome hour at the gym, lunch at Pita Pit, with a felafel and 5 berry drink plus protein, I did 50 lengths in 40 minutes in the pool.
At the gym I did some new things and I did the old things bigger and better.

Shoulder press, 70 X 10, chest press, 80 X 10, upper back 120 X 20, lower back, 200 X 30, ab crunch, 100 X 30, leg curl 60 X 30, all X2.
Abductor, 150 X 30, lats, 100 X 20, toe lifts, 54 pounds in each hand X 50, flexibility posterior, 53%, ball with arm curls, 10 X 10 X 3, then 15 X 10 X 3.
Ball, back extensions, 20, ball leg raise, 10 each leg, skating, 5 min., ball wall squats, 10 X 60, cat stretch X30, and dead Bug X 30, followed by 28 stairs.

Yakupov is on the first line with Nugent-Hopkins and Eberle, with Taylor Hall suspended, and he set up the Oilers' first goal!

140 Sunday
Woke up early, did some writing. Got ready for a morning at the gym but drove to the lake and then the ski hill instead and didn't walk or do anything physical. Came home and slept and wrote.
Swam for an hour in the afternoon but felt like quitting after half an hour. - Did a lot of water walking.

On the way out, I almost hit a guy walking behind my car in the parking lot. A military guy with a hockey bag glared at me. He was at my right rear bumper. I didn't see him.
He looked like he wanted to fight.
I ignored him and drove away slowly and under control.
It made me feel angry for a moment and then sad.

Drove up to the lake again but the Eatery was closed, so I came home and made a sandwich.
And now I feel sleepy again.
Did I overdo it yesterday?

The Ending

Shortly after that, I went back to work. First I worked half time, for a week, and then I went back full-time, and I was very happy.

I kept thinking about the time the time the doctor said, You're going to die, you'll never work again, you'll never walk again, activate your short and long term disability, get your things in order, get ready to say good-bye.

But I proved him wrong, and I was happy about that.

However, after a couple of months at work, something horrible happened: At work, one teacher's aid after another and then one teacher after another got the word that were not wanted back the next school year.

I got a letter from the principal saying the same thing. So did many of my colleagues.

As word got out that so many staff members were leaving, many of the students made plans to go to other schools the next year.

It looked as though our school was closing, or shrinking fast, and the high school half of the building might close so it would only be open as a little elementary school.

To make a long story short, there was a lot of sadness, followed by the various stages of the grieving process, as our little school appeared to be about to die.

First I was told I was going to die, or never work again, but I made it back to work just in time to see the school go down. It looked like it was going to die.

They say that when one door closes, another door opens, because that's what doors do.

I looked around for a new job, but there weren't many positions to apply for in Northern Alberta, or the rest of Alberta, or Ontario, or the rest of Canada, either. So, I applied for a job in China.

I got hired to teach in a school in the north-east part of China, in a school called Maple Leaf, in the city of Dalian.

I found out Dalian had the best air quality in all of Asia.

When I got there, in the fall, one of the first things I did was find a doctor, for acupuncture and Traditional Chinese Medicine.

I wrote a book about that, called Intro To Acupuncture.

After a few months, my new doctor gave me some good news: She said, "You have very good jing-chi-shen." She told me what that meant: "You are now in good health, good spirits, and it looks as though you will a long time."

A Canadian doctor told me I was going to die soon but Dr. Sofia Guan Wang said, "You have a biological age of around twenty-four. You are as strong and healthy as a twenty-four year old. And you will live a long time."

She worked in a clinic that combined the best of the East and the West, in terms of medical practice, and I came to appreciate the integrative approach to healing in a very deep way. But that's another story for another day, as they say, and you can read about it in my book called An Intro To TCM.

About the author ....

Martin Avery, author and educator, indie publisher and teacher, drama coach and basketball coach, ATC, OTC, Hons. B.A., B.Ed., MFA in Writing, AQ Drama, Hons. Spec. English, also has a Diploma In Spiritual Healing and is a Reikimaster, Qigong leader, and Zen meditation instructor.

He has taught Writing at York University, Georgian College, The Ginger Press Bookstore, Centauri Arts Retreat, at high schools across Ontario for The Writers Union of Canada, Bracebridge Public Library, Pickering Public Library, Meta4 Art Gallery, etc. He has also taught at the Halifax School For The Blind, Frontier College at Bannock Point Rehab Camp in the Whiteshell Provincial Park, Manitoba, and the Kennedy Youth Facility school inside a maximum security prison, Ajax, Ontario, as well as Grey Highlands, The OSCVI, Gravenhurst High School, and the high schools of Durham Region.

He is the founder of A Novel Marathon, in Owen Sound, and the Muskoka Novel Marathon at the Huntsville Festival of the Arts, and the author of over 100 books available on Amazon.

(Amazon? one of my students said. Well, we don't live in the Amazon, so what do I care?!)

He was the English teacher and Department Head at the French school, Ecole Voyageur, in Cold Lake, for two years, and is now the English Department Head of Maple Leaf International School in Dalian, China.

He was recently awarded a Certificate of Achievement for Expert Authors from Ezine @articles and a Gold Medal for sports writing from Bleacher Report. He is also the winner of the Balzac award for poetry and four of his books have been nominated for the Leacock Award for Humour. He was the Writer in Residence at the Bracebridge Public Library, the Pickering Public Library, and the Zen Forest.

He recently recovered from an idiopathic neuropathy emergency and severe nerve damage, leading to the loss of his legs and feet, or Guillain-Barré Syndrome. Or something.

www.ingramcontent.com/pod-product-compliance
Lightning Source LLC
Chambersburg PA
CBHW031810170526

45157CB00001B/29